Disaster Risk Reduction

Methods, Approaches and Practices

Series Editor

Rajib Shaw, Keio University, Shonan Fujisawa Campus, Fujisawa, Japan

Disaster risk reduction is a process that leads to the safety of communities and nations. After the 2005 World Conference on Disaster Reduction, held in Kobe, Japan, the Hyogo Framework for Action (HFA) was adopted as a framework for risk reduction. The academic research and higher education in disaster risk reduction has made, and continues to make, a gradual shift from pure basic research to applied, implementation-oriented research. More emphasis is being given to multi-stakeholder collaboration and multi-disciplinary research. Emerging university networks in Asia, Europe, Africa, and the Americas have urged process-oriented research in the disaster risk reduction field. With this in mind, this new series will promote the output of action research on disaster risk reduction, which will be useful for a wide range of stakeholders including academicians, professionals, practitioners, and students and researchers in related fields. The series will focus on emerging needs in the risk reduction field, starting from climate change adaptation, urban ecosystem, coastal risk reduction, education for sustainable development, community-based practices, risk communication, and human security, among other areas. Through academic review, this series will encourage young researchers and practitioners to analyze field practices and link them to theory and policies with logic, data, and evidence. In this way, the series will emphasize evidence-based risk reduction methods, approaches, and practices.

All submitted proposals are carefully reviewed by the series editor.

Publication Ethics and Publication Malpractice Statement: https://www.springernature.com/gp/policies/book-publishing-policies

Ethical, Conflict-of-Interest, and Editorial Policy Statement: https://www.springer.com/us/authors-editors/journal-author/journal-author-helpdesk/before-you-start/before-you-start/1330#c14214

Qingyue Du · Rajib Shaw

Social Organizations in Disaster Response and Recovery

Development and Potentials in China

Springer

Qingyue Du
School of Political Science and Public Administration
Shandong University
Qingdao, China

Rajib Shaw
Graduate School of Media and Governance
Keio University
Fujisawa, Kanagawa, Japan

ISSN 2196-4106 ISSN 2196-4114 (electronic)
Disaster Risk Reduction
ISBN 978-981-92-2845-4 ISBN 978-981-92-2846-1 (eBook)
https://doi.org/10.1007/978-981-92-2846-1

This Springer imprint is published by the registered company Springer Nature Singapore Pte Ltd.
The registered company address is: 152 Beach Road, #21-01/04 Gateway East, Singapore 189721, Singapore

Preface

Disaster response and recovery increasingly rely on the capacity of social organizations to operate across institutional, community, and transnational boundaries. As disasters become more frequent, interconnected, and differentiated in impact, state-centered systems often struggle to manage the cascading risks and governance blind spots they generate. In this evolving landscape, social organizations—including NGOs, volunteer groups, and grassroots networks—have become crucial actors in resource mobilization, coordination, and long-term recovery. Many begin with spontaneous relief efforts, but over time, transition into more organized roles. Understanding their evolution, modes of collaboration, and pathways to institutionalization is vital to rethinking disaster governance in an age of complexity and uncertainty.

This book explores the changing roles of social organizations in disaster response and recovery, with a particular focus on China and its engagement with international humanitarian systems. Through theoretical frameworks and in-depth empirical analysis, the book examines how civic actors have moved from reactive, fragmented efforts toward more professionalized, digitally enabled, and structurally embedded participation. It also analyzes how governance settings, technological infrastructures, and evolving state–society relations influence collaboration during crises. China offers a dynamic site of transformation: from the Wenchuan Earthquake to the Zhengzhou floods, civic engagement has undergone cycles of innovation and constraint. Meanwhile, the growing presence of Chinese social organizations in overseas responses—such as in Nepal, Ecuador, Indonesia, Türkiye, and Myanmar—has tested their capacity to navigate complex geopolitical, legal, and operational terrains.

Drawing from both domestic and international experiences, this book presents selected cases to highlight key shifts and tensions in disaster governance involving social organizations. Though not exhaustive, we believe it will help broaden the future scope of research and practice on collaborative and resilient disaster governance. We will be glad if the readers find the book useful.

Qingdao, China
Fujisawa, Japan

Qingyue Du
Rajib Shaw

About This Book

Social organizations are playing an increasingly vital role in disaster response and recovery as disasters grow more frequent, complex, and transboundary. Understanding how these actors mobilize, collaborate, and become institutionally embedded is key to rethinking disaster governance. China offers a particularly dynamic case, where evolving state–society relations, digital innovation, and international outreach converge. This book focuses on the experiences of Chinese social organizations within both domestic and cross-border contexts, analyzing their roles through governance, network, and co-production perspectives. It explores how civic actors transition from spontaneous relief to institutionalized, professionalized, and digitally supported engagement, and how coordination, legitimacy, and resilience are built under crisis conditions. Drawing on global frameworks—such as the UN Cluster System and VOAD networks—and rich Chinese cases from Wenchuan to Zhengzhou, as well as Nepal, Ecuador, Indonesia, Türkiye, and Myanmar, the book bridges theoretical insight with practice-based learning. This book presents recent advances, trends, challenges, and potentials in disaster governance involving social organizations through 20 chapters across four thematic parts.

Contents

About the Authors

Qingyue Du is a researcher at the School of Political Science and Public Administration, Shandong University, China. Her research focuses on risk communication, crisis communication, and disaster risk reduction. She holds a Ph.D. in Journalism from Sichuan University, where her dissertation examined the dissemination of information about the participation of social organizations in disaster relief. She has published several academic articles and has been engaged in volunteer efforts aimed at improving disaster information services and coordination.

Rajib Shaw is a professor in Graduate School of Media and Governance of Keio University, Japan. He is the co-founder of a Delhi (India) based social entrepreneur startup Resilience Innovation Knowledge Academy (RIKA), and RIKA Institute. He is the Co-chair of the United Nations Asia Pacific Science Technology Advisory Group (AP-STAG) and was CLA for IPCC's 6th Assessment Report. Professor Shaw is the recipient of prestigious "Pravasi Bharatiya Samman Award (PBSA)" in 2021 for his contribution in education sector. He also received United Nations Sasakawa Award in 2022 for his lifetime contributions in the field of disaster risk reduction. He published 89 books and over 450 research papers. More about his work can be found in: www.rajibshaw.org.

Part I
Overview and Theory

Chapter 1
Global Risk Landscape: Rising Disasters and the Role of Social Organizations

Abstract In recent decades, the global landscape of disasters has undergone a dramatic transformation—marked not only by rising frequency and intensity but also by increasing complexity, regional disparities, and governance challenges. As climate change accelerates and urban vulnerabilities deepen, disasters are no longer isolated shocks but systemic events that cascade across sectors and borders. This chapter maps the contours of this evolving risk environment, beginning with long-term trends in hazard escalation and climate-driven shifts in disaster typologies. It then explores the uneven geography of exposure and vulnerability, highlighting how structural inequalities amplify risks in the Global South and among marginalized populations. The discussion further examines persistent data gaps that obscure invisible risks and limit anticipatory action. These dynamics reveal the limitations of conventional disaster management and underscore the growing relevance of decentralized, inclusive models—where social organizations play an increasingly vital role in navigating uncertainty and building resilience from the ground up.

Keywords Disaster trends · Climate risk · Social vulnerability · Disaster governance · Civil society organizations

1.1 Global Escalation of Disasters: A Systemic Risk Landscape

The past two decades have witnessed a sharp escalation in the frequency, intensity, and complexity of disasters, with climate change, urban vulnerability, and data deficiencies amplifying systemic risks across regions (Seneviratne et al., 2021; UNDRR, 2020). Far from isolated shocks, disasters are increasingly interwoven with structural drivers such as global warming, socio-economic inequalities, and fragile governance systems. Between 2000 and 2019, over 7300 major disaster events were recorded globally, resulting in 1.23 million deaths and economic losses nearing US $3 trillion—figures that reflect not only more frequent hazards but also deepening exposure and vulnerability (UNDRR, 2020). In 2024 alone, disasters caused economic losses

Q. Du and R. Shaw, *Social Organizations in Disaster Response and Recovery*, Disaster Risk Reduction, https://doi.org/10.1007/978-981-92-2846-1_1

49% higher than the 30-year average, despite advancements in early warning systems (UNDRR, 2025). Meanwhile, high-confidence climate science confirms a marked rise in hot extremes, heavy precipitation, and ecological droughts since the mid-twentieth century, with risks projected to intensify further under warming scenarios. These converging dynamics underscore a critical shift: disasters can no longer be understood in isolation, but as part of an evolving, interconnected global risk system. This section synthesizes recent insights to map the contours of this transformation and its implications for disaster risk governance.

1.1.1 Overall Trends in the Frequency and Intensity of Disasters

Over the past two decades, there has been a marked escalation in both the frequency and intensity of disasters globally. This trend is no longer anecdotal but substantiated by longitudinal data, climate science projections, and institutional disaster monitoring systems. Drawing on empirical evidence and authoritative climate models, this section explores how disasters have intensified over time and how future risk trajectories suggest even greater severity in the coming decades. The analysis brings together retrospective data assessments from disaster risk reports and forward-looking projections from climate science to offer a comprehensive understanding of the shifting global risk landscape.

Between 2000 and 2019, the world experienced 7348 major recorded disaster events, which resulted in 1.23 million deaths, affected over 4 billion people, and caused approximately US $2.97 trillion in economic losses (UNDRR, 2020). This represents a sharp increase compared to the preceding two decades (1980–1999), during which 4212 disasters were recorded. The vast majority of disasters recorded between 2000 and 2019—91%—were climate-related, including floods, storms, droughts, heatwaves, and wildfires (UNDRR, 2020). These figures demonstrate not only an increased frequency of hazard occurrence but also a systemic shift in the nature of disasters, with hydro-meteorological and climate-related events becoming dominant.

This upward trajectory continued into the most recent years. According to the 2024 Global Natural Disaster Assessment Report, the year 2024 alone recorded 318 major disasters, exceeding the average of the preceding three decades (1994–2023) (UNDRR, 2025). While the number of deaths slightly declined due to improvements in early warning systems and emergency response, the economic loss increased by 49% compared to the 30-year average, revealing a troubling trend: global assets and critical infrastructure are becoming more exposed and vulnerable to natural hazards (UNDRR, 2025). Storms and floods remained the most frequent and costly hazard types, accounting for the largest shares of both human and economic losses.

From a scientific standpoint, the Intergovernmental Panel on Climate Change (IPCC) confirms the causal relationship between anthropogenic climate change and the increase in extreme weather and climate events. According to the Sixth Assessment Report, high-confidence evidence indicates that hot extremes, heavy precipitation, and agricultural and ecological droughts have increased in both frequency and intensity across most regions since 1950. The report further projects that with global warming of 1.5 °C, the world will experience "statistically significant increases in extreme heat, heavy rainfall, and tropical cyclone intensification," with even more dramatic effects under higher warming scenarios (Seneviratne et al., 2021).

The IPCC's climate scenario modeling introduces a nonlinear dimension to future disaster trends: as global temperatures rise, the frequency of extreme events increases disproportionately. For example, a 1-in-50-year heatwave in the pre-industrial period is now a 1-in-10-year event under current warming levels, and it may occur as often as once every five years under a 2 °C scenario (Seneviratne et al., 2021). Similarly, extreme rainfall events are projected to intensify by approximately 7% per 1 °C of warming, significantly increasing flood risks even in regions not traditionally vulnerable.

The UNDRR Special Report 2024 also points to the growing complexity of disasters, with compound and cascading events emerging as a new normal (UNDRR, 2024). Examples include concurrent drought-heatwave-wildfire occurrences, which strain emergency response systems and disrupt interconnected supply chains. These complex events are not only more destructive but also harder to predict and manage using traditional risk models. They reveal the limitations of linear planning frameworks in the face of systemic, cross-sectoral risk, further underscoring the need for integrated governance approaches.

Moreover, while hazard frequency is increasing globally, the intensity of disasters is not equally distributed. As reported by UNDRR, Asia remains the most disaster-affected region, accounting for 42% of global events in 2024 alone (UNDRR, 2025). Smaller island developing states and low-income countries disproportionately suffer higher death rates and economic damage relative to GDP, highlighting global risk asymmetries. This is further complicated by data gaps in vulnerable areas, which prevent comprehensive assessments and timely response (UNDRR, 2023).

The data challenge is particularly relevant in the context of growing disaster intensity. As emphasized in UNDRR's 2023 report, limited access to disaggregated, timely, and localized data has hindered efforts to understand and manage the evolving risk landscape. The report states that "disaster data remains fragmented, outdated, or missing in many high-risk areas," particularly among marginalized populations such as informal settlers, women, and people with disabilities (UNDRR, 2023). These data gaps obscure the full scope of disaster impact and constrain the design of anticipatory strategies.

1.1.2 The Evolution of Disaster Types Driven by Climate Change

Climate change is not only intensifying the frequency of natural hazards but is also reshaping their typology, behavior, and systemic impacts. Over the past two decades, extreme weather events—particularly extreme heat, intense tropical cyclones, prolonged droughts, and compound disasters—have become more prevalent, with future climate scenarios suggesting a continued escalation in both severity and complexity. This section explores how climate change is altering the landscape of disaster types, drawing on key findings from observational data and climate projections presented in authoritative global reports.

According to high-confidence assessments from the Sixth Assessment Report of the IPCC, there is clear evidence that hot extremes, heavy precipitation, and both agricultural and ecological droughts have increased in frequency and intensity across most regions since 1950 (Seneviratne et al., 2021). The report underscores that the likelihood and magnitude of these events are strongly linked to anthropogenic climate change. In particular, extreme heat events, which were once rare, are now becoming commonplace, with some regions experiencing record-breaking temperatures more frequently than at any point in recorded history. The probability of events such as the European heatwave of 2003 or the Pacific Northwest heat dome of 2021 is projected to rise dramatically with every increment of warming (Seneviratne et al., 2021).

Similarly, droughts have shown intensifying patterns in both duration and geographic spread. The probability of drought conditions affecting agriculture and ecosystems has increased significantly, particularly in semi-arid and arid regions, due to shifts in rainfall patterns and rising evapotranspiration rates. These changes are already manifesting in Southern Africa, Central Asia, and parts of South America—regions experiencing prolonged dry spells with direct implications for food security, migration, and conflict.

Beyond isolated hazard types, recent years have seen the emergence of compound and cascading disasters, where multiple hazards interact, overlap, or follow in rapid succession. This includes events such as simultaneous heatwaves, wildfires, and droughts, which overwhelm response systems and generate disproportionately high impacts. For example, the UNDRR Special Report 2024 highlights the increasing frequency of compound events, noting that disasters "can no longer be treated as discrete and isolated incidents" (UNDRR, 2024). Instead, climate change is acting as a risk multiplier, creating interdependencies across hazard systems, infrastructures, and vulnerable populations.

One of the most critical shifts in disaster typology is the observed and projected increase in the intensity of tropical cyclones. Based on scientific modeling and trend analysis, the IPCC concludes that while the overall number of tropical cyclones may not increase, the proportion of intense storms is rising under climate change scenarios. Warmer sea surface temperatures provide more energy for cyclones, leading to higher wind speeds, increased rainfall, and slower storm movement—all factors that amplify destructive potential. Events such as Cyclone Idai (2019) and Typhoon Haiyan (2013) exemplify how intensified storms devastate critical infrastructure and result in high casualty rates, especially in coastal regions with high population density.

Supporting this trend, the 2024 Global Natural Disaster Assessment Report notes that storms accounted for 32.8% of total economic losses globally in 2024, making them the most financially destructive hazard type for that year (UNDRR, 2025). Additionally, floods—often linked to storm surges or extreme precipitation—remained the most frequent hazard, responsible for 41% of fatalities and affecting over one-third of the global disaster-impacted population in 2024 alone (UNDRR, 2025). These statistics reinforce the dual concern: not only are climate-linked hazards becoming more frequent but they are also among the most deadly and economically disruptive.

The UNDRR 2020 report further provides long-term data to corroborate this trend. Between 2000 and 2019, floods and storms represented the two most prevalent disaster types globally, with floods accounting for 44% of all recorded events and storms for 28% (UNDRR, 2020). Together, they caused over 700,000 deaths and impacted billions of people. Importantly, these hazards are also the most sensitive to climatic variability, making them likely candidates for further escalation under future warming trajectories.

Another emerging concern is the increasing prevalence of wildfires, particularly in regions previously considered low-risk. As detailed in recent climate-risk scenarios, rising temperatures and prolonged dry periods have extended fire seasons and increased ignition likelihood. While not as immediately deadly as cyclones or floods, wildfires contribute to air pollution, displacement, biodiversity loss, and public health crises, particularly when occurring near densely populated areas.

Across all these disaster types, one consistent pattern is that climate change is intensifying their physical characteristics while also transforming their social and economic impacts. As shown in the GAR Special Report 2024, the growing complexity of disasters challenges traditional emergency management frameworks, which are often designed for single-hazard events (UNDRR, 2024). The report calls for a paradigm shift toward multi-hazard risk governance, where climate data, infrastructure planning, and social vulnerability assessments are integrated into a unified response strategy.

Finally, it is important to emphasize that these evolving disaster types do not affect all regions equally. Asia continues to bear the brunt, with 42% of all disasters occurring in the region in 2024 (UNDRR, 2025). Meanwhile, small island developing states (SIDS) and parts of Africa suffer disproportionately high mortality and loss rates relative to their population and GDP. These trends highlight the importance of localized risk assessments and tailored adaptation measures to address the unique exposure and vulnerability profiles across regions.

In conclusion, climate change is not merely increasing the number of disasters but is transforming the very nature of disasters themselves. From more powerful tropical cyclones to slow-onset droughts and multi-hazard compound events, the evolution of disaster types underscores the urgent need for systemic, anticipatory, and inclusive disaster risk governance.

1.1.3 Regional Disparities and Unequal Risk Exposure in the Global Landscape of Disasters

While the global frequency and intensity of disasters are rising, their impacts are neither uniformly distributed nor equally experienced. Certain regions, particularly in the Global South, are disproportionately exposed and vulnerable to hazard-related losses due to a confluence of geographical, socio-economic, and governance-related factors. The pattern of disaster impacts reflects deep-rooted structural inequalities in global risk exposure, adaptive capacity, and data access. Understanding these regional disparities is crucial for shaping equitable and inclusive disaster risk governance systems.

The most disaster-affected region globally over the past two decades has been Asia. According to long-term data compiled by the Centre for Research on the Epidemiology of Disasters and published by UNDRR, Asia accounted for 44% of all recorded disasters between 2000 and 2019, far surpassing other regions in terms of event frequency (UNDRR, 2020). This high incidence is largely attributed to the region's geographic exposure to multiple hazard types—such as floods, typhoons, droughts, and earthquakes—as well as its dense populations and rapidly urbanizing environments. Countries like China, India, the Philippines, and Indonesia consistently rank among the top disaster-affected nations globally, both in terms of event frequency and number of people impacted (UNDRR, 2020).

While Asia experiences the highest number of disasters, Small Island Developing States (SIDS) and Sub-Saharan African countries endure the most severe consequences relative to their size, population, and economic output. The 2024 Global Natural Disaster Assessment Report notes that although SIDS represent only a small fraction of global territory and population, they suffer from exceptionally high mortality rates and economic losses as a percentage of GDP (UNDRR, 2025). This is particularly evident during high-impact tropical cyclones and tsunamis, where limited land area, fragile infrastructure, and economic dependency on climate-sensitive sectors—such as tourism and fisheries—leave these states acutely exposed.

In Sub-Saharan Africa, vulnerability is exacerbated by widespread poverty, weak governance, and chronic underinvestment in risk-reduction systems. Droughts, floods, and epidemics have had recurrent and devastating impacts on livelihoods, especially among rural populations dependent on rain-fed agriculture. The compound effects of climate variability and limited response capacity have made disasters in African nations disproportionately lethal. The 2025 report highlights that disaster mortality rates in low-income African countries remain among the highest globally, despite relatively fewer reported disaster events compared to other regions (UNDRR, 2025).

Another region showing increased exposure is the Middle East and South Asia, particularly in relation to extreme heat events. The 2024 and 2025 UNDRR reports warn of a sharp uptick in heatwave frequency and intensity in these regions, compounded by urban heat island effects, high population density, and inadequate infrastructure for heat mitigation (UNDRR, 2024, 2025). Cities such as Karachi, Dhaka, and Cairo are already experiencing climatic thresholds that threaten human survivability without cooling interventions. The problem is made more urgent by demographic trends: many of the world's fastest-growing cities are located in regions with limited adaptive capacity to manage chronic and extreme heat.

These regional differences in exposure are closely linked to global inequality in risk governance capacity. Wealthier nations typically have stronger infrastructure, early warning systems, and social protection networks that buffer disaster impacts, even when exposed to high-intensity events. In contrast, vulnerable countries often lack the resources, institutions, or data systems necessary to prepare for, respond to, or recover from disasters in a timely and equitable manner.

One of the key limitations in addressing regional disparities lies in data availability and quality. As highlighted in the 2023 UNDRR report on data gaps, many high-risk regions in the Global South suffer from fragmented, outdated, or entirely missing disaster and climate risk data (UNDRR, 2023). This includes critical gaps in disaggregated data related to gender, disability, informal settlements, and small-scale localized hazards. Without such data, it becomes nearly impossible to design effective and inclusive disaster risk reduction strategies. The report notes, for example, that many countries do not report subnational disaster loss data, thereby masking intra-national disparities and vulnerable groups within regions (UNDRR, 2023).

The absence of reliable data not only limits disaster preparedness but also distorts global policy priorities. Global risk models, funding mechanisms, and insurance schemes are often based on datasets biased toward regions with better monitoring capacity. This can lead to the systematic underestimation of risk in poorer regions, resulting in lower investment in adaptation and recovery, and perpetuating a vicious cycle of vulnerability.

Moreover, the combination of high exposure, weak adaptive capacity, and institutional marginalization makes certain populations more vulnerable within regions. For example, informal urban settlements in Southeast Asia or the Horn of Africa are highly susceptible to flooding but are often excluded from official risk assessments and emergency planning. The same applies to pastoralist communities in the Sahel, who face slow-onset disasters like drought and desertification, yet are rarely captured in rapid-onset disaster databases (UNDRR, 2023).

These findings point to the need for place-based, equity-oriented disaster governance strategies. Disaster risk is deeply entangled with broader issues of development inequality, historical marginalization, and institutional capacity gaps. Any meaningful global response must therefore address not only the physical exposure to hazards but also the socio-political and economic structures that determine who bears the brunt of disaster impacts.

In conclusion, regional disparities in disaster risk exposure and outcomes reflect profound global inequalities. While disasters are becoming more frequent and severe everywhere, the ability to cope, adapt, and recover is unequally distributed, with vulnerable regions facing exponentially higher risks. Addressing these asymmetries requires a fundamental shift in disaster risk governance—one that incorporates localized data, inclusive planning, and resource redistribution to support those most at risk. This uneven geography of risk also lays the groundwork for the next section, which turns to the governance challenges in managing increasingly complex and unequal disaster landscapes.

1.1.4 Data Gaps and Monitoring Challenges: Invisible Risks

As disasters grow more frequent and complex in the context of accelerating climate change, the demand for reliable, inclusive, and anticipatory data becomes increasingly urgent. Many global and national disaster monitoring systems remain unprepared to meet this demand. Critical information is missing, fragmented, or inaccessible—particularly in regions and among communities most exposed to risk. These gaps are not merely technical oversights; they reflect deep-seated inequalities in governance, representation, and resource allocation. In many parts of the world, disasters are not only underestimated—they are undocumented, making them all the more difficult to prevent or respond to.

One of the most persistent challenges across disaster risk systems is the lack of disaggregated and locally relevant data. Information about how disasters affect different social groups—such as women, persons with disabilities, informal settlement residents, or ethnic minorities—is often incomplete or entirely absent. As a result, those most vulnerable to disasters are frequently excluded from risk assessments, early warning systems, and recovery efforts, rendering their needs invisible within formal governance structures (UNDRR, 2023).

This invisibility is compounded by the way disaster data is collected and managed. In many national contexts, data flows remain highly centralized, with little integration of insights from subnational governments, local actors, or non-state responders. Local knowledge—often held by municipalities, civil society groups, or frontline workers—is rarely incorporated into national disaster databases or global monitoring platforms. Instead, these systems tend to rely on aggregated, post-event statistics that fail to reflect the lived realities of risk at the community level (UNDRR, 2023).

The consequences of these gaps are especially visible during compound or emerging risks, such as heatwaves linked to health crises or droughts contributing to migration. These events require new types of data, including indicators that span climate, health, infrastructure, and socio-economic vulnerability. Most existing systems are built around single-hazard, short-term models that do not account for cascading impacts or slow-onset risks. For example, while flood damage is typically recorded through infrastructure losses, the long-term social and public health consequences of extreme heat or chronic drought often go unmeasured and unmonitored (UNDRR, 2023, 2024).

Another challenge lies in the fragmentation of data systems. Risk information is often scattered across government departments, international organizations, and scientific agencies—each using different formats, taxonomies, and timeframes. Without interoperability between these systems, it becomes difficult to assemble a comprehensive picture of who is at risk, what assets are exposed, and how multiple threats may interact. This siloed structure hampers cross-sectoral coordination and delays disaster response efforts, especially when multiple hazards occur simultaneously (UNDRR, 2023).

The problem is particularly acute in low-income and high-risk countries, where institutional and technical capacity to generate risk data is limited. In many cases, data is outdated, non-digital, or collected manually. Governments may depend on external partners for assessment and analysis, creating bottlenecks in access and ownership. This reliance undermines local decision-making and delays critical interventions, especially in fast-moving crises that require real-time situational awareness.

Even where data exists, barriers to access and usability remain. Many datasets are stored in non-standard formats, hidden behind institutional restrictions, or designed without consideration for local decision-makers. Social organizations and frontline responders often lack the tools or permissions needed to access national or international datasets, even when those datasets are directly relevant to their operations. In some cases, data literacy and digital infrastructure constraints prevent local actors from making full use of available information (UNDRR, 2023).

These data exclusions have far-reaching consequences. Risks that are not tracked or measured tend to be ignored in planning processes, left out of funding mechanisms, and deprioritized in public discourse. This is especially true for slow-onset or non-spectacular disasters—like prolonged heatwaves, environmental degradation, or food system disruptions—that lack the visual immediacy of floods or earthquakes but often have equal or greater long-term impacts (Seneviratne et al., 2021; UNDRR, 2024).

Furthermore, many predictive models used in disaster planning are based on historical trends and linear assumptions. With the planet warming, the likelihood of non-linear, high-impact events—including mega-storms, tipping points, and cascading failures—is growing. Without data systems that can anticipate these shifts, disaster governance remains fundamentally reactive. The gap between what we know and what we need to know is widening at precisely the moment when risk is becoming more systemic and less predictable.

Recent assessments show that when early warning systems fail, the issue is often not technological. In many cases, risk messages do not reach the people who need them—because those populations were never identified in the system design. Residents of informal settlements, linguistic minorities, people with disabilities, and others are routinely overlooked in communication strategies, emergency drills, and evacuation planning (UNDRR, 2025). Their exclusion is not coincidental; it is the result of being uncounted in the underlying data architecture.

At its core, disaster data is not neutral. It reflects underlying questions of whose risks are visible, whose knowledge is valued, and whose safety is prioritized. Risk information is shaped by access to power, resources, and political voice. Without a deliberate effort to correct these asymmetries, current data systems will continue to replicate patterns of marginalization, leaving the most vulnerable communities out of sight—and out of policy.

Creating more equitable and effective disaster risk governance means confronting these structural issues head-on. Improving data quality alone is not enough; it must be paired with investments in participatory data collection, transparent governance, and institutional accountability. This shift also opens the door to new kinds of partnerships: with social organizations, local governments, and informal networks who hold vital knowledge about their communities—knowledge that is too often ignored.

1.2 Global Challenges and the Governance Imperative in a Shifting Risk Landscape

The accelerating complexity of the global risk landscape has rendered traditional disaster governance frameworks increasingly inadequate. Today's disasters are rarely isolated; instead, they unfold as systemic, compound, and transboundary events that span across environmental, economic, and social domains. These evolving dynamics—such as cascading infrastructure failures, multi-hazard interactions, and climate-health crises—demand governance approaches capable of anticipating interconnected disruptions (UNDRR, 2025). Current institutional models, often designed around linear, sector-specific, and reactive logics, struggle to respond to these cross-cutting risks (Schweizer & Renn, 2019; UNDRR, 2024). Moreover, vulnerability and response capacities are distributed highly unevenly across geographies and populations, compounding the effects of disaster exposure in marginalized or under-resourced communities (Manatsa et al., 2025; Mark et al., 2024). In this context, centralized, top-down approaches are increasingly misaligned with the realities of a polycentric risk environment. There is thus a growing need to transition toward inclusive, adaptive, and multi-actor governance systems that embed collaboration, local capacity, and cross-scale coordination. This section addresses four interconnected challenges driving this transformation: the rise of systemic risks, unequal capacities, institutional limitations, and the imperative for inclusive risk governance.

1.2.1 The Rise of Systemic and Compound Risks: Challenging Traditional Governance Logics

The global risk landscape is increasingly shaped by systemic, interconnected, and compound disasters that challenge the linear logic of traditional risk governance. These risks no longer emerge in isolation but are part of tightly interwoven systems, where disturbances in one domain can cascade into multiple others. The conventional hazard-specific response model is increasingly inadequate for dealing with this complexity, as it lacks the capacity to account for cross-sectoral feedbacks and transboundary implications.

Systemic risks are defined by their capacity to propagate through interdependent infrastructures, economies, and ecosystems. These risks often manifest in nonlinear ways, where small triggers can lead to disproportionately large consequences. Theoretical frameworks of systemic risk governance stress that such risks are difficult to predict and manage because they operate across spatial, temporal, and institutional boundaries. Unlike isolated shocks, systemic risks accumulate silently within the system and reveal themselves during moments of stress, often as polycrises—simultaneous, interacting crises that strain governance capacities beyond their limits (Liu & Renn, 2025; Schweizer & Renn, 2019).

Recent disasters highlight how these dynamics unfold in practice. Global urban centers, for example, increasingly experience cascading disruptions, where climatic events like heatwaves or flash floods trigger failures in energy grids, transport systems, and public health infrastructures. These interlinked failures are often intensified by underlying structural inequalities and uncoordinated institutional responses. In such contexts, a localized event can escalate into a systemic crisis, particularly when it intersects with vulnerabilities such as informal housing, aging infrastructure, or under-resourced local governance (Mark et al., 2024; Tagliacozzo et al., 2024).

The compound nature of contemporary disasters is especially evident in climate-related events. Increasingly frequent combinations of drought, heatwaves, and air pollution are being recorded across regions such as South Asia and the Mediterranean. These overlapping hazards produce amplified impacts on human health, agriculture, and energy systems, with vulnerable populations bearing the brunt of exposure and limited access to coping mechanisms. Research highlights how such compound climate-health-social risks not only raise the scale and duration of disasters but also undermine traditional early warning systems, which are often designed for single-event scenarios (Higuera Roa et al., 2025; Senevirathne, 2024).

Environmental risks are also becoming increasingly entangled. Wildfires, for example, no longer result solely from dry conditions, but from a complex interaction of climatic stressors, land management failures, and socio-economic pressures. Studies on fire regimes in Southern Europe reveal how drought, depopulation of rural areas, and institutional fragmentation contribute to the emergence of uncontrollable fire events. Similarly, in the urban water sector, recurring droughts are no longer solely a function of rainfall deficit but are connected to governance failures, infrastructure limitations, and competing water demands—illustrating the compound nature of slow-onset and sudden-onset risks (Bacciu et al., 2022; Grasham et al., 2019).

These examples underscore a critical governance gap. Current institutional architectures remain largely fragmented, designed to handle discrete risks within specific domains—environmental, health, or economic—rather than the complex entanglements now observed. Without a shift in perspective, responses are likely to remain reactive, misaligned, and resource-inefficient. What is required instead is an integrated governance logic capable of anticipating interdependencies, sharing knowledge across sectors, and fostering adaptive capacity at multiple levels. Recent proposals advocate for iterative frameworks that embed systems thinking into risk assessment, scenario planning, and decision-making processes (Parviainen et al., 2025).

1.2.2 Uneven Distribution of Resources and Capacities: Spatial and Social Inequalities in Risk Exposure

Disasters may be global in reach, but their impacts are profoundly unequal. The intersection of geography, governance capacity, and social vulnerability produces asymmetric patterns of disaster exposure and resilience. These disparities are neither incidental nor temporary; they are deeply embedded in historical development trajectories, structural inequalities, and uneven state capacity.

Regions such as the Global South, and particularly small island developing states (SIDS) and least developed countries (LDCs), face disproportionately high exposure to natural hazards despite contributing the least to climate change and global environmental degradation. The 2025 global risk assessment underscores that nations with the lowest adaptive capacities are often located in the highest-risk zones, facing recurrent tropical cyclones, sea-level rise, and water stress (UNDRR, 2025). Moreover, as emphasized in recent research, humanitarian frameworks continue to operate under colonial legacies that restrict the agency of these nations in global governance processes, thereby perpetuating their vulnerability (Mena, 2025).

Within countries, urban–rural divides further exacerbate inequities in risk governance. Disaster preparedness and recovery capacities are often concentrated in major cities, leaving rural and peri-urban communities with limited infrastructure, fewer early warning systems, and weaker institutional presence. A national-scale study of China revealed significant gaps in vulnerability mapping between urban and rural regions, highlighting how resource allocation and public service provision systematically favor urban cores (Li et al., 2025). These geographic disparities directly translate into increased disaster mortality and slower recovery in under-resourced areas.

Social stratification compounds spatial inequality. Gender, poverty, and social marginalization intersect to shape vulnerability in highly differentiated ways. Across many low-income agrarian communities, disaster risk is entangled with historical patterns of disenfranchisement. In Zimbabwe, for instance, women in rural farming regions experience overlapping vulnerabilities stemming from limited access to land, credit, and early warning systems—conditions that severely constrain their ability to respond to droughts or floods (Manatsa et al., 2025). Likewise, in disaster-prone regions of the Global South, systemic exclusion of Indigenous, disabled, and migrant populations continues to limit the reach and legitimacy of state-led disaster preparedness efforts (Eriksen et al., 2025).

This layered vulnerability is often mirrored in the distribution of institutional capacity. While national governments may set policy, the actual burden of response falls disproportionately on local actors, many of whom are underfunded and poorly coordinated. Municipal authorities in parts of Southern Europe, for example, have struggled to implement civil protection protocols due to limited fiscal autonomy and fragmented vertical coordination with central government agencies (Papadopoulou et al., 2025). Similarly, in Pacific island states, community-based disaster risk reduction programs suffer from inconsistent support and a lack of trained personnel, leaving them unable to implement long-term resilience measures (Fraser, 2023).

Addressing these disparities requires more than technocratic solutions. Risk governance must confront its own embedded asymmetries in capacity, voice, and accountability. Overreliance on centralized command-and-control models neglects the differentiated needs of communities and undermines horizontal coordination across sectors. Central governments alone cannot manage the scale and diversity of contemporary risks. Instead, collaborative governance frameworks that enable local participation, redistribute resources more equitably, and leverage social capital are increasingly recognized as essential to building resilience.

To this end, a growing body of literature advocates for integrated approaches that combine top-down support with bottom-up empowerment. Enhancing adaptive capacity in underserved areas—rural, coastal, marginalized—depends not only on financial investment but also on shifting governance practices toward greater inclusion, transparency, and local agency. Institutional pluralism, resource-sharing mechanisms, and participatory planning processes are not ancillary tools; they are fundamental to overcoming the entrenched geographies of risk exposure.

In short, disaster risk is not merely a product of natural hazard intensity but of who is exposed, where, and under what conditions. Without directly addressing the uneven distribution of capacity and vulnerability, efforts at resilience will continue to fall short—exacerbating, rather than mitigating, the inequalities that disasters so starkly reveal.

1.2.3 The Growing Limitations of Traditional Disaster Governance Models

As disaster risks grow increasingly complex, the prevailing governance models—centered on reactive, hierarchical, and sectoral approaches—have revealed significant limitations. The traditional frameworks, often designed for single-hazard events and linear response cycles, are ill-equipped to manage the evolving realities of cascading, compound, and systemic risks. These limitations manifest in three interrelated dimensions: response lag, vertical fragmentation, and declining public trust.

One of the most pressing issues is the reactive orientation embedded in most governance systems. Disaster risk management (DRM) frameworks still predominantly emphasize post-disaster response and recovery, rather than anticipatory planning and systemic resilience. Such designs are misaligned with today's risk landscape, which demands predictive capacity and long-term adaptation strategies. Research on governance failures in South and Southeast Asia has shown how institutional inertia and rigid bureaucratic routines limit early action and adaptive learning, especially in response to multi-hazard crises like pandemics overlapping with climate shocks (Mitra & Shaw, 2023). Forensic disaster investigations further reveal how root causes of disaster impacts—such as poor planning and knowledge silos—are repeatedly overlooked, reinforcing a cycle of reactive governance (McDermott et al., 2022).

Another structural challenge lies in the coordination failures between different levels of government. National agencies typically dominate strategic decision-making, while sub-national governments are tasked with operationalizing plans—often without adequate authority, funding, or capacity. This disconnect impairs timely and context-sensitive responses. For instance, evidence from Latin America highlights how municipal governments operate within limited legal mandates, lacking the flexibility to adapt global frameworks like the Sendai Framework to local needs (Alcántara-Ayala et al., 2025). In the Nordic context, regional preparedness systems remain fragmented, as local actors are neither sufficiently integrated into national planning nor empowered to act autonomously in crisis situations (Steinsund et al., 2025). These vertical coordination gaps not only slow down implementation but also create ambiguities in accountability.

Equally concerning is the erosion of public trust in formal disaster governance structures. Centralized decision-making, when decoupled from community realities, often results in misaligned priorities, ineffective communication, and low legitimacy. In urban resilience policies, the dominant focus on infrastructural robustness has sidelined citizen knowledge and neighborhood-level dynamics. This "top-down technocratic logic" reinforces a narrative of control while underestimating the potential of grassroots adaptation (Meriläinen, 2020). Similarly, competency-building frameworks that fail to consider social capital, local histories, and culturally grounded risk perceptions have proven insufficient in building resilient communities. Community-based actors frequently report feeling excluded from planning processes, which diminishes trust in institutions and reduces willingness to participate in formal preparedness programs (Clark et al., 2025).

These shortcomings point to an urgent need to shift from reactive and hierarchical governance toward more anticipatory, inclusive, and horizontally integrated systems. The increasing complexity and velocity of disasters demand governance models that embrace uncertainty, foster learning across sectors, and bridge the gap between institutional mandates and community agency. A transition to adaptive governance—one that incorporates plural knowledge systems, invests in local capacities, and facilitates co-produced risk assessments—is now widely recognized as essential.

1.2.4 From Reactive Response to Multi-actor Collaboration: The Imperative of Inclusive Risk Governance

The increasing complexity of global disaster risks, marked by their systemic, interconnected, and cross-scalar nature, has exposed the inadequacy of traditional emergency-centric governance models. Effective disaster risk reduction now demands a shift from command-and-control responses to governance approaches grounded in multi-actor collaboration, knowledge co-production, and adaptive, decentralized coordination. Rather than concentrating authority within state structures, current challenges call for inclusive systems that engage actors across government levels, sectors, and communities.

The rationale for such a shift lies in the structural limitations of hierarchical governance systems, which often isolate decision-making from social realities. Collaborative governance theory provides a compelling framework for addressing these shortcomings, particularly through the integration of diverse knowledge systems and the institutionalization of co-decision-making across sectors (Schweizer & Renn, 2019; Tiwari, 2024). This orientation enables more nuanced, anticipatory risk management attuned to local conditions and systemic interdependencies.

Community-science partnerships exemplify the potential of knowledge co-production in disaster governance. For instance, collaborative models that link citizen-generated data with scientific monitoring have improved early warning accuracy and context-sensitive response, especially in regions with limited formal infrastructure (Freschi et al., 2024; Sarker et al., 2020). Such participatory initiatives have helped bridge knowledge gaps while also fostering local ownership of risk governance processes.

Equally critical is the role of cross-sector collaboration, particularly between public institutions and private actors. In several risk governance contexts, public–private partnerships (PPPs) have been deployed to mobilize financial resources, technical expertise, and innovation capacity. Research into urban resilience planning has demonstrated how integrated frameworks that include infrastructure providers, insurance agencies, and municipal governments can enable shared accountability and proactive planning (Lin et al., 2020; Tagliacozzo et al., 2024).

At the urban governance level, horizontal collaboration between municipalities and local communities is emerging as a key mechanism for addressing vulnerabilities. Case studies from European cities reveal that inclusive planning processes, in which neighborhood groups actively participate in resilience strategies, lead to improved risk awareness and more legitimate policy outcomes (Fraser, 2023; Meriläinen, 2020). In particular, joint scenario planning, risk mapping, and budgeting exercises between city officials and residents have shown promise in aligning top-down strategies with bottom-up priorities.

Participatory disaster prevention is also gaining traction in pre-crisis contexts. For example, capacity-building networks that link civil society organizations (CSOs) with disaster management authorities have enhanced local readiness by embedding risk knowledge within community institutions (Iao-Jörgensen, 2024). These approaches emphasize not only procedural inclusion but also the cultivation of local leadership and long-term social capital.

Empirical evidence further underscores the value of municipal–civil society partnerships in formal institutional settings. The example of civil protection mechanisms in Greek cities illustrates how co-managed emergency response structures and citizen-led monitoring initiatives can be incorporated into formal governance frameworks, improving both reactivity and legitimacy (Papadopoulou et al., 2025). Similarly, in China, integrated rural–urban resilience programs have demonstrated the feasibility of bridging administrative divides through co-governance platforms that prioritize context-specific needs and capacities (Li et al., 2025).

Ultimately, the transition toward multi-actor governance requires institutional support, regulatory flexibility, and sustained investment in social infrastructure. Risk governance can no longer rely on ad hoc participation but must instead embed collaborative arrangements into formal systems of planning, response, and accountability. Institutionalizing such practices will be critical to building what has been described as "all-of-society competence"—the collective capacity to anticipate, absorb, and adapt to cascading risks (Clark et al., 2025).

1.3 The Rising Role and Importance of Civil Society Organizations in Disaster Risk Governance

As disaster risks become increasingly complex, transboundary, and frequent, traditional state-centered governance models often struggle to deliver timely, inclusive, and context-sensitive responses. In this shifting global landscape, civil society organizations have risen to prominence not as peripheral actors, but as essential partners embedded across all stages of disaster risk governance. Their ability to operate at the intersection of state systems and local communities enables them to bridge governance gaps, especially in contexts where formal institutions are constrained or absent. This section explores the evolving role and growing importance of CSOs by examining five interrelated dimensions of their contribution. First, CSOs act as intermediaries between governments and communities, facilitating information flow and ensuring local relevance. Second, their organizational agility and adaptive capacity allow for rapid mobilization and effective response during crises. Third, CSOs increasingly serve as central coordinators within multi-actor governance systems, aligning efforts across sectors and scales. Fourth, their long-term embeddedness in communities positions them as vital components of social infrastructure, capable of mobilizing trust, resources, and local agency. Finally, the institutionalization of CSOs into formal governance frameworks reflects a paradigm shift from ad hoc engagement to strategic collaboration. These dimensions underscore the necessity of integrating civil society into resilient, decentralized disaster governance systems.

1.3.1 Bridging the Governance Gap: Social Organizations as Intermediaries Between States and Communities

In the face of increasingly complex disaster risks, the capacity of centralized governance systems to deliver effective and timely responses at the community level remains limited. Fragmented institutional architectures and vertical hierarchies often produce a disconnect between top-down policy frameworks and local realities. Within this governance gap, social organizations—ranging from grassroots non-governmental organizations (NGOs) to community-based associations—have emerged as critical intermediaries that bridge the state–society divide. Their embeddedness in local contexts, coupled with operational flexibility, allows them to translate policy into practice, facilitate community engagement, and ensure bidirectional information flows.

A central function of social organizations in disaster governance lies in their ability to enhance downward accountability and localized communication. During the Brumadinho mining disaster in Brazil, institutional paralysis and delayed state communication hindered public access to risk information. In contrast, civil society organizations not only mobilized immediate relief but also served as crucial "information intermediaries," linking affected residents with technical experts and governmental aid (Melo et al., 2024). This capacity to interpret, contextualize, and circulate risk information tailored to local needs underscores the communicative role of civil society in disaster scenarios.

Evidence from rural China further illustrates how community organizations assist in disseminating risk knowledge and promoting preparedness. In areas with limited state presence, these groups often conduct early warning campaigns, organize mock drills, and facilitate knowledge exchange among residents—functions that would otherwise be neglected by formal institutions (Fu, 2019). Rather than merely filling a service gap, such organizations embed resilience within the social fabric, transforming abstract risk policy into meaningful, actionable practices for communities.

In sub-Saharan Africa, particularly in Cameroon, similar patterns have been observed. Local NGOs operate as the primary channels for disseminating disaster warnings in remote villages, where physical and institutional access is constrained. Their longstanding relationships within these communities position them as trusted messengers, enhancing both the reception and credibility of alerts (Aka et al., 2017). This bridging actor role is especially significant in regions where state legitimacy or capacity is contested, allowing social organizations to mediate not only information but also trust.

Conceptually, this intermediary function can be situated within the "connected community approach," which emphasizes the importance of integrating community networks into broader governance architectures. In the Australian context, research on localized disaster preparedness found that when community-based organizations are actively engaged in both planning and response, levels of risk awareness and civic trust increase substantially (Poland et al., 2021). By acting as conduits between citizens and authorities, these organizations enable more inclusive and effective governance mechanisms.

Moreover, the presence of intermediary actors contributes to social learning by facilitating feedback loops that inform adaptive policy. The grassroots transmission of lived experiences and community needs back to decision-makers allows for more responsive and context-sensitive governance. This dynamic, however, requires institutional openness and mechanisms that formally recognize the role of civil society in disaster policy frameworks.

Ultimately, social organizations do not merely supplement state functions—they reshape the contours of disaster governance by embedding participatory practices and fostering co-production of knowledge. Their intermediary position between formal systems and informal networks enhances both the legitimacy and effectiveness of risk governance, making them indispensable partners in managing the systemic and localized dimensions of contemporary disaster risks.

1.3.2 Organizational Agility and Rapid Responsiveness: Demonstrating Resilience in Times of Crisis

In an era characterized by increasingly frequent and unpredictable disasters, the capacity for rapid response and flexible coordination has become central to effective risk governance. Social organizations, particularly those embedded within local communities, have demonstrated distinctive advantages over state bureaucracies in terms of organizational agility, speed of mobilization, and responsiveness to evolving needs. These traits enable them to operate as frontline actors during crises, responding not only faster than many formal institutions but also with a deeper sensitivity to local priorities.

During repeated flooding events in Myanmar between 2015 and 2018, subnational NGOs played a pivotal role in initiating early response efforts before formal government interventions arrived. These organizations leveraged their proximity to affected communities and pre-existing networks to distribute relief materials, assess needs, and coordinate shelter operations with greater efficiency than state agencies constrained by hierarchical protocols and limited access (Aung & Lim, 2021). Their actions underscore the advantage of decentralized response structures during acute emergencies.

Organizational agility is further enhanced by the social capital and trust that many community-based organizations cultivate over time. In the Philippine context, purpose-driven nonprofit networks were found to occupy central positions in emergency response systems, functioning as information hubs and coordination nodes across fragmented actors. These networks—characterized by loose structures and adaptive leadership—enabled flexible resource reallocation and rapid improvisation in volatile environments (Tang & Chen, 2025). Their structure allowed them to balance centralized coordination with decentralized action, reducing communication delays and bureaucratic friction.

Empirical evidence from Bangladesh offers additional support for the comparative speed of NGOs. In flood-prone rural areas, local organizations were able to complete damage assessments and deliver immediate assistance within 24 h—far outpacing government units that often took several days to mobilize due to procedural bottlenecks (Rahman & Tasnim, 2023). This capability stems not only from physical proximity but also from embedded knowledge of community vulnerabilities and capacities, allowing for more targeted and need-based interventions.

Beyond individual cases, the concept of organizational agility has been advanced to explain how civil society actors adapt under uncertainty. A study of post-disaster governance in Indonesia demonstrates that agility is not simply a matter of speed but also of adaptive capacity—responding to changing circumstances with minimal friction and leveraging social networks for quick mobilization. Here, agility is closely tied to the density of horizontal linkages and the degree of autonomy from rigid command structures (Gonzaga & Pratama, 2025).

These findings collectively highlight the structural and relational assets that enable civil society organizations to respond swiftly and adaptively in disaster settings. Their capacity to act ahead of formal state interventions, recalibrate strategies in real time, and maintain continuity of services makes them indispensable in managing complex crises. Moreover, their responsiveness strengthens community trust and resilience, providing not only immediate relief but also reinforcing long-term risk governance through iterative learning and adaptation.

1.3.3 Collaborative Mechanisms: Civil Society Organizations as Central Nodes in Multi-actor Governance Systems

As disaster risks become increasingly systemic and cross-scalar, collaborative governance has emerged as a necessary response to overcome the limitations of state-centered disaster management. In this context, civil society organizations have evolved from peripheral actors into key nodes within multi-level, cross-sectoral governance architectures. Their ability to navigate between local needs and institutional mandates allows them to broker partnerships, synchronize efforts across domains, and mediate competing interests.

Collaborative disaster governance is increasingly structured around hybrid alliances involving public, private, and civil actors. For instance, during flood relief operations in India, a public–private partnership (PPP) model enabled CSOs to work alongside government logistics agencies, significantly improving the efficiency of aid delivery. These arrangements allowed for rapid deployment of financial and material resources while ensuring community feedback mechanisms were maintained throughout the response cycle (Syal et al., 2021).

Beyond operational complementarity, CSOs also serve as coordination platforms in complex governance environments. In Kenya, a multi-level disaster governance framework has positioned civil society actors as crucial intermediaries linking climate adaptation, disaster risk reduction, and land restoration strategies. Rather than operating in isolation, these organizations facilitate horizontal integration across sectors and vertical coordination between national policies and local practices (Smucker et al., 2020). This "nexus-based" approach strengthens institutional coherence and resource alignment in fragmented policy landscapes.

International cooperation further amplifies the collaborative capacities of civil society. In Mexico, tri-sectoral partnerships—comprising the Red Cross, government institutions, and universities—have demonstrated high levels of functional coordination during post-disaster reconstruction phases. These arrangements not only improved the allocation of responsibilities and technical expertise but also institutionalized knowledge co-production, enhancing long-term resilience planning (Alcántara-Ayala et al., 2020).

Transnational collaboration has also empowered advocacy-based NGOs to bridge local and global governance spaces. Following Brazil's Mariana mining disaster, CSOs combined grassroots mobilization with international legal and media advocacy to hold private corporations accountable. This dual-scale approach forced state and private actors to engage with risk governance as a matter of both justice and policy reform (Carlos, 2020).

Public preference further reinforces the legitimacy of CSO-centered models. In Finland, survey-based research indicates a higher degree of public trust and willingness to support humanitarian aid initiatives led by civil society rather than government-alone schemes. Citizens perceived CSO-led collaborations as more transparent, participatory, and responsive to local needs—attributes often lacking in top-down state mechanisms (Jäpölä et al., 2024).

In sum, the governance of disaster risk today increasingly relies on inter-organizational coordination, where CSOs function not only as service providers but as architects of multi-actor cooperation. Their embeddedness in local contexts, coupled with institutional fluency across sectors and scales, allows them to serve as vital connectors in an era where systemic risks demand integrated and adaptive governance approaches.

1.3.4 Organizational Embeddedness and Community Mobilization: Civil Society as Social Infrastructure

In the context of increasingly frequent and complex disasters, the effectiveness of disaster governance depends not only on formal institutions but also on the strength and activation of what scholars have termed "social infrastructure." Civil society organizations, particularly those embedded in local networks, play a pivotal role as both carriers and catalysts of this infrastructure. Their pre-existing relationships, cultural proximity, and everyday service delivery mechanisms enable them to mobilize resources and trust rapidly when crises occur.

Rather than functioning merely as external responders, CSOs are deeply integrated within the social fabric of many communities. This embeddedness allows them to act as connectors between informal networks and formal response systems. For example, in the Philippines, local organizations operating in disaster-prone areas have been identified as both "connectors" and "brokers" within complex, multi-layered networks. Social network analysis revealed that these organizations held central positions in communication and coordination flows, enabling faster response and more inclusive recovery efforts (Trias et al., 2019).

Beyond logistical coordination, CSOs contribute emotional and psychological resilience by fostering solidarity and community belonging. In post-earthquake Nepal, mutual aid networks facilitated by community-based organizations provided not only material support but also psychosocial healing. These organizations became spaces of collective care, particularly for marginalized groups, creating routines and rituals that supported emotional recovery (Maly et al., 2022). Similarly, research in Latin America and the Caribbean has highlighted the role of CSO mobilization networks as essential vehicles for implementing the Sendai Framework at the local level. These networks are not just operational arms but essential social systems that embed disaster governance in community routines (Cabral-Ramírez et al., 2025).

Volunteer-based systems are another critical dimension of civil society's infrastructural function. International NGOs, when effectively localized, have demonstrated strong capacities to leverage volunteer networks for both rapid deployment and long-term engagement. These volunteers often possess local knowledge, cultural fluency, and relational trust that are difficult for state or international actors to replicate. For instance, in humanitarian settings, leadership models rooted in "emotional resonance" and relational engagement have proven more effective in sustaining volunteer commitment and mobilizing grassroots participation (Hermann & Pagé, 2016).

Importantly, the infrastructure provided by civil society is often informal yet remarkably robust. As noted in recent scholarship, "informal social infrastructures" such as neighborhood associations, religious groups, and community-based NGOs act as parallel systems of support during disasters. These structures are particularly vital in low-resource or governance-fragile settings, where state responses may be delayed or insufficient (Russell et al., 2023).

Thus, CSOs serve not only as service providers but as embedded nodes of resilience, capable of activating trust-based networks, channeling community capacities, and sustaining collective agency. Their role in building and maintaining social infrastructure positions them as indispensable actors in both preparedness and long-term recovery strategies, especially as global disaster governance seeks more localized, participatory, and adaptive models.

1.3.5 *Institutionalizing Civil Society and Advancing Strategic Collaboration*

The evolving global risk landscape underscores a critical transition in the role of civil society organizations—from provisional actors supplementing state responses to formal collaborators embedded within disaster governance architectures. As demonstrated across diverse contexts, the growing relevance of CSOs reflects not only their adaptability and proximity to communities but also the emergence of multi-scalar governance systems that rely on decentralized and inclusive structures.

In post-disaster settings such as Sichuan, China, the integration of CSOs into formal recovery frameworks signaled a strategic shift from ad hoc engagement to policy-guided coordination. Research on the region's recovery process shows that local authorities increasingly adopted a strategy of "absorption," wherein nongovernmental organizations were selectively incorporated into the administrative disaster response framework—both to expand reach and to stabilize legitimacy (Kang, 2020; Paltemaa, 2017). This inclusion was not merely operational but also political, shaping how accountability and trust were reconstructed in the wake of crisis.

Beyond emergency response, civil society's engagement in disaster risk reduction (DRR) is now contributing to measurable improvements in policy implementation. Empirical findings based on spatial mapping and project databases indicate that nonprofit actors involved in pre-disaster mitigation efforts have played a key role in enhancing local-level preparedness and risk communication. A recent study analyzing disaster mitigation initiatives in China observed a significant rise in CSO participation, especially in urban risk zoning and public education, thereby increasing the efficiency and granularity of state-led planning mechanisms (Zhao et al., 2025).

In fragile or authoritarian contexts, CSOs often operate within constrained political environments but nonetheless fill crucial governance gaps. Evidence from conflict-affected settings demonstrates that CSOs may serve as functional surrogates for the state, organizing relief and managing services in the absence—or failure—of formal institutions. In such cases, civil society becomes a "parallel governance apparatus," not only responding to disasters but also redefining who governs and how (Aldoughli, 2025). These dynamics challenge conventional binaries of state versus non-state actors and suggest a need to reconceptualize resilience governance as inherently hybrid.

To realize the full potential of CSOs in disaster governance, institutional frameworks must evolve from reactive inclusion to proactive integration. This entails embedding civil society participation in every phase of disaster management—preparedness, response, and recovery—and aligning coordination mechanisms with long-term resilience planning. It also requires building formal channels for knowledge co-production, participatory decision-making, and shared accountability, thereby transitioning from partnership-as-practice to partnership-as-policy.

Strategic collaboration between state and civil society actors is no longer optional; it is a structural imperative for effective disaster governance. As global risks intensify and diversify, institutionalizing civil society's roles—through legal recognition, funding mechanisms, and inter-organizational platforms—will be critical to ensuring inclusive, flexible, and adaptive responses that meet the demands of a rapidly changing world.

1.4 Confronting Complexity: Risks, Inequality, and the Way Forward

The global landscape of disasters is undergoing a profound transformation. Intensifying climate change, coupled with deepening social and geographic inequalities, has turned disasters into systemic and increasingly complex events. From the rising frequency of heatwaves and floods to the emergence of compound crises that strain governance capacities, the nature of risk is no longer linear nor predictable. Vulnerability is unevenly distributed, with regions such as the Global South and small island states disproportionately affected. At the same time, persistent data gaps—especially in marginalized communities—limit the effectiveness of early warning systems and inclusive preparedness. This chapter traces these interconnected dynamics, revealing not only the shifting contours of global risk but also the structural blind spots that exacerbate them. It also begins to highlight the critical importance of localized, socially embedded responses—laying the foundation for understanding the role of civil society in navigating today's uncertain risk terrain.

1.5 About the Book

This book offers a comprehensive examination of the evolving roles of social organizations in disaster response and recovery, with a primary focus on China and comparative global insights. Structured into four parts and spanning 21 chapters, it integrates theoretical frameworks, empirical case studies, and international practices to advance an interdisciplinary understanding of collaborative disaster governance.

The first part lays the theoretical foundation for understanding the transformation of social organizations in disaster contexts. It comprises two chapters that frame the analytical lens used throughout the volume.

This chapter focuses on the shifting global risk landscape, emphasizing the rising frequency, intensity, and systemic complexity of disasters. It maps long-term hazard trends and the expanding influence of climate change, showing how disasters increasingly manifest as cascading, transboundary events. The chapter highlights how structural inequalities and governance deficits deepen vulnerabilities, particularly in the Global South, while also pointing to data blind spots that hinder early warning and risk-informed planning. In this context, the chapter argues for the growing importance of social organizations as frontline actors capable of addressing localized risks, fostering adaptive capacity, and enhancing community resilience beyond the limits of state-centered disaster management.

Chapter 2 focuses on the theoretical foundations underpinning disaster relief collaboration. It integrates three analytical lenses—collaborative governance, network governance, and co-production—to explain how diverse actors engage in structured, institutionalized cooperation under high-stakes, time-sensitive conditions. The chapter discusses how rules, norms, and shared leadership sustain collaboration; how networks facilitate cross-organizational coordination; and how co-production enables the joint creation of public value. By merging these perspectives, it proposes a multi-dimensional framework for understanding disaster governance as a relational, evolving, and institutionally embedded process.

The second part focuses on international frameworks and models of network-based humanitarian coordination, offering comparative insights into how social organizations collaborate across diverse governance systems. Through five chapters, it explores both top-down institutional mechanisms and bottom-up regional initiatives that demonstrate the evolution and professionalization of global disaster response networks.

Chapter 3 focuses on the UN Cluster System, examining how it institutionalizes humanitarian coordination through modular structures and platform governance. It analyzes the transformation from ad hoc collaboration to a structured system with distributed leadership, professionalized roles, and multi-layered accountability mechanisms. The chapter further explores how the system adapts to shifting humanitarian contexts by transitioning from global oversight to localized ownership and responsiveness.

Chapter 4 focuses on the Sphere Project, detailing how it evolved from a technical handbook into a norm-generating framework for ethical and professional standards in humanitarian response. The chapter unpacks how Sphere's global standards are developed, negotiated, and adapted across local contexts. It emphasizes how civil society actors do not merely follow these guidelines but actively reinterpret and shape them within varied humanitarian settings, highlighting the pluralistic and co-governed nature of standard diffusion.

Chapter 5 focuses on ADRRN (Asian Disaster Reduction and Response Network), showcasing how regional networks in Asia enable localized capacity-building, peer exchange, and bottom-up norm creation. It highlights ADRRN's unique structure based on thematic hubs and its role in promoting context-sensitive, trust-based collaboration among Asian NGOs. The chapter illustrates how regional solidarity and embedded innovation empower southern-led governance models in humanitarian response.

Chapter 6 focuses on the U.S. VOAD (Voluntary Organizations Active in Disaster) system, analyzing how volunteer networks have evolved into institutionalized actors embedded in national and local disaster governance. It emphasizes the operationalization of the "4Cs" principle—Communication, Coordination, Collaboration, and Cooperation—and how these have become procedural mechanisms. The chapter also explores how VOADs achieve legitimacy through professionalization and how structured volunteerism promotes equity and long-term recovery.

Chapter 7 focuses on Japan's VOAD, which represents a rare case of formal civil society leadership in disaster response. The chapter traces JVOAD's institutionalization following repeated coordination failures and analyzes its legally grounded, decentralized structure. It discusses how JVOAD facilitates cross-sectoral cooperation during crises without central command authority, and how peacetime preparedness—via training systems, regional hubs, and financial support—enables operational resilience.

The third part shifts the focus from global models to the empirical evolution of social organizations in disaster governance within China. This section, comprising eight chapters, traces how non-state actors have increasingly asserted relevance across varying disaster events and institutional contexts—moving from marginal, improvised participation to embedded, recognized roles in emergency governance. Through case studies of major disasters from Wenchuan (2008) to Zhengzhou (2021), it captures the evolving modalities, constraints, and breakthroughs in the governance landscape.

Chapter 8 focuses on the historical and institutional trajectories that have shaped Chinese social organizations' capacities in disaster relief. It traces the shift from imperial communal practices to the sidelining of civic actors in the socialist era, and finally to post-reform resurgence. This foundational analysis foregrounds how capacity-building efforts—technical, organizational, and collaborative—have emerged under asymmetrical governance structures, establishing foundational pathways for understanding later case-specific transformations.

Chapter 9 focuses on the 2008 Wenchuan Earthquake, which marked a historic inflection point for disaster governance in China. The chapter chronicles how grassroots NGOs and citizen volunteers mobilized spontaneously, then negotiated operational space within a state-dominated system. It details how informal initiatives evolved into more structured collaborations, resulting in the nascent yet formalizing institutional inclusion of civic actors in disaster response. This transformation redefined the landscape for future engagement by social organizations.

Chapter 10 focuses on the Yushu Earthquake, exploring the complexities of disaster governance in ethnically diverse, high-altitude, and peripheral regions. It reveals how the governance vacuum enabled rapid civic engagement from local monasteries and volunteer networks. The chapter highlights adaptive strategies such as cultural embedding and local trust-building, while also identifying governance dilemmas and persistent exclusions that challenge replicability in more centralized contexts.

Chapter 11 focuses on the Lushan Earthquake, a turning point that illustrated maturing government–society collaboration. The case underscores incremental shifts: improved government receptiveness, enhanced NGO professionalism, and early forms of coordination mechanisms. It also points to continuing inequalities in service access and the fragility of informal arrangements. This chapter illustrates both the possibilities and the institutional constraints of collaborative governance under pressure.

Chapter 12 focuses on the Ludian Earthquake, where digital coordination and spontaneous networks of volunteers significantly reshaped the disaster response. It demonstrates how emergent grassroots actors initiated action ahead of formal mechanisms, fostering new forms of hybrid governance. The case illustrates how civil society gained partial institutional recognition, laying the foundation for procedural and normative change in China's disaster architecture.

Chapter 13 focuses on the Jiuzhaigou Earthquake, which tested the implementation of institutional learning from prior disasters. The chapter highlights how technological upgrades, legal reforms, and multi-level collaboration improved coordination capacity, while simultaneously exposing ongoing systemic constraints. It presents Jiuzhaigou as a governance stress test—demonstrating how adaptive, polycentric governance is emerging but still limited in reach and resilience.

Chapter 14 focuses on the Changning Earthquake, situated within China's fourth-generation emergency management reform. It showcases how newly established institutional actors like the Ministry of Emergency Management and the Sichuan Coordination Center facilitated real-time coordination through digital platforms and formalized civil society integration. The chapter also explores how narrative governance and participatory media contributed to legitimacy-building, emphasizing the role of public trust in hybrid governance models.

Chapter 15 focuses on the Zhengzhou 7.20 extreme rainfall disaster, which revealed both infrastructural fragility and the transformative power of digital mobilization. The chapter maps the bottom-up emergence of volunteer networks, analyzing how civil society entered governance through functionality rather than formal inclusion. It also unpacks the institutional frictions—legal, spatial, and informational—that constrained inclusive governance, underscoring the complexity of civic-state dynamics in compound urban crises.

The fourth part focuses on the evolving transnational engagement of Chinese social organizations in cross-border humanitarian response, analyzing their strategic adaptation to diverse political, cultural, and operational landscapes. Comprising five chapters, this section captures their learning curve across multiple disaster sites—ranging from Nepal, Ecuador, and Indonesia to Türkiye and Myanmar—highlighting patterns of improvisation, adaptation, institutional integration, and normative alignment.

Chapter 16 focuses on the 2015 Nepal Earthquake as a foundational moment for the overseas engagement of Chinese social organizations. It examines how actors mobilized under uncertainty, operated with limited institutional experience, and engaged in grassroots trust-building amidst cultural unfamiliarity. The chapter highlights how these early efforts catalyzed organizational learning and identity transformation, laying the groundwork for a shift from domestic relief providers to emerging transnational humanitarian actors.

Chapter 17 focuses on the 2016 Ecuador Earthquake, exploring how Chinese social organizations collaborated with both local Chinese communities and international partners within a multi-layered disaster governance environment. By leveraging diasporic networks as intermediary infrastructures, they embedded themselves in health, shelter, and relief operations. This chapter illustrates a shift from spontaneous emergency actions toward more structured—yet still evolving—forms of overseas engagement grounded in practical experience.

Chapter 18 focuses on the 2018 Sulawesi earthquake and tsunami in Indonesia, where Chinese civil society organizations expanded their roles in a sovereignty-sensitive context. It discusses how they accessed disaster zones through diaspora mediation and relational diplomacy, adapted to operational challenges, and began shaping regionally anchored strategies. The chapter emphasizes the difficulties of institutionalizing these efforts but also points to a growing strategic orientation toward Southeast Asia.

Chapter 19 focuses on the 2023 Türkiye–Syria Earthquake as a milestone for Chinese social organizations' integration into international humanitarian architectures. It analyzes the transition from fragmented mobilization to structured collaboration through platforms, the localization of global humanitarian norms, and the internal capacity-building sparked by crisis engagement. This case illustrates the deepening of professional practices and reflexive reform, positioning Chinese actors as credible, albeit still maturing, contributors to transnational disaster response.

Chapter 20 focuses on the 2025 Myanmar Earthquake, a complex and politically volatile crisis that exposed the institutional and operational limits of conventional humanitarian systems. Chinese social organizations mounted rapid, digitally enabled responses, but faced deep frictions including legal ambiguity and burnout. The chapter concludes by analyzing their pivot toward long-term strategies of governance embedding, resilience programming, and regional partnerships—signaling a pivotal shift from reactive, short-term interventions toward more sustained, structurally embedded humanitarian engagement.

Building on the integrated theoretical, comparative, and empirical analyses developed across the four parts of the book, the final chapter draws together the overarching themes and insights presented across the volume. Chapter 21 focuses on synthesizing the book's major insights across its four thematic parts—ranging from theoretical foundations and global networks to China's institutional evolution and overseas practices of social organizations. It highlights how disaster governance has progressively shifted from centralized, state-led models toward more pluralistic, adaptive, and digitally mediated systems. Drawing on comparative case studies, the chapter emphasizes the rising interdependence between civil society actors, government institutions, and technological infrastructures in building disaster resilience. It proposes an integrated framework grounded in collaborative governance, networked coordination, and co-production, offering both analytical clarity and practical relevance. The chapter concludes by identifying future directions for research and policy—focusing on deepened institutional embedding, transnational knowledge transfer, inclusive governance mechanisms, and digital innovation as critical levers for building resilient, equitable disaster response systems worldwide.

References

Aka, F. T., Buh, G. W., Fantong, W. Y., Issa, Zouh, I. T., Djomou, S. L. B., Ghogomu, R. T., Gibson, T., Marmol Del, M.-A., Sigha, L. N., Ohba, T., Kusakabe, M., Yoshida, Y., Tanyileke, G., Nnange, J. M., & Hell, J. V. (2017). Disaster prevention, disaster preparedness and local community resilience within the context of disaster risk management in Cameroon. *Natural Hazards, 86*(1), 57–88. https://doi.org/10.1007/s11069-016-2674-5

Alcántara-Ayala, I., Rodríguez-Velázquez, D., Garnica-Peña, R. J., & Maldonado-Martínez, A. (2020). Multi-sectoral reflections and efforts in strengthening partnerships to reduce disaster risk in Mexico: The first MuSe-IDRiM conference. *International Journal of Disaster Risk Science, 11*(5), 686–691. https://doi.org/10.1007/s13753-020-00302-w

Alcántara-Ayala, I., Velásquez-Espinoza, G., & De Jesús, A. M. (2025). From mandates to mechanisms: Institutional vulnerability, decentralized governance, and the challenges of local disaster risk reduction implementation. *International Journal of Disaster Risk Science, 16*(5), 709–723. https://doi.org/10.1007/s13753-025-00673-y

Aldoughli, R. (2025). Authoritarianism and the emergence of parallel state dynamics: Evidence from the Syrian earthquake. *Digest of Middle East Studies, 34*(1), e12347. https://doi.org/10.1111/dome.12347

Aung, T. M., & Lim, S. (2021). Evolution of collaborative governance in the 2015, 2016, and 2018 Myanmar flood disaster responses: A longitudinal approach to a network analysis. *International Journal of Disaster Risk Science, 12*(2), 267–280. https://doi.org/10.1007/s13753-021-00332-y

Bacciu, V., Sirca, C., & Spano, D. (2022). Towards a systemic approach to fire risk management. *Environmental Science & Policy, 129*, 37–44. https://doi.org/10.1016/j.envsci.2021.12.015

Cabral-Ramírez, M., Niño-Barrero, Y., & DiBella, J. (2025). Lessons from the implementation of the Sendai Framework for disaster risk reduction from Latin America and the Caribbean. *International Journal of Disaster Risk Science, 16*(1), 72–83. https://doi.org/10.1007/s13753-025-00613-w

Carlos, E. (2020). Civil society and social mobilizations in the context of the Rio Doce socioenvironmental disaster. *Integrated Environmental Assessment and Management, 16*(5), 681–690. https://doi.org/10.1002/ieam.4282

Clark, N., Boersma, K., Raju, E., Opromolla, A., Orru, K., Hansson, S., Russo, R., Gargiulo, M. V., Duca, G., Capuano, P., Schweizer, P.-J., Cumiskey, L., Steinhausen, M., Branlat, M., Olson, A., Andersen, N. B., Larruina, R., Atun, F., Van Westen, C., et al. (2025). Strengthening all-of-society approaches for disaster resilient societies through competency building: A European research agenda. *International Journal of Disaster Risk Reduction, 121*, 105345. https://doi.org/10.1016/j.ijdrr.2025.105345

Eriksen, C., Kirschner, J., Simon, G. L., O'Grady, N., Uyttewaal, K., Lüthi, S., Prior, T., Zeffiri, F., Emmenegger, R., Ay, D., Chmutina, K., Raju, E., & Grove, K. (2025). From rigidity traps towards reparative disaster governance and management. *International Journal of Disaster Risk Reduction, 125*, 105603. https://doi.org/10.1016/j.ijdrr.2025.105603

Fraser, A. (2023). Reflections from PEARL for an era of systemic risks, uncertain futures and complex governance. *Jàmbá Journal of Disaster Risk Studies, 15*(1). https://doi.org/10.4102/jamba.v15i1.1485

Freschi, G., Menegatto, M., & Zamperini, A. (2024). Conceptualising the link between citizen science and climate governance: A systematic review. *Climate, 12*(5), 60. https://doi.org/10.3390/cli12050060

Fu, J. S. (2019). Unpacking the influence of informational, organizational, and structural factors on the longitudinal change of the NPO follower-followee network on Twitter. *International Journal of Communication.*

Gonzaga, G. L., & Pratama, A. B. (2025). Social capital and prospects for collaborative governance: An ethnographic case study of Sama-Bajaus' post-disaster experiences. *Public Administration and Development, 45*(3), 201–212. https://doi.org/10.1002/pad.2096

Grasham, C. F., Korzenevica, M., & Charles, K. J. (2019). On considering climate resilience in urban water security: A review of the vulnerability of the urban poor in sub-Saharan Africa. *WIREs Water, 6*(3), e1344. https://doi.org/10.1002/wat2.1344

Hermann, M. G., & Pagé, C. (2016). Leadership and behavior in humanitarian and development transnational non-governmental organizations. *Politics and Governance, 4*(2), 127–137. https://doi.org/10.17645/pag.v4i2.569

Higuera Roa, O., Bachmann, M., Mechler, R., Šakić Trogrlić, R., Reimann, L., Mazzoleni, M., Aerts, J. C. J. H., Buskop, F. E., Pirani, A., & Mysiak, J. (2025). Challenges and opportunities in climate risk assessment: Future directions for assessing complex climate risks. *Environmental Research Letters, 20*(5), 053003. https://doi.org/10.1088/1748-9326/adc756

Iao-Jörgensen, J. (2024). Networking in action: Taking collaborative capacity development seriously for disaster risk management. *Progress in Disaster Science, 21*, 100311. https://doi.org/10.1016/j.pdisas.2024.100311

Jäpölä, J.-P., Van Schoubroeck, S., & Van Passel, S. (2024). Preferences on funding humanitarian aid and disaster management under climatic losses and damages: A multinational Delphi panel. *Climatic Change, 177*(7), 113. https://doi.org/10.1007/s10584-024-03741-2

Kang, Y. (2020). Dispersed domination through patron-clientelism: The evolution of the local state–NGO relationship in post-disaster Sichuan. *Journal of Contemporary China, 29*(124), 598–613. https://doi.org/10.1080/10670564.2019.1677367

Li, Y., Cheng, W., Zhang, Y., Stanny, M., & Du, G. (2025). Mapping urban-rural divide of disaster vulnerability and implications for community resilience building in China. *Habitat International, 165*, 103580. https://doi.org/10.1016/j.habitatint.2025.103580

Lin, X., Zhang, H., Wu, H., & Cui, D. (2020). Mapping the knowledge development and frontier areas of public risk governance research. *International Journal of Disaster Risk Reduction, 43*, 101365. https://doi.org/10.1016/j.ijdrr.2019.101365

Liu, H., & Renn, O. (2025). Polycrisis and systemic risk: Assessment, governance, and communication. *International Journal of Disaster Risk Science, 16*(4), 526–549. https://doi.org/10.1007/s13753-025-00636-3

Maly, E., Vahanvati, M., & Sararit, T. (2022). People-centered disaster recovery: A comparison of long-term outcomes of housing reconstruction in Thailand, India, and Japan. *International Journal of Disaster Risk Reduction, 81*, 103234. https://doi.org/10.1016/j.ijdrr.2022.103234

Manatsa, D., Bowora, J., Mukwenyi, P., Pedzisai, E., & Gomo, M. (2025). Intersectional vulnerabilities in disaster resilience: Gender, geography, and poverty in Zimbabwe's agrarian communities. *International Journal of Disaster Risk Reduction, 129*, 105771. https://doi.org/10.1016/j.ijdrr.2025.105771

Mark, S., Holder, S., Hoyer, D., Schoonover, R., & Aldrich, D. P. (2024). Understanding polycrisis: Definitions, applications, and responses. *SSRN Electronic Journal*. https://doi.org/10.2139/ssrn.4593383

McDermott, R., Fraser, A., Ensor, J., & Seddighi, H. (2022). The role of forensic investigation in systemic risk enquiry: Reflections from case studies of disasters in Istanbul, Kathmandu, Nairobi, and Quito. *Progress in Disaster Science, 16*, 100262. https://doi.org/10.1016/j.pdisas.2022.100262

Melo, T. L. D., Guimarães, L. D. O., & Philippe Cortese, T. T. P. (2024). Disaster governance and the role of civil society: A study of the mining disaster in Brumadinho, Brazil. *International Journal of Disaster Risk Reduction, 114*, 104945. https://doi.org/10.1016/j.ijdrr.2024.104945

Mena, R. (2025). Humanitarianism and the Sendai Framework: A 10-year review of converging and diverging paths. *International Journal of Disaster Risk Science, 16*(1), 20–32. https://doi.org/10.1007/s13753-024-00595-1

Meriläinen, E. (2020). The dual discourse of urban resilience: Robust city and self-organised neighbourhoods. *Disasters, 44*(1), 125–151. https://doi.org/10.1111/disa.12367

Mitra, A., & Shaw, R. (2023). Systemic risk from a disaster management perspective: A review of current research. *Environmental Science & Policy, 140*, 122–133. https://doi.org/10.1016/j.envsci.2022.11.022

Paltemaa, L. (2017). China's changing disaster management regime. *China Information, 31*(3), 284–303. https://doi.org/10.1177/0920203X17726611

Papadopoulou, C.-I., Kalogiannidis, S., Kalfas, D., Konteos, G., & Kapageridis, I. (2025). Civil protection in Greece's cities and regions: Multi-hazard performance, systemic gaps, and a roadmap to integrated urban resilience. *Urban Science, 9*(9), 362. https://doi.org/10.3390/urbansci9090362

Parviainen, J., Hochrainer-Stigler, S., Cumiskey, L., Bharwani, S., Schweizer, P.-J., Hofbauer, B., & Cubie, D. (2025). The risk-tandem framework: An iterative framework for combining risk governance and knowledge co-production toward integrated disaster risk management and climate change adaptation. *International Journal of Disaster Risk Reduction, 116*, 105070. https://doi.org/10.1016/j.ijdrr.2024.105070

Poland, B., Gloger, A., Morgan, G. T., Lach, N., Jackson, S. F., Urban, R., & Rolston, I. (2021). A connected community approach: Citizens and formal institutions working together to build community-centred resilience. *International Journal of Environmental Research and Public Health, 18*(19), 10175. https://doi.org/10.3390/ijerph181910175

Rahman, S., & Tasnim, F. (2023). The role of NGOs in ensuring local governance in Bangladesh: From the perception of other actors of governance. *Asia-Pacific Journal of Regional Science, 7*(3), 1007–1034. https://doi.org/10.1007/s41685-023-00283-w

Russell, C., Gyawali, D., Linnerooth-Bayer, J., & Thompson, M. (2023). Disaster risk reduction reconsidered. *International Journal of Disaster Risk Reduction, 96*, 103895. https://doi.org/10.1016/j.ijdrr.2023.103895

Sarker, M. N. I., Peng, Y., Yiran, C., & Shouse, R. C. (2020). Disaster resilience through big data: Way to environmental sustainability. *International Journal of Disaster Risk Reduction, 51*, 101769. https://doi.org/10.1016/j.ijdrr.2020.101769

Schweizer, P.-J., & Renn, O. (2019). Governance of systemic risks for disaster prevention and mitigation. *Disaster Prevention and Management: An International Journal, 28*(6), 862–874. https://doi.org/10.1108/DPM-09-2019-0282

Senevirathne, M. (2024). *Enhancing systemic risk management to strengthen community resilience: Key recommendations from state-of-the-art practices of past complex incidents.*

Seneviratne, S. I., Zhang, X., Adnan, M., Badi, W., Dereczynski, C., Di Luca, A., Ghosh, S., Iskandar, I., Kossin, J., Lewis, S., Otto, F., Pinto, I., Satoh, M., Vicente-Serrano, S. M., Wehner, M., & Zhou, B. (2021). Weather and climate extreme events in a changing climate. In V. Masson-Delmotte, P. Zhai, A. Pirani, S. L. Connors, C. Péan, S. Berger, N. Caud, Y. Chen, L. Goldfarb, M. I. Gomis, M. Huang, K. Leitzell, E. Lonnoy, J. B. R. Matthews, T. K. Maycock, T. Waterfield, Ö. Yelekçi, R. Yu, & B. Zhou (Eds.), *Climate change 2021: The physical science basis. Contribution of working group I to the sixth assessment report of the Intergovernmental Panel on Climate Change* (pp. 1513–1766). Cambridge University Press. https://doi.org/10.1017/9781009157896.001

Smucker, T. A., Oulu, M., & Nijbroek, R. (2020). Foundations for convergence: Sub-national collaboration at the nexus of climate change adaptation, disaster risk reduction, and land restoration under multi-level governance in Kenya. *International Journal of Disaster Risk Reduction, 51*, 101834. https://doi.org/10.1016/j.ijdrr.2020.101834

Steinsund, S., Kelman, I., Pescaroli, G., & Eid, J. (2025). Crisis preparedness and systemic risk: The role of municipal leaders in disaster risk governance in Norway. *Scandinavian Journal of Psychology*. https://doi.org/10.1111/sjop.70015

Syal, R., Van Wessel, M., & Sahoo, S. (2021). Collaboration, co-optation or navigation? The role of civil society in disaster governance in India. *VOLUNTAS: International Journal of Voluntary and Nonprofit Organizations, 32*(4), 795–808. https://doi.org/10.1007/s11266-021-00344-8

Tagliacozzo, S., Ayeb-Karlsson, S., & Ahmed, B. (2024). An integrated governance framework to map out and act on the interrelationships between human mobility and disaster risk. *Journal of Ethnic and Migration Studies, 50*(5), 1277–1298. https://doi.org/10.1080/1369183X.2023.2279756

Tang, G., & Chen, B. (2025). Unpacking the dynamics of nonprofit purpose-oriented networks in disaster relief: A tale of two cases. *Public Administration and Development*. https://doi.org/10.1002/pad.70044

Tiwari, A. (2024). Collaborative governance and integrated risk management framework of natural disasters. *Journal of Asia Business Studies, 18*(6), 1668–1686. https://doi.org/10.1108/JABS-10-2022-0342

Trias, A. P. L., Lassa, J., & Surjan, A. (2019). Connecting the actors, discovering the ties: Exploring disaster risk governance network in Asia and the Pacific. *International Journal of Disaster Risk Reduction, 33*, 217–228. https://doi.org/10.1016/j.ijdrr.2018.10.007

UNDRR. (2020). *The human cost of disasters: An overview of the last 20 years (2000–2019).* UNDRR. https://www.undrr.org/publication/human-cost-disasters-overview-last-20-years-2000-2019

UNDRR. (2023). *Closing climate and disaster data gaps—New challenges, new thinking.* United Nations Office for Disaster Risk Reduction.

UNDRR. (2024). *GAR special report 2024 forensic insights for future resilience: Learning from past disasters.* https://www.undrr.org/gar/gar2024-special-report

UNDRR. (2025). *2024 global natural disaster assessment report.* https://www.preventionweb.net/publication/documents-and-publications/2024-global-natural-disaster-assessment-report

Zhao, X., Wang, H., Morikawa, S., Wang, S., Tang, J., Sugiyama, S., & Sriwarnasinghe, S. M. (2025). Mapping non-profits participation in disaster mitigation: A data-driven study of functional diversity, spatial patterns and driving factors of China's emergency management nonprofit organizations. *International Journal of Disaster Risk Reduction, 118*, 105252. https://doi.org/10.1016/j.ijdrr.2025.105252

Chapter 2
Theoretical Foundations of Disaster Relief Collaboration: Governance, Networks, and Co-production

Abstract Disaster relief collaboration unfolds at the intersection of governance arrangements, relational structures, and everyday practices of collective action. Addressing complex, time-sensitive, and uncertain disaster situations requires more than isolated organizational capacity or improvised coordination; it demands governance frameworks capable of mobilizing diverse actors, structuring interaction, and enabling joint problem-solving across institutional boundaries. This chapter develops the theoretical foundations for analyzing disaster relief collaboration by bringing together three complementary perspectives: collaborative governance, network governance, and co-production. Collaborative governance highlights how cooperation is institutionally embedded and sustained through rules, leadership, and interactive processes. Network governance foregrounds the relational architectures through which actors, resources, and power are connected under pressure. Co-production shifts attention to how public value and response capacity are jointly created through multi-actor interaction. These perspectives provide an integrated analytical lens for understanding disaster relief collaboration as a dynamic, relational, and institutionally grounded mode of governance rather than a series of ad hoc responses.

Keywords Collaborative governance · Network governance · Institutional embedding · Public value co-creation · Institutional embedding

2.1 Collaborative Governance Perspective: Institutional Embedding and Collaboration Frameworks

Disaster relief collaboration unfolds in contexts of uncertainty, interdependence, and fragmented authority, where effective action depends on more than hierarchical control or ad hoc coordination. A collaborative governance perspective offers a way to understand how such cooperation is structured, enabled, and sustained through institutional design and interactive processes. By situating disaster relief within broader

Q. Du and R. Shaw, *Social Organizations in Disaster Response and Recovery*, Disaster Risk Reduction, https://doi.org/10.1007/978-981-92-2846-1_2

transformations in public governance, this chapter examines why collaborative governance is necessary, how collaboration becomes institutionally embedded, and how it operates through consensus formation, trust-building, learning, and collective leadership. These elements provide an integrated framework for analyzing disaster relief collaboration as an institutionally grounded mode of governance.

2.1.1 Why Is "Collaborative Governance" Needed to Understand Disaster Relief Collaboration?

Disaster relief collaboration cannot be adequately understood through traditional models of hierarchical administration or single-organization management. Disaster situations are typically marked by high uncertainty, rapidly changing conditions, and complex causal relationships, while responsibilities, information, and resources are dispersed across multiple organizations and sectors. These characteristics closely align disaster relief with the notion of "wicked problems," where problems are difficult to define, solutions are contested, and no single actor possesses sufficient authority or knowledge to act independently (Weber & Khademian, 2008). Under such conditions, unilateral decision-making and command-and-control approaches are structurally constrained.

Research in public administration and emergency management demonstrates that disaster response and recovery increasingly depend on interorganizational collaboration. Governments rely on nonprofit organizations, private firms, and community actors to provide specialized services, local knowledge, logistical capacity, and social support that public agencies alone cannot supply (McGuire, 2006). These interdependencies mean that disaster relief outcomes are shaped not only by managerial efficiency, but by how relationships among actors are governed, coordinated, and sustained. Collaboration thus emerges as a necessity rather than a discretionary choice, particularly in situations where time pressure and uncertainty intensify coordination challenges (O'Leary & Vij, 2012).

Within this context, collaborative governance offers a more appropriate theoretical lens for analyzing disaster relief cooperation. Collaborative governance conceptualizes collaboration as a governance arrangement in which public authorities engage non-state actors in collective decision-making processes structured by formal or semi-formal rules and procedures (Ansell & Gash, 2008). This perspective moves beyond viewing collaboration as ad hoc coordination and instead highlights the institutional conditions that enable joint action, such as shared decision arenas, clear role allocation, and mechanisms for managing power asymmetries. Moreover, collaborative governance scholarship emphasizes trust-building, facilitative leadership, and consensus-oriented processes, which are particularly critical in disaster settings characterized by urgency and risk (Sørensen & Torfing, 2015).

2.1.2 *The Core Meaning of Collaborative Governance: From Collaboration to Institutionalized Collaboration*

Collaborative governance provides a conceptual framework for understanding collaboration as a structured mode of governance rather than a series of temporary cooperative acts. In contrast to generic notions of collaboration, collaborative governance refers to governance arrangements in which public authorities and non-governmental actors jointly engage in decision-making processes aimed at addressing public problems through direct interaction, shared problem definition, and consensus-oriented deliberation within formally or semi-formally recognized arenas (Ansell & Gash, 2008). Collaboration in this sense is not peripheral to public action but constitutes a central mechanism through which authority, responsibility, and collective goals are negotiated across organizational boundaries.

This understanding has been further developed by conceptualizing collaborative governance as a regime composed of institutional arrangements, system context, and interactive dynamics that together enable sustained collective action over time (Emerson et al., 2012). Collaboration is therefore embedded in rules, norms, and procedures that structure participation, define decision-making processes, and allocate responsibilities among actors. Such institutional embedding distinguishes collaborative governance from episodic or informal cooperation, which may arise quickly in response to immediate needs but often lacks durability, clarity of accountability, and coordination capacity once situational pressures subside (McGuire, 2006; O'Leary & Vij, 2012).

The institutionalization of collaboration is particularly significant in policy domains characterized by uncertainty and interdependence. By embedding collaboration within governance structures, collaborative governance arrangements organize how knowledge, authority, and responsibility are shared and integrated, enabling collective problem-solving while maintaining accountability (Weber & Khademian, 2008). At the same time, institutional design plays a critical role in managing the inherent tension between control and collaboration, preventing both excessive centralization and unstructured decentralization that can undermine collective action (Sundaramurthy & Lewis, 2003). Collaboration thus functions not merely as a mechanism for resource pooling, but as a governance logic for managing complexity.

The rise of collaborative governance is closely linked to broader transformations in public governance. Within the paradigm of New Public Governance, governing capacity is increasingly distributed across networks of public, private, and civil society actors, while governments assume roles as conveners, facilitators, and metagovernors rather than sole decision-makers (Sørensen & Torfing, 2015). Collaborative governance represents a concrete institutional expression of this shift, translating networked interaction into structured governance arrangements. Importantly, this transformation does not imply a retreat of public authority. Governments remain central in designing institutional frameworks, defining participation boundaries, and safeguarding legitimacy and accountability, while enabling non-state actors to contribute complementary resources and expertise (Meijer & Bolívar, 2016). In this sense, collaborative governance captures the transition from collaboration as temporary coordination to collaboration as an institutionally embedded mode of governance.

2.1.3 Institutional Embedding: The Key Mechanism Enabling Collaborative Governance

The effectiveness of collaborative governance in disaster relief depends fundamentally on institutional embedding. Collaboration among diverse actors does not emerge or sustain itself automatically; rather, it requires governance structures that stabilize expectations, coordinate action, and allocate authority and responsibility across organizational boundaries. Institutional embedding refers to the process through which collaboration is anchored in formal rules, organizational mandates, coordination platforms, and accountability arrangements, transforming cooperation from episodic interaction into a durable governance mechanism.

Within collaborative governance arrangements, governments no longer function as sole decision-makers or command-and-control authorities. Instead, they assume a central institutional role as designers and stewards of collaboration frameworks. Public authorities define participation rules, establish coordination arenas, and create incentive structures that enable non-state actors to contribute resources, expertise, and operational capacity without undermining public accountability (McGuire, 2006). This role is particularly critical in disaster relief, where fragmented authority and time pressure make unstructured collaboration highly vulnerable to coordination failure.

Institutional embedding also addresses the knowledge and coordination challenges inherent in complex and uncertain environments. By formalizing channels for information sharing and joint problem-solving, governments enable the integration of diverse forms of expertise while maintaining collective responsibility for outcomes (Weber & Khademian, 2008). At the same time, institutional design helps manage the persistent tension between control and collaboration, ensuring that collaboration does not devolve into either excessive centralization or uncoordinated decentralization (Sundaramurthy & Lewis, 2003).

Beyond internal governance design, institutional embedding shapes the broader relationship between state and society. Empirical research shows that when governments provide transparent rules, data access, and coordination platforms, collaboration with social organizations, firms, and citizens becomes more effective and outcome-oriented (Wu & Xie, 2024). Similarly, institutional arrangements that clarify role boundaries and incentive mechanisms strengthen the integration of non-state actors into public governance processes without eroding state capacity (Liu et al., 2024). From a New Public Governance perspective, these arrangements position governments as meta-governors who enable collaboration while safeguarding legitimacy and accountability (Sørensen & Torfing, 2015).

2.1.4 *The Operating Mechanisms of Collaborative Governance: How Collaboration Actually Happens*

Institutional embedding alone does not guarantee effective collaboration. Collaborative governance becomes operational through a set of interactive processes that translate institutional arrangements into coordinated action. These processes shape how actors engage with one another, build shared understanding, and adjust behavior over time, thereby determining whether collaboration produces sustained collective outcomes.

A central operating mechanism of collaborative governance is consensus-oriented interaction. Collaboration unfolds through repeated deliberation in which actors jointly define problems, negotiate priorities, and align expectations. Such interaction enables participants to move beyond fragmented perspectives and develop shared frames of reference, which is particularly critical in complex and uncertain environments. Interactive governance frameworks emphasize that collaboration is sustained not by formal authority alone, but by ongoing communication, mutual responsiveness, and iterative negotiation embedded within institutional settings (Emerson et al., 2012). Through these interactions, collaboration becomes a process of joint sense-making rather than mere coordination.

Closely related to consensus formation is trust-building, which functions as both an outcome and a condition of collaboration. Trust develops gradually through transparency, reliability, and repeated interaction, allowing actors to manage risk and uncertainty without excessive reliance on formal control. Participatory and co-productive governance research demonstrates that trust is reinforced when actors perceive decision processes as inclusive, fair, and meaningful, rather than symbolic or instrumental (Jagosh et al., 2015). Experimental and practice-oriented studies further show that trust is strengthened through concrete collaboration experiences that generate learning and visible progress, rather than through abstract commitments alone (Puerari et al., 2018).

Learning and feedback constitute a third core mechanism through which collaboration evolves. Collaborative governance arrangements create spaces for collective learning by enabling actors to reflect on outcomes, revise strategies, and adapt roles over time. Such feedback loops allow collaboration to remain flexible while remaining anchored in institutional frameworks.

These interactive processes are enabled and coordinated through collective leadership rather than individual authority. Leadership in collaborative governance is distributed across actors who facilitate interaction, bridge organizational boundaries, and integrate diverse forms of knowledge. Governance research emphasizes that effective collaboration depends on boundary-spanning roles and shared leadership practices that connect expertise across institutional divides (Weber & Khademian, 2008). Leadership thus emerges as a relational capacity embedded in collaborative processes rather than a positional attribute of any single actor (Ospina, 2017).

2.1.5 *Disaster Relief Collaboration as a Governance Problem: Integrating Collaborative Governance Theory*

Research on disaster relief collaboration has increasingly converged on the recognition that disasters constitute complex governance challenges rather than merely technical or operational failures. Across diverse hazard types and institutional contexts, disaster situations are characterized by high uncertainty, cascading impacts, fragmented authority, and dispersed resources. These conditions align closely with the concept of "wicked problems," in which no single actor possesses sufficient knowledge, capacity, or legitimacy to respond independently, making collaboration a structural necessity rather than a strategic choice (Parker et al., 2020; Tierney, 2012; Weber & Khademian, 2008).

Within this context, collaborative governance theory provides a foundational framework for understanding how disaster relief cooperation is organized, stabilized, and sustained. Rather than treating collaboration as ad hoc coordination activated during emergencies, collaborative governance conceptualizes disaster response as an institutionally embedded arrangement in which public authorities engage non-state actors in structured, consensus-oriented processes under formal or semi-formal rules (Ansell & Gash, 2008; Emerson et al., 2012). This perspective has been widely adopted in disaster governance research to explain why hierarchical command-and-control approaches routinely encounter coordination breakdowns in large-scale crises (McGuire, 2006; O'Leary & Vij, 2012).

A core insight emerging from the disaster-focused literature is that collaboration effectiveness depends fundamentally on institutional embedding. Empirical studies across national and local contexts show that disaster relief networks perform more effectively when collaboration is anchored in legal mandates, coordination platforms, shared information infrastructures, and predefined roles, rather than relying solely on informal relationships or improvisation (Downey & Myers, 2020; Kapucu & Demiroz, 2011; Kapucu & Garayev, 2013). Where such embedding is weak, collaboration tends to remain episodic, uneven, and vulnerable to power asymmetries or political contestation (Hermansson, 2019; Ku et al., 2024).

At the same time, institutional embedding does not eliminate the need for dynamic interaction. A substantial body of research highlights that disaster collaboration operates through ongoing processes of consensus formation, trust-building, learning, and adaptation. These interaction mechanisms enable actors to negotiate problem definitions, integrate diverse forms of expertise, and adjust coordination as conditions evolve (Emerson et al., 2012; Jagosh et al., 2015; Puerari et al., 2018). Studies of post-disaster recovery, emergency logistics, and climate-related disasters consistently demonstrate that collaboration becomes durable only when interactive processes are supported by institutionalized arenas for repeated engagement (Jayasinghe et al., 2020; Quarshie & Leuschner, 2018; Wang et al., 2025).

Another major theme concerns the evolving role of government within collaborative disaster governance. Rather than acting as sole decision-makers, governments increasingly function as meta-governors, responsible for designing coordination frameworks, enabling participation, and safeguarding accountability while mobilizing non-state capacities (Meijer & Bolívar, 2016; Sørensen & Torfing, 2015). Empirical evidence from both centralized and federal systems shows that disaster outcomes improve when governments successfully balance authority with facilitation, particularly in integrating community organizations, nonprofits, private firms, and digitally enabled grassroots initiatives (Aung & Lim, 2021; Calyx, 2020; Tuncalp & Cetindamar, 2026).

Finally, recent scholarship extends collaborative governance analysis beyond immediate response to encompass preparedness, recovery, and resilience building. Disaster collaboration is increasingly understood as a longitudinal governance process shaped by historical relationships, institutional legacies, and structural inequalities (Bello Carvalho et al., 2024; Brondizio et al., 2016; Simon et al., 2025). From this perspective, collaborative governance provides not only a mechanism for coordinating action during crises but also a framework for understanding how disaster governance systems evolve over time.

2.2 Network Governance Theory: Actors, Structures, and Pathways of Collaboration

Disaster relief increasingly unfolds through complex interorganizational arrangements that cannot be adequately captured by hierarchical or market-based governance models. Network governance provides a relational and structural lens for understanding how diverse actors coordinate under conditions of uncertainty, time pressure, and interdependence. This section develops a network governance perspective by first clarifying why disasters expose the limits of traditional governance dichotomies, and then introducing an analytical framework centered on actors, ties, and network structures. It further examines the distinctive forms, role differentiation, and evolutionary dynamics of disaster response networks, laying the conceptual groundwork for analyzing collaboration pathways and governance capacity in crisis settings.

2.2.1 From Hierarchy and Market to Network: The Problem Awareness of Network Governance

Traditional governance theories have long been structured around a dichotomy between hierarchical authority and market-based coordination. Hierarchical governance presumes centralized decision-making, stable authority relations, and command-and-control mechanisms, while market governance relies on contractual exchange, price signals, and competition to coordinate action. However, the governance of disasters consistently reveals the limitations of this dichotomy. Disaster situations are marked by profound uncertainty, rapidly evolving conditions, and strong interdependencies among actors whose resources, information, and operational capacities are distributed across organizational and sectoral boundaries. Under such conditions, neither hierarchical control nor market exchange can independently ensure effective coordination.

Disaster response and recovery require the simultaneous mobilization of heterogeneous capabilities, including public authority, professional expertise, logistical resources, community knowledge, and voluntary labor. These tasks are highly interdependent and temporally compressed, making it impossible for a single organization to exercise comprehensive control. At the same time, market mechanisms are ill-suited to disaster contexts, where urgency, moral imperatives, and public responsibility constrain contractual negotiation and price-based coordination. As a result, governance increasingly occurs through relational arrangements that depend on repeated interaction, mutual adjustment, and negotiated coordination among multiple actors. Within this context, networks have been conceptualized as a distinct governance form—neither market nor hierarchy, but a third mode of coordination grounded in interorganizational relationships (Powell, 1990; Powell & Smith-Doerr, 1994).

Network governance shifts analytical attention from formal authority structures to patterns of interaction among interdependent actors. Coordination is achieved not primarily through commands or contracts, but through stable relationships, shared norms, and reciprocal exchanges that evolve over time. This perspective has significant political and institutional implications. Governance is no longer equated with direct government control but increasingly involves steering through facilitation, negotiation, and indirect influence. Public authority operates within, rather than above, networks of organizations, leading to governance arrangements often described as "governing without government" (Rhodes, 1996).

Disaster governance research demonstrates that these dynamics are particularly pronounced in crisis situations. Emergency response networks combine formal mandates with informal collaboration, linking governmental agencies with nonprofits, private firms, and citizen groups. Authority, resources, and responsibility are dispersed across the network, and coordination outcomes depend on relational positioning rather than hierarchical rank alone (Tierney, 2012). Empirical studies of disaster response networks further show that performance is shaped by how actors are connected, how information flows, and how dependencies are managed under conditions of stress and uncertainty (Bodin & Nohrstedt, 2016; Nowell et al., 2018).

Understanding disaster governance through a network lens therefore provides a problem-oriented rationale for moving beyond hierarchical and market-based explanations. It establishes the theoretical foundation for examining the internal composition of governance networks—specifically the actors involved, the ties that connect them, and the structural configurations through which collective action becomes possible—topics that are addressed in the subsequent sections.

2.2.2 *Core Components of Network Governance: Actors, Ties, and Structure*

Network governance conceptualizes collective action not as the product of isolated organizations, but as the outcome of patterned relations among interdependent actors embedded in a shared structure. To move beyond abstract invocations of "networks," this perspective requires a set of analytical components that make governance dynamics empirically observable and analytically tractable. These components include actors positioned within networks, the ties that connect them, the structural configurations that shape interaction, and the flows of resources and trust that circulate through these arrangements.

At the most basic level, network governance focuses on actors as positionally embedded entities rather than autonomous units. Governments, nonprofit organizations, firms, volunteer groups, and community actors participate in governance not simply by virtue of their formal mandates, but through the positions they occupy within interorganizational networks. Action capacity is therefore shaped by access to others, control over key connections, and proximity to decision-making arenas. From this perspective, organizational behavior cannot be inferred from organizational type alone; similar actors may exercise very different levels of influence depending on their structural location within the network (Bodin et al., 2022; Klijn et al., 1995). Disaster governance research confirms that actors positioned as coordinators, brokers, or hubs often exert disproportionate influence over information flows and resource allocation, regardless of their formal authority (Bodin & Nohrstedt, 2016; Kapucu & Garayev, 2016).

The second analytical component concerns the ties that connect actors. Ties represent the channels through which collaboration is enacted, including information exchange, resource sharing, joint operations, and trust-based relationships. Network governance distinguishes between different types of ties, such as strong and weak connections, as well as formal and informal relations. Strong ties often support intensive coordination and reliability under pressure, while weak ties expand access to diverse information and external resources, particularly in complex and uncertain environments (Powell & Smith-Doerr, 1994). Disaster response studies highlight that effective coordination frequently depends on a combination of both types, as formal command relationships coexist with informal communication and personal trust developed through prior interaction (Krogh & Lo, 2023; Steelman et al., 2014).

Beyond individual actors and dyadic ties, network governance emphasizes the importance of overall network structure. Structural properties such as density, centrality, core–periphery configurations, and brokerage positions shape how power, efficiency, and vulnerability are distributed across the network. Dense networks may facilitate rapid information diffusion but can also reinforce redundancy and groupthink, while sparse networks may enhance flexibility at the cost of coordination. Centralized structures can improve speed and clarity in emergency response but also create bottlenecks and single points of failure (Klijn et al., 2015; Van Der Weert et al., 2022). Empirical analyses of disaster networks demonstrate that highly centralized core–periphery structures often emerge during crises, concentrating coordination authority in a small set of actors while relegating others to peripheral roles (Ku et al., 2024; Sun et al., 2025).

Finally, network governance adopts a dynamic, flow-oriented perspective. Networks are not static maps of relationships but evolving processes through which resources, information, authority, and trust circulate over time. Governance outcomes depend on how these flows are enabled, constrained, or redirected by institutional arrangements and relational practices. From this view, power is exercised not only through formal decisions but through control over information pathways and coordination channels (Rhodes, 1996). Disaster governance scholarship further shows that trust and legitimacy flow unevenly across networks, shaping participation, compliance, and learning during and after crises (Nowell et al., 2018; Tierney, 2012).

2.2.3 *Network Forms in Disaster Relief: Temporality, Pressure, and Unstable Equilibria*

Disaster relief networks differ fundamentally from governance networks operating under routine conditions. Extreme time pressure, high uncertainty, and rapidly evolving task demands generate network forms that are inherently provisional, adaptive, and unstable. Rather than emerging through gradual institutionalization, disaster governance networks are often assembled through rapid recombination of existing organizations and relationships, producing governance arrangements that are effective precisely because they are temporary.

A defining characteristic of disaster relief networks is their high degree of temporality. Empirical studies consistently show that disaster networks are "swiftly assembled" structures that bring together public agencies, nonprofit organizations, private actors, and volunteers in response to urgent needs (Nowell et al., 2018). These networks often rely on pre-existing relationships and latent capacities but are activated and reconfigured in compressed timeframes, resulting in governance structures that differ markedly from routine interorganizational collaboration. Research on disaster responses in China and Southeast Asia demonstrates that such networks frequently emerge through ad hoc coordination mechanisms that evolve in parallel with unfolding events rather than through predefined organizational charts (Aung & Lim, 2021; Tao & Zhang, 2023).

At the same time, disaster relief networks are shaped by the coexistence of cooperation and competition. While collaboration is necessary for effective response, networks are also arenas of strategic behavior. Competition over information control, public visibility, funding, and reputational credit is commonly observed, particularly among organizations operating in overlapping functional domains (Moynihan, 2012). Network analyses of disaster response reveal that some actors strategically hoard information or position themselves as indispensable intermediaries, thereby gaining influence while increasing coordination risks for the network as a whole (Boersma et al., 2021; Ku et al., 2024). These dynamics highlight that disaster networks cannot be understood as purely cooperative systems, but as politically charged structures in which power and accountability are continuously negotiated.

Network structure in disaster relief also exhibits pronounced phase-dependent variation. During the immediate response phase, governance networks tend to become more centralized, with a limited number of core actors coordinating information flows and operational decisions to reduce ambiguity and accelerate action (Bodin & Nohrstedt, 2016). As recovery and reconstruction progress, networks typically decentralize, allowing for broader participation, specialization, and community engagement. Longitudinal analyses of earthquake and flood responses demonstrate that these structural shifts are closely aligned with changing task requirements and resource dependencies over time (Sun et al., 2025; Wu et al., 2021).

Taken together, disaster relief networks can be understood as systems operating in a state of unstable equilibrium. They must balance speed with inclusiveness, coordination with flexibility, and authority with collaboration under conditions of extreme stress. This instability does not represent a failure of governance, but rather a structural feature of disaster response networks. Recognizing these network forms is essential for understanding how different actors come to occupy distinct roles within disaster governance systems, a question addressed in the following section.

2.2.4 Role Differentiation and Power Structures: Who Does What in Disaster Governance Networks?

Disaster relief networks are characterized not only by interdependence but also by pronounced role differentiation and asymmetrical power relations. Network governance research emphasizes that actors occupy functionally distinct positions shaped by their structural location, resource endowments, and institutional mandates, rather than by formal authority alone (Bodin et al., 2022; Klijn et al., 1995). In disaster contexts, these differences become especially salient as urgent coordination needs amplify both cooperation and inequality within networks.

Governments typically occupy central positions in disaster governance networks, functioning as hub actors and institutional anchors. Public authorities often control legal mandates, emergency plans, and access to critical information infrastructures, enabling them to convene actors and stabilize coordination under crisis conditions (Aung & Lim, 2021). However, network centrality does not automatically translate into unilateral control. Empirical studies show that while governments often sit at the structural core, their effectiveness depends on the ability to mobilize non-state capacities and manage horizontal relationships rather than impose hierarchical commands (Kapucu & Garayev, 2016; Zhang et al., 2023). This distinction highlights that centrality reflects coordination responsibility more than dominance.

Non-governmental organizations and civil society actors frequently occupy brokerage positions that connect otherwise fragmented parts of disaster networks.

These actors translate information across organizational, sectoral, and community boundaries, facilitating coordination between local populations, government agencies, and external donors (Johnson et al., 2021). Brokerage roles are particularly critical in cross-regional and multi-level disaster responses, where institutional fragmentation limits direct interaction among formal authorities (Ku et al., 2024). Studies of disaster governance in the Global South further show that NGOs often compensate for state capacity gaps while simultaneously reshaping power relations through agenda-setting and resource mobilization (Melo et al., 2024).

Volunteers and grassroots groups typically operate at the network periphery but play indispensable functional roles. While structurally less central, these actors contribute local knowledge, rapid response capacity, and innovative problem-solving, particularly in the early phases of disaster response (Krogh & Lo, 2023; Schmidt, 2019). Their peripheral position does not imply marginality; rather, it reflects a division of labor in which execution and adaptation are decoupled from formal coordination.

Private firms occupy a distinct position as providers of specialized resources, logistics, and technological solutions. Their influence depends less on network centrality than on the criticality of the resources they control, which can temporarily shift power balances during response and recovery phases (Malone & Kinnear, 2015).

Taken together, disaster governance networks exhibit layered power structures in which authority, influence, and functionality are unevenly distributed. Centrality does not equate to control, and peripheral actors may exercise disproportionate influence through brokerage, expertise, or innovation. Understanding these differentiated roles is essential for explaining how disaster networks function—and malfunction—under pressure, and provides a foundation for analyzing collaboration pathways in subsequent sections.

2.2.5 Collaboration Pathways and Network Evolution: From Emergency Coordination to Governance Capacity

Network governance in disaster relief cannot be fully understood through static descriptions of actors and structures alone. What distinguishes effective disaster networks is their capacity to evolve through interaction, learning, and institutional memory. From this perspective, collaboration is not a one-off response mechanism but a dynamic pathway through which short-term emergency coordination can gradually translate into longer-term governance capacity.

At the operational level, collaboration in disaster networks typically begins with intensive information exchange. In the early response phase, rapid sharing of situational data, needs assessments, and operational updates enables actors to align actions under conditions of extreme uncertainty (Nowell et al., 2018; Steelman et al., 2014). These initial exchanges are often pragmatic and task-oriented, driven by urgency rather than by established trust. However, repeated interaction around information sharing creates the conditions for relational stabilization, as actors develop expectations about reliability, responsiveness, and competence.

Over time, sustained information exchange supports the accumulation of trust, which functions as a critical coordination resource in network governance. Trust reduces transaction costs, mitigates fears of opportunism, and allows actors to rely less on formal control mechanisms when coordinating complex tasks (Fan et al., 2022; Krogh & Lo, 2023). Empirical studies of interagency disaster networks show that trust is reinforced through joint problem-solving and visible performance outcomes, rather than through formal agreements alone (Quarshie & Leuschner, 2020). As trust deepens, coordination shifts from parallel action toward more integrated forms of collaboration, including joint planning and resource pooling.

This progression from information exchange to trust-based coordination underpins the evolutionary nature of disaster governance networks. Network governance capacity emerges through learning and feedback processes, as actors reflect on past responses, adjust roles, and modify coordination practices (Steelman et al., 2014). Such learning enhances network resilience by improving adaptability to future shocks and reducing reliance on improvisation alone (Sun et al., 2025). Importantly, learning is unevenly distributed across networks, often concentrating among core and brokerage actors who participate most frequently in coordination arenas.

At the same time, network evolution is shaped by strong path dependencies. Historical relationships, prior collaboration experiences, and institutional legacies influence how networks form, whom actors trust, and which coordination routines are reactivated during subsequent disasters (Hilhorst et al., 2020; Lin et al., 2020). Networks that have previously invested in coordination platforms and cross-sector ties are more likely to transition from ad hoc collaboration to sustained governance arrangements, whereas fragmented networks often revert to short-lived coordination cycles (Kapucu & Garayev, 2016).

In this sense, disaster networks evolve not simply through structural reconfiguration but through cumulative interaction processes that convert emergency collaboration into enduring governance capacity. Understanding these collaboration pathways is essential for explaining why some disaster networks become more resilient over time, while others repeatedly struggle with coordination failures.

2.3 Co-production Theory: Multi-actor Co-creation of Public Value

In disaster governance, public value is rarely produced through unilateral state action. Uncertainty, time pressure, fragmented authority, and socially embedded risk conditions expose the limits of traditional public service delivery models and foreground the role of non-state actors in shaping governance outcomes. Co-production theory offers a conceptual framework for understanding how disaster response, recovery, and resilience are jointly produced through sustained interaction among governments, communities, civil society organizations, and other stakeholders. By emphasizing multi-actor involvement, interactive processes, and public value creation, co-production reframes disaster governance as a relational and processual form of collective action.

2.3.1 From "Public Service Delivery" to "Public Value Co-creation": The Problem Awareness of Co-production

Conventional public management has long relied on a supply-oriented understanding of public service provision, in which governments are assumed to design, deliver, and control services, while citizens occupy the role of recipients. This model presumes that public value can be generated primarily through administrative capacity, professional expertise, and organizational efficiency. Such assumptions become increasingly untenable in the face of complex public problems, where outcomes depend on dispersed knowledge, contextual interpretation, and the behavior of service users themselves. The limits of unilateral provision are therefore not merely technical but structural, reflecting a fundamental mismatch between governance models and problem characteristics.

Disaster governance exposes these limits in particularly stark ways. Disaster response and recovery are shaped by uncertainty, urgency, and fragmented authority, while formal institutions often lack timely information, local access, and implementation capacity. Empirical research on disaster contexts shows that official interventions routinely depend on community action, volunteer labor, informal coordination, and local adaptation to be effective (Fraser, 2017; Hilhorst et al., 2020). Under such conditions, public authorities cannot simply "deliver" safety, relief, or recovery as finished outputs. Instead, outcomes emerge through interaction with affected populations whose choices, knowledge, and practices directly shape response trajectories.

Co-production theory develops from this problem awareness by rejecting the assumption that public services—and the public value associated with them—can be produced without user involvement. Early contributions demonstrated that many public services are inherently dependent on user inputs, making citizens indispensable participants in production rather than optional add-ons (Parks et al., 1981). This insight was later institutionalized through the argument that co-production is not an ad hoc practice but a structural feature of public service systems, embedded in the rules, routines, and capacities that govern collective action (Ostrom, 1996). In this view, governance outcomes are jointly produced through sustained interaction between public organizations and non-state actors.

Subsequent scholarship refined this perspective by examining why citizens engage in co-production and under what conditions such engagement is sustained. Participation is shaped by a combination of material incentives, normative commitments, perceived efficacy, and trust in public institutions, rather than by altruism alone (Alford, 2002). These insights paved the way for a broader conceptual shift from participation as consultation toward co-production as public value co-creation. Rather than focusing narrowly on service delivery, later work emphasized interactive processes through which value is jointly created by governments, communities, and other stakeholders over time (Bovaird, 2007; Bovaird & Loeffler, 2012).

Within disaster governance, this shift has profound implications. Research on community-led response, grassroots innovation, and knowledge co-production demonstrates that public value in disasters is often generated through collective sense-making, mutual support, and adaptive practices that cannot be orchestrated solely through hierarchical control (De Sisto et al., 2024; Tuncalp & Cetindamar, 2026). Co-production thus provides a distinct analytical lens for understanding disaster response as a process of multi-actor value creation, establishing the theoretical foundation for examining interaction mechanisms, role configurations, and learning dynamics in subsequent sections.

2.3.2 The Core Meaning of Co-production: Multi-actor Involvement, Interactive Processes, and Public Value

Co-production refers to a mode of public action in which public value is generated through sustained interaction among multiple actors rather than through unilateral provision by public authorities. Its analytical strength lies not merely in acknowledging participation, but in specifying how diverse actors, interactive processes, and value orientations jointly shape governance outcomes. Clarifying these core dimensions is essential to avoid reducing co-production to volunteerism or symbolic engagement.

A first defining feature of co-production is multi-actor involvement. Public value creation involves the coordinated contribution of governments, communities, nonprofit organizations, volunteers, private firms, and experts, each mobilizing distinct resources such as authority, local knowledge, labor, networks, expertise, and legitimacy. Rather than treating the state as the sole producer and non-state actors as auxiliary supporters, co-production conceptualizes public services as inherently relational and interdependent. Research on the role of the third sector highlights how civil society organizations act not only as service deliverers but as co-designers and co-implementers of public action, particularly where public capacity is fragmented or uneven (Brandsen & Pestoff, 2006). Subsequent work further demonstrates that co-production frequently operates across organizational, sectoral, and governance levels, producing hybrid arrangements that blur boundaries between public and private responsibility (Brandsen et al., 2018). In disaster contexts, such multi-actor configurations are especially prominent, as no single actor possesses sufficient information or capacity to manage response and recovery independently.

Second, co-production is fundamentally interactive and processual. It does not occur as a single event or discrete intervention but unfolds through ongoing communication, negotiation, contestation, learning, and adaptation. This process orientation distinguishes co-production from episodic participation mechanisms, such as consultations or one-off engagements. Empirical research emphasizes that co-production should be understood as a dynamic process shaped by institutional conditions, power asymmetries, and practical constraints, rather than as a guaranteed pathway to positive outcomes (Verschuere et al., 2012). Interaction is often uneven, conflictual, and contingent, requiring continuous adjustment by participating actors. Public service research further shows that co-production processes are embedded within broader service systems, where relationships evolve over time and outcomes depend on cumulative interaction rather than isolated contributions (Osborne et al., 2013, 2016).

Third, co-production is oriented toward public value creation rather than narrow efficiency gains. While efficiency may be one outcome, co-production explicitly foregrounds broader values such as equity, resilience, trust, inclusion, and capacity building. Within the public service logic perspective, value is not delivered but co-created through interaction between service systems and users, emphasizing experiential and relational dimensions of public outcomes (Osborne, 2018, 2020). This perspective has been extended through the concept of public service ecosystems, which situates co-production within complex constellations of actors, institutions, and socio-material conditions that jointly shape value creation over time (Osborne et al., 2021, 2022).

2.3.3 *Why Disaster Governance Relies Heavily on Co-production*

Disaster governance constitutes a governance domain in which the limitations of unilateral public action are most acutely exposed. Extreme uncertainty, severe time pressure, fragmented information, and rapidly shifting risk conditions fundamentally constrain the capacity of governments to act alone. Hazards unfold faster than formal procedures can adapt, while needs emerge unevenly across space and social groups. Under such conditions, disaster governance cannot rely solely on centralized authority or standardized administrative routines, but instead depends on the active involvement of multiple societal actors in the production of response, recovery, and resilience.

A defining structural characteristic of disasters is uncertainty combined with urgency. Decision-makers must act with incomplete information, while delayed action can significantly amplify human and material losses. Information relevant to risk exposure, vulnerability, and immediate needs is often highly localized and socially embedded, making it inaccessible to centralized authorities in real time. Disaster research consistently shows that such information asymmetries cannot be resolved through technical systems alone, but require engagement with communities, local organizations, and informal actors who possess situational knowledge derived from everyday experience (Fraser, 2017; Hilhorst et al., 2020). Co-production provides a conceptual framework for understanding how this dispersed knowledge becomes an integral input into governance rather than an external supplement.

In addition, disasters frequently expose moments of partial or temporary state incapacity. Formal emergency systems may be overwhelmed, infrastructure disrupted, and administrative reach uneven, particularly in peripheral or marginalized areas. In these contexts, governance outcomes depend on the capacity of communities and local actors to mobilize resources, organize collective action, and adapt response strategies on the ground. Empirical studies of post-disaster recovery demonstrate that housing reconstruction, livelihood restoration, and risk reduction practices are routinely co-produced through interactions between public agencies and affected populations, rather than delivered through top-down programs alone (Ojeda et al., 2018; Schaer & Hanonou, 2017). These dynamics highlight that co-production is not an optional enhancement but a functional necessity.

Community behavior further reinforces the indispensability of co-production in disaster governance. Risk exposure is directly shaped by everyday practices related to land use, housing construction, evacuation decisions, and mutual aid. Without the active participation of residents, even well-designed policies may fail to reduce vulnerability. Research on wildfire response and flood adaptation illustrates that local initiative, informal coordination, and collective problem-solving often determine whether response efforts succeed or collapse under pressure (Arifeen & Eriksen, 2020; Goldstein, 2008). Co-production captures these processes by recognizing communities as necessary producers of safety and resilience rather than passive beneficiaries.

2.3.4 Co-production Mechanisms in Disaster Contexts: How Response Capacity Is Jointly Produced

In disaster settings, co-production becomes operational through a set of concrete mechanisms that translate multi-actor engagement into collective response capacity. Rather than functioning as an abstract principle, co-production unfolds through recurrent practices that structure how knowledge is generated, resources are mobilized, decisions are made, and learning is accumulated over time. Four interrelated mechanisms are particularly central to understanding how disaster response capacity is jointly produced.

A first mechanism concerns the co-production of information and knowledge. Disaster governance requires the integration of scientific expertise, technical models, and professional assessments with localized experience and everyday practices.

Formal risk models and early-warning systems often lack fine-grained sensitivity to social vulnerability, informal settlements, or adaptive behaviors, while community-based knowledge alone may not capture broader hazard dynamics. Co-production enables these knowledge forms to be combined through iterative interaction, producing more actionable and context-sensitive understandings of risk and response (Aguilar-Barajas et al., 2019; Yumagulova & Vertinsky, 2019). Recent work on disaster resilience further highlights the importance of linking data infrastructures, modeling tools, and governance processes, demonstrating that knowledge becomes effective only when it is jointly interpreted and translated into collective action (Schröter et al., 2024).

A second mechanism operates through resource integration and capacity complementarity. Disaster response systems routinely face gaps in manpower, logistics, and institutional reach, particularly during large-scale or cascading crises. Co-production allows volunteers, community groups, and civil society organizations to complement formal response structures by filling operational gaps and extending coverage. Citizen science initiatives and digitally enabled participation illustrate how non-state actors contribute sustained monitoring, situational awareness, and operational support beyond the initial emergency phase (Calyx, 2020). Similarly, grassroots digital humanitarianism demonstrates how decentralized networks mobilize technical skills, platforms, and social capital to augment official response capacities under extreme conditions (Tuncalp & Cetindamar, 2026).

A third mechanism involves joint decision-making and collective action. In co-produced disaster governance, affected communities and non-state actors are not confined to consultative roles but increasingly participate in shaping priorities, strategies, and implementation pathways. This shift from consultation to shared decision-making inevitably introduces contestation, as different forms of expertise, values, and interests intersect. Research on post-disaster reconstruction and climate adaptation shows that co-production processes are inherently political, requiring negotiation over authority, responsibility, and the legitimacy of knowledge claims (Borie & Fraser, 2023; Visconti, 2023). These dynamics are not pathologies but integral features of co-produced governance.

Finally, learning, feedback, and capacity accumulation form a fourth mechanism through which co-production strengthens disaster response over time. Repeated interaction enables experiential learning, the refinement of practices, and the gradual institutionalization of effective collaboration patterns. Participatory research and evaluation demonstrate that learning emerges through feedback loops linking action, reflection, and adaptation (Jagosh et al., 2012). Such learning processes support long-term resilience by embedding knowledge and coordination capacity within both institutions and communities (Cumiskey et al., 2025). Social learning frameworks further show that co-production often follows dual pathways, combining formal institutional learning with informal community-based knowledge accumulation (Chen & Xu, 2025).

2.3.5 Tensions and Boundaries of Co-production: Not a Universal Remedy

While co-production offers a compelling framework for understanding how disaster response capacity is jointly created, it is neither normatively neutral nor universally beneficial. Treating co-production as an unqualified solution risks obscuring the structural constraints, power asymmetries, and unintended consequences that often shape participatory arrangements in disaster governance. A critical examination of its tensions and boundaries is therefore essential for maintaining analytical rigor and empirical credibility.

A first and fundamental tension concerns power asymmetry. Co-production takes place within pre-existing social, political, and economic hierarchies that shape whose knowledge is recognized, whose voices are heard, and whose priorities ultimately prevail. In disaster contexts marked by inequality, marginalized communities may formally participate in co-production processes while remaining structurally constrained in their ability to influence outcomes. Empirical studies of disaster risk governance demonstrate that co-produced arrangements can reproduce existing patterns of exclusion when unequal access to resources, expertise, and institutional channels is left unaddressed (Ruiz-Cortés & Alcántara-Ayala, 2020). Under such conditions, participation does not automatically translate into empowerment.

A second tension relates to the uneven distribution of participation burdens. Co-production relies heavily on the time, labor, emotional energy, and local knowledge of citizens and community actors. These contributions are often taken for granted, particularly in prolonged disaster recovery processes. Research on post-disaster participation shows that expectations of sustained engagement can disproportionately burden already vulnerable groups, exacerbating fatigue and social stress (Fraser, 2017). The capacity to participate is therefore unevenly distributed, raising questions about fairness and sustainability in co-produced disaster governance.

A third boundary emerges from the risk of instrumentalization and symbolic participation. Co-production may be adopted as a governance label without meaningful shifts in decision-making authority or resource control. In such cases, participatory mechanisms serve primarily to legitimize predetermined policies rather than to reshape them. Critical analyses of disaster recovery and reconstruction processes highlight that participation can be reduced to consultation exercises that mask persistent top-down control (Tuhkanen, 2023). This symbolic use of co-production undermines trust and may ultimately weaken governance legitimacy rather than strengthen it.

These tensions underscore that co-production is best understood as a conditional and context-dependent governance strategy rather than a universally applicable model. Its effectiveness depends on institutional design, attention to power relations, and the capacity to align participation with real influence. Recognizing these boundaries does not diminish the value of co-production; instead, it clarifies the conditions under which it can meaningfully contribute to disaster governance. This critical stance creates conceptual space for integrating co-production with complementary perspectives—such as network governance and collaborative governance—that explicitly address coordination structures, power distribution, and institutional embedding in complex disaster response systems.

2.4 Conclusion

This chapter has established an integrated theoretical foundation for understanding disaster relief collaboration as a governance phenomenon shaped by institutions, networks, and multi-actor interaction. By examining collaborative governance, the chapter has shown how disaster relief cooperation is institutionally embedded through rules, leadership, and interactive processes that enable collective action under uncertainty. The network governance perspective has further illuminated how collaboration unfolds through relational structures, role differentiation, and evolving coordination pathways that shape power, information flows, and capacity over time. Building on these insights, co-production theory has reframed disaster relief not as a unidirectional delivery process but as the joint creation of public value through sustained interaction among governments, communities, and other actors, while also recognizing its tensions and limits. Taken together, these perspectives conceptualize disaster relief collaboration as a dynamic, relational, and context-dependent mode of governance, providing a coherent analytical framework for examining how collaborative capacity is produced, contested, and transformed in disaster settings.

References

Aguilar-Barajas, I., Sisto, N. P., Ramirez, A. I., & Magaña-Rueda, V. (2019). Building urban resilience and knowledge co-production in the face of weather hazards: Flash floods in the Monterrey Metropolitan Area (Mexico). *Environmental Science & Policy, 99*, 37–47. https://doi.org/10.1016/j.envsci.2019.05.021

Alford, J. (2002). Why do public-sector clients coproduce?: Toward a contingency theory. *Administration & Society, 34*(1), 32–56. https://doi.org/10.1177/0095399702034001004

Ansell, C., & Gash, A. (2008). Collaborative governance in theory and practice. *Journal of Public Administration Research and Theory, 18*(4), 543–571. https://doi.org/10.1093/jopart/mum032

Arifeen, A., & Eriksen, S. (2020). The politics of disaster vulnerability: Flooding, post-disaster interventions and water governance in Baltistan, Pakistan. *Environment and Planning E: Nature and Space, 3*(4), 1137–1157. https://doi.org/10.1177/2514848619880899

Aung, T. M., & Lim, S. (2021). Evolution of collaborative governance in the 2015, 2016, and 2018 Myanmar flood disaster responses: A longitudinal approach to a network analysis. *International Journal of Disaster Risk Science, 12*(2), 267–280. https://doi.org/10.1007/s13753-021-00332-y

Bello Carvalho, R., Oliveras Menor, I., Belloni Schmidt, I., Berlinck, C. N., Genes, L., & Dirzo, R. (2024). Brazil on fire: Igniting awareness of the 2024 wildfire crisis. *SSRN*. https://doi.org/10.2139/ssrn.5049664

Bodin, Ö., Guerrero, A. M., Nohrstedt, D., Baird, J., Summers, R., Plummer, R., & Jasny, L. (2022). Choose your collaborators wisely: Addressing interdependent tasks through collaboration in responding to wildfire disasters. *Public Administration Review, 82*(6), 1154–1167. https://doi.org/10.1111/puar.13518

Bodin, Ö., & Nohrstedt, D. (2016). Formation and performance of collaborative disaster management networks: Evidence from a Swedish wildfire response. *Global Environmental Change, 41*, 183–194. https://doi.org/10.1016/j.gloenvcha.2016.10.004

Boersma, K., Ferguson, J., Groenewegen, P., & Wolbers, J. (2021). The dynamics of power in disaster response networks. *Risk, Hazards & Crisis in Public Policy, 12*(4), 418–433. https://doi.org/10.1002/rhc3.12218

Borie, M., & Fraser, A. (2023). The politics of expertise in building back better: Contrasting the co-production of reconstruction post-Irma in the Dutch and French Caribbean. *Geoforum, 145*, 103813. https://doi.org/10.1016/j.geoforum.2023.103813

Bovaird, T. (2007). Beyond engagement and participation: User and community coproduction of public services. *Public Administration Review, 67*(5), 846–860. https://doi.org/10.1111/j.1540-6210.2007.00773.x

Bovaird, T., & Loeffler, E. (2012). From engagement to co-production: The contribution of users and communities to outcomes and public value. *VOLUNTAS: International Journal of Voluntary and Nonprofit Organizations, 23*(4), 1119–1138. https://doi.org/10.1007/s11266-012-9309-6

Brandsen, T., & Pestoff, V. (2006). Co-production, the third sector and the delivery of public services: An introduction. *Public Management Review, 8*(4), 493–501. https://doi.org/10.1080/14719030601022874

Brandsen, T., Steen, T., & Verschuere, B. (2018). *Co-production and co-creation: Engaging citizens in public services* (1st ed.). Routledge. https://doi.org/10.4324/9781315204956

Brondizio, E. S., O'Brien, K., Bai, X., Biermann, F., Steffen, W., Berkhout, F., Cudennec, C., Lemos, M. C., Wolfe, A., Palma-Oliveira, J., & Chen, C.-T.A. (2016). Re-conceptualizing the Anthropocene: A call for collaboration. *Global Environmental Change, 39*, 318–327. https://doi.org/10.1016/j.gloenvcha.2016.02.006

Calyx, C. (2020). Sustaining citizen science beyond an emergency. *Sustainability, 12*(11), 4522. https://doi.org/10.3390/su12114522

Chen, F., & Xu, H. (2025). Social learning for disaster-resilient urban destinations: Dual-path knowledge co-production and bridging organizations. *Tourism Geographies, 27*(6), 1232–1257. https://doi.org/10.1080/14616688.2025.2546536

Cumiskey, L., Parviainen, J., Bharwani, S., Ng, N., Bagli, S., Drews, M., Genillard, C., Hedderich, D., Hochrainer-Stigler, S., Hofbauer, B., Huszti, L., Kropf, C. M., Löhrlein, J., Pou, A. M., Mazzoli, P., Pedersen, J., Rosa, A., Schweizer, P.-J., Steinhausen, M., et al. (2025). Capacity development for locally-led knowledge co-production processes in Real World Labs for managing climate and disaster risk. *International Journal of Disaster Risk Reduction, 125*, 105398. https://doi.org/10.1016/j.ijdrr.2025.105398

De Sisto, M., Shearing, C., Heffernan, T., & Sanderson, D. (2024). Reshaping disaster management: An integrated community-led approach. *Australian Journal of Public Administration.* https://doi.org/10.1111/1467-8500.12668

Downey, D. C., & Myers, W. M. (2020). Federalism, intergovernmental relationships, and emergency response: A comparison of Australia and the United States. *The American Review of Public Administration, 50*(6–7), 526–535. https://doi.org/10.1177/0275074020941696

Emerson, K., Nabatchi, T., & Balogh, S. (2012). An integrative framework for collaborative governance. *Journal of Public Administration Research and Theory, 22*(1), 1–29. https://doi.org/10.1093/jopart/mur011

Fan, B., Li, Z., & Desouza, K. C. (2022). Interagency collaboration within the city emergency management network: A case study of Super Ministry Reform in China. *Disasters, 46*(2), 371–400. https://doi.org/10.1111/disa.12495

Fraser, A. (2017). The missing politics of urban vulnerability: The state and the co-production of climate risk. *Environment and Planning A: Economy and Space, 49*(12), 2835–2852. https://doi.org/10.1177/0308518X17732341

Goldstein, B. E. (2008). Skunkworks in the embers of the cedar fire: Enhancing resilience in the aftermath of disaster. *Human Ecology, 36*(1), 15–28. https://doi.org/10.1007/s10745-007-9133-6

Hermansson, H. (2019). Disaster response in Turkey: Conditions promoting cross-sectoral collaboration and implications for effectiveness. *Administration & Society, 51*(7), 1051–1078. https://doi.org/10.1177/0095399716680058

Hilhorst, D., Boersma, K., & Raju, E. (2020). Research on politics of disaster risk governance: Where are we headed? *Politics and Governance, 8*(4), 214–219. https://doi.org/10.17645/pag.v8i4.3843

Jagosh, J., Bush, P. L., Salsberg, J., Macaulay, A. C., Greenhalgh, T., Wong, G., Cargo, M., Green, L. W., Herbert, C. P., & Pluye, P. (2015). A realist evaluation of community-based participatory research: Partnership synergy, trust building and related ripple effects. *BMC Public Health, 15*(1), 725. https://doi.org/10.1186/s12889-015-1949-1

Jagosh, J., Macaulay, A. C., Pluye, P., Salsberg, J., Bush, P. L., Henderson, J., Sirett, E., Wong, G., Cargo, M., Herbert, C. P., Seifer, S. D., Green, L. W., & Greenhalgh, T. (2012). Uncovering the benefits of participatory research: Implications of a realist review for health research and practice. *The Milbank Quarterly, 90*(2), 311–346. https://doi.org/10.1111/j.1468-0009.2012.00665.x

Jayasinghe, K., Kenney, C. M., Prasanna, R., & Velasquez, J. (2020). Enacting "accountability in collaborative governance": Lessons in emergency management and earthquake recovery from the 2010–2011 Canterbury earthquakes. *Journal of Public Budgeting, Accounting & Financial Management, 32*(3), 439–459. https://doi.org/10.1108/JPBAFM-09-2019-0143

Johnson, L., Swapan, M. S. H., & Ashikuzzaman, M. (2021). Disaster risk governance in Bangladesh: How do NGOs find themselves in the network? *Environment and Urbanization ASIA, 12*(2), 240–256. https://doi.org/10.1177/09754253211040205

Kapucu, N., & Demiroz, F. (2011). Measuring performance for collaborative public management using network analysis methods and tools. *Public Performance & Management Review, 34*(4), 549–579. https://doi.org/10.2753/PMR1530-9576340406

Kapucu, N., & Garayev, V. (2013). Designing, managing, and sustaining functionally collaborative emergency management networks. *The American Review of Public Administration, 43*(3), 312–330. https://doi.org/10.1177/0275074012444719

Kapucu, N., & Garayev, V. (2016). Structure and network performance: Horizontal and vertical networks in emergency management. *Administration & Society, 48*(8), 931–961. https://doi.org/10.1177/0095399714541270

Klijn, E., Koppenjan, J., & Termeer, K. (1995). Managing networks in the public sector: A theoretical study of management strategies in policy networks. *Public Administration, 73*(3), 437–454. https://doi.org/10.1111/j.1467-9299.1995.tb00837.x

Klijn, E.-H., Ysa, T., Sierra, V., Berman, E., Edelenbos, J., & Chen, D. Y. (2015). The influence of network management and complexity on network performance in Taiwan, Spain and the Netherlands. *Public Management Review, 17*(5), 736–764. https://doi.org/10.1080/14719037.2014.957340

Krogh, A. H., & Lo, C. (2023). Managing spontaneous volunteers during emergency response: The role of trust. *International Journal of Disaster Risk Reduction, 97*, 104010. https://doi.org/10.1016/j.ijdrr.2023.104010

Ku, M., Kim, M. S., & Oh, S. S. (2024). Core-periphery structure and power imbalance in disaster management networks: A Bayesian approach to interorganizational response to a public health disaster. *Public Management Review, 26*(2), 488–513. https://doi.org/10.1080/14719037.2022.2103171

Lin, X., Zhang, H., Wu, H., & Cui, D. (2020). Mapping the knowledge development and frontier areas of public risk governance research. *International Journal of Disaster Risk Reduction, 43*, 101365. https://doi.org/10.1016/j.ijdrr.2019.101365

Liu, J., Su, X., Liu, Y., & Shui, W. (2024). A review of research on progress in the theory and practice of eco-product value realization. *Land, 13*(3), 316. https://doi.org/10.3390/land13030316

Malone, E. L., & Kinnear, S. (2015). How and why: Complementary analyses of social network structures and cultural values: Improving flood response networks in Queensland, Australia. *Quality & Quantity, 49*(1), 203–220. https://doi.org/10.1007/s11135-013-9982-6

McGuire, M. (2006). Collaborative public management: Assessing what we know and how we know it. *Public Administration Review, 66*(s1), 33–43. https://doi.org/10.1111/j.1540-6210.2006.00664.x

Meijer, A., & Bolívar, M. P. R. (2016). Governing the smart city: A review of the literature on smart urban governance. *International Review of Administrative Sciences, 82*(2), 392–408. https://doi.org/10.1177/0020852314564308

Melo, T. L. D., Guimarães, L. D. O., & Philippe Cortese, T. T. P. (2024). Disaster governance and the role of civil society: A study of the mining disaster in Brumadinho, Brazil. *International Journal of Disaster Risk Reduction, 114*, 104945. https://doi.org/10.1016/j.ijdrr.2024.104945

Moynihan, D. P. (2012). Extra-network organizational reputation and blame avoidance in networks: The Hurricane Katrina example. *Governance, 25*(4), 567–588. https://doi.org/10.1111/j.1468-0491.2012.01593.x

Nowell, B., Steelman, T., Velez, A.-L.K., & Yang, Z. (2018). The structure of effective governance of disaster response networks: Insights from the field. *The American Review of Public Administration, 48*(7), 699–715. https://doi.org/10.1177/0275074017724225

Ojeda, L., Bacigalupe, G., & Pino, A. (2018). Co-production after an urban forest fire: Post-disaster reconstruction of an informal settlement in Chile. *Environment and Urbanization, 30*(2), 537–556. https://doi.org/10.1177/0956247818790731

O'Leary, R., & Vij, N. (2012). Collaborative public management: Where have we been and where are we going? *The American Review of Public Administration, 42*(5), 507–522. https://doi.org/10.1177/0275074012445780

Osborne, S. P. (2018). From public service-dominant logic to public service logic: Are public service organizations capable of co-production and value co-creation? *Public Management Review, 20*(2), 225–231. https://doi.org/10.1080/14719037.2017.1350461

Osborne, S. P. (2020). *Public service logic: Creating value for public service users, citizens, and society through public service delivery* (1st ed.). Routledge. https://doi.org/10.4324/9781003009153

Osborne, S. P., Nasi, G., & Powell, M. (2021). Beyond co-production: Value creation and public services. *Public Administration, 99*(4), 641–657. https://doi.org/10.1111/padm.12718

Osborne, S. P., Powell, M., Cui, T., & Strokosch, K. (2022). Value creation in the public service ecosystem: An integrative framework. *Public Administration Review, 82*(4), 634–645. https://doi.org/10.1111/puar.13474

Osborne, S. P., Radnor, Z., & Nasi, G. (2013). A new theory for public service management? Toward a (public) service-dominant approach. *The American Review of Public Administration, 43*(2), 135–158. https://doi.org/10.1177/0275074012466935

Osborne, S. P., Radnor, Z., & Strokosch, K. (2016). Co-production and the co-creation of value in public services: A suitable case for treatment? *Public Management Review, 18*(5), 639–653. https://doi.org/10.1080/14719037.2015.1111927

Ospina, S. M. (2017). Collective leadership and context in public administration: Bridging public leadership research and leadership studies. *Public Administration Review, 77*(2), 275–287. https://doi.org/10.1111/puar.12706

Ostrom, E. (1996). Crossing the great divide: Coproduction, synergy, and development. *World Development, 24*(6), 1073–1087. https://doi.org/10.1016/0305-750X(96)00023-X

Parker, C. F., Nohrstedt, D., Baird, J., Hermansson, H., Rubin, O., & Baekkeskov, E. (2020). Collaborative crisis management: A plausibility probe of core assumptions. *Policy and Society, 39*(4), 510–529. https://doi.org/10.1080/14494035.2020.1767337

Parks, R. B., Baker, P. C., Kiser, L., Oakerson, R., Ostrom, E., Ostrom, V., Percy, S. L., Vandivort, M. B., Whitaker, G. P., & Wilson, R. (1981). Consumers as coproducers of public services: Some economic and institutional considerations. *Policy Studies Journal, 9*(7), 1001–1011. https://doi.org/10.1111/j.1541-0072.1981.tb01208.x

Powell, W. (1990). Neither market nor hierarchy: Network forms of organization. *Research in Organizational Behaviour, 12*, 295–336.

Powell, W., & Smith-Doerr, L. (1994). Networks and economic life. In *Networks and economic life* (Vol. 4). https://doi.org/10.17323/1726-3247-2003-3-61-105

Puerari, E., De Koning, J. I. J. C., Von Wirth, T., Karré, P. M., Mulder, I. J., & Loorbach, D. A. (2018). Co-creation dynamics in urban living labs. *Sustainability, 10*(6), 1893. https://doi.org/10.3390/su10061893

Quarshie, A. M., & Leuschner, R. (2018). Cross-sector social interactions and systemic change in disaster response: A qualitative study. *Journal of Business Ethics, 150*(2), 357–384. https://doi.org/10.1007/s10551-018-3860-z

Quarshie, A. M., & Leuschner, R. (2020). Interorganizational interaction in disaster response networks: A government perspective. *Journal of Supply Chain Management, 56*(3), 3–25. https://doi.org/10.1111/jscm.12225

Rhodes, R. A. W. (1996). The new governance: Governing without government. *Political Studies, 44*(4), 652–667. https://doi.org/10.1111/j.1467-9248.1996.tb01747.x

Ruiz-Cortés, N. S., & Alcántara-Ayala, I. (2020). Landslide exposure awareness: A community-based approach towards the engagement of children. *Landslides, 17*(6), 1501–1514. https://doi.org/10.1007/s10346-020-01391-w

Schaer, C., & Hanonou, E. K. (2017). The real governance of disaster risk management in peri-urban Senegal: Delivering flood response services through co-production. *Progress in Development Studies, 17*(1), 38–53. https://doi.org/10.1177/1464993416674301

Schmidt, A. (2019). Tensions and dilemmas in crisis governance: Responding to citizen volunteers. *Administration & Society, 51*(7), 1171–1195. https://doi.org/10.1177/0095399719836734

Schröter, K., Schweizer, P.-J., Gräler, B., Cumiskey, L., Bharwani, S., Parviainen, J., Kropf, C., Hakansson, V. W., Drews, M., Irvine, T., Dondi, C., Apel, H., Löhrlein, J., Hochrainer-Stigler, S., Bagli, S., Huszti, L., Genillard, C., Unguendoli, S., & Steinhausen, M. (2024). Invited perspectives: Fostering interoperability of data, models, communication and governance for disaster resilience through transdisciplinary knowledge co-production. *Risk Assessment, Mitigation and Adaptation Strategies, Socioeconomic and Management Aspects.* https://doi.org/10.5194/nhess-2024-135

Simon, G. L., O'Grady, N., Grove, K., Eriksen, C., Chmutina, K., Emmenegger, R., Raju, E., Ay, D., Lüthi, S., Prior, T., Uyttewaal, K., & Zeffiri, F. (2025). A reparative paradigm for thinking with disasters. *The Geographical Journal*, e70057. https://doi.org/10.1111/geoj.70057

Sørensen, E., & Torfing, J. (2015). Enhancing public innovation through collaboration, leadership and new public governance. In A. Nicholls, J. Simon, & M. Gabriel (Eds.), *New frontiers in social innovation research* (pp. 145–169). Palgrave Macmillan UK. https://doi.org/10.1057/9781137506801_8

Steelman, T. A., Nowell, B., Bayoumi, D., & McCaffrey, S. (2014). Understanding information exchange during disaster response: Methodological insights from infocentric analysis. *Administration & Society, 46*(6), 707–743. https://doi.org/10.1177/0095399712469198

Sun, F., Zhou, J., Hu, S., Zhang, R., & Xing, H. (2025). Dynamically examining emergency response network resilience: A case study of a typical earthquake in China. *Safety Science, 184*, 106766. https://doi.org/10.1016/j.ssci.2024.106766

Sundaramurthy, C., & Lewis, M. (2003). Control and collaboration: Paradoxes of governance. *The Academy of Management Review, 28*(3), 397–415. https://doi.org/10.2307/30040729

Tao, Z., & Zhang, H. (2023). The emergence of complex adaptive response networks in China: A case study of four disasters. *Risk Analysis, 43*(11), 2223–2240. https://doi.org/10.1111/risa.14121

Tierney, K. (2012). Disaster governance: Social, political, and economic dimensions. *Annual Review of Environment and Resources, 37*(1), 341–363. https://doi.org/10.1146/annurev-environ-020911-095618

Tuhkanen, H. (2023). Shifting power through participation in post-disaster recovery: A scoping review. *International Journal of Disaster Risk Reduction, 97*, 104041. https://doi.org/10.1016/j.ijdrr.2023.104041

Tuncalp, D., & Cetindamar, D. (2026). Grassroots digital humanitarianism and regenerative synergy in the 2023 Türkiye earthquake. *Technology in Society, 85*, 103174. https://doi.org/10.1016/j.techsoc.2025.103174

Van Der Weert, G., Burzynska, K., & Knoben, J. (2022). An integrative perspective on interorganizational multilevel healthcare networks: A systematic literature review. *BMC Health Services Research, 22*(1), 923. https://doi.org/10.1186/s12913-022-08314-6

Verschuere, B., Brandsen, T., & Pestoff, V. (2012). Co-production: The state of the art in research and the future agenda. *VOLUNTAS: International Journal of Voluntary and Nonprofit Organizations, 23*(4), 1083–1101. https://doi.org/10.1007/s11266-012-9307-8

Visconti, C. (2023). Co-production of knowledge for climate-resilient design and planning in Naples, Italy. *Habitat International, 135*, 102748. https://doi.org/10.1016/j.habitatint.2023.102748

Wang, D., Liu, G., Wang, H., Huang, J., & Wang, Y. (2025). Characteristics and adaptive structure of urban flood governance network: The 'July 20th' flooding event in Zhengzhou, China. *International Journal of Disaster Risk Reduction, 122*, 105465. https://doi.org/10.1016/j.ijdrr.2025.105465

Weber, E. P., & Khademian, A. M. (2008). Wicked problems, knowledge challenges, and collaborative capacity builders in network settings. *Public Administration Review, 68*(2), 334–349. https://doi.org/10.1111/j.1540-6210.2007.00866.x

Wu, D., & Xie, Y. (2024). Unveiling the impact of public data access on collaborative reduction of pollutants and carbon emissions: Evidence from open government data policy. *Energy Economics, 138*, 107822. https://doi.org/10.1016/j.eneco.2024.107822

Wu, M., Gao, X., Cao, M., Papa, E., & Zhu, X. (2021). The changes of intergovernmental collaboration dynamic in post-disaster destination management: Network analysis. *Journal of Hospitality and Tourism Management, 48*, 32–45. https://doi.org/10.1016/j.jhtm.2021.05.006

Yumagulova, L., & Vertinsky, I. (2019). Moving beyond engineering supremacy: Knowledge systems for urban resilience in Canada's Metro Vancouver region. *Environmental Science & Policy, 100*, 66–73. https://doi.org/10.1016/j.envsci.2019.05.022

Zhang, X., Wang, C., Wang, W., & Zhang, H. (2023). Inter-organizational collaboration after institutional reform in China: A perspective based on the revision of the emergency plan. *International Journal of Disaster Risk Reduction, 98*, 104084. https://doi.org/10.1016/j.ijdrr.2023.104084

Part II
Networks of Social Organizations in Disaster Relief: Global Overview

Chapter 3
UN Cluster System: Network Logic and Platform-Based Humanitarian Coordination

Abstract The UN Cluster System represents a paradigmatic shift in humanitarian governance—from ad hoc coordination to institutionalized, network-based collaboration. Built on principles of modularity, sectoral specialization, and distributed leadership, the system enables multi-actor engagement across volatile and resource-constrained environments. This chapter unpacks the structural and functional foundations of the Cluster System, including its modular architecture, formalized division of labor, and professionalized coordination roles. It further explores the evolution toward dynamic, platform-based operations, emphasizing data infrastructures, activation mechanisms, and community-integrated feedback loops. Accountability, both upward to institutional actors and downward to affected populations, is examined as a critical governance mechanism. Finally, the chapter analyzes the transition from internationally driven coordination to localized ownership, highlighting the system's adaptive capacity and potential for institutional embedding. Collectively, these elements demonstrate how the Cluster System operationalizes humanitarian coordination as a multi-layered platform for ethical, responsive, and sustainable action.

Keywords Modular architecture · Platform governance · Multi-directional accountability · Network governance

3.1 Functional Modularity and Division of Labor: Structural Foundations of the Cluster System

The Cluster System is architected around a modular logic that enables humanitarian response to be functionally specialized, operationally coherent, and ethically accountable. Through sector-specific clusters led by designated agencies, the system institutionalizes division of labor, professional roles, and coordination mandates. This structure allows for scalable activation, technical focus, and inter-agency synchronization in complex emergencies. At its core, the system embeds ethical safeguards such as the "Provider of Last Resort," while relying on professionalized roles like

Q. Du and R. Shaw, *Social Organizations in Disaster Response and Recovery*, Disaster Risk Reduction, https://doi.org/10.1007/978-981-92-2846-1_3

Cluster Coordinators and IM Officers to operationalize strategy. This section analyzes the structural principles that underlie the system's modular coordination framework.

3.1.1 *Functional Modularity of the Cluster System*

The United Nations Cluster System operationalizes humanitarian coordination through a functionally modular architecture. This structure divides responsibilities across sector-specific clusters—such as Protection, Shelter, Camp Coordination and Camp Management (CCCM), and others—each of which is led by designated agencies and guided by standardized terms of reference. This modular division enables focused specialization while maintaining alignment with an overarching coordination framework, thus addressing the operational complexity of humanitarian crises (UNHCR, 2025b).

Each cluster acts as a semi-autonomous coordination node with its own leadership, objectives, and technical standards. For instance, the Protection Cluster is led globally by UNHCR and focuses on ensuring rights-based humanitarian assistance in contexts of displacement, armed conflict, and protection risks (UNHCR, 2025e). Similarly, the CCCM Cluster, co-led by UNHCR and IOM, addresses displacement site management and coordination among service providers (UNHCR, 2025a). These structures reflect what organizational theory identifies as modular systems, in which semi-independent units function within a larger architecture, allowing for both specialization and adaptability (Altay & Pal, 2014).

The system's modularity enhances responsiveness by allowing each cluster to develop technical expertise, methodologies, and context-specific tools. For example, the Shelter Cluster has developed detailed coordination guidelines adapted for conflict and disaster settings, while still interfacing with the CCCM and WASH clusters to ensure comprehensive site-based services (UNHCR, 2024). These parallel yet interlinked units facilitate horizontal coordination, reducing operational overlap and information silos (Clarke & Campbell, 2018).

Furthermore, the modular design allows clusters to be activated selectively based on contextual needs and scale, as outlined in the IASC Scale-Up Activation protocols (UNHCR, 2018). This flexibility is particularly valuable in multi-crisis environments, where only certain sectors may require formal coordination structures. For instance, during complex emergencies, the activation of the Protection and Health Clusters may precede other clusters, depending on severity assessments (UNHCR, 2025d).

However, the system is not without limitations. Field studies from South Sudan and the Democratic Republic of Congo show that while modularity enhances technical focus, it may also create segmentation, leading to coordination gaps at the inter-cluster level (Grigoli et al., 2024). Such findings underscore the importance of integrated platforms—like the Inter-Cluster Coordination Group (ICCG)—to harmonize efforts across clusters and maintain coherence in strategy and resource allocation.

3.1.2 Division of Labor and Sectoral Leadership

A defining feature of the Cluster System is its formalized division of labor through the Sector Lead Agency model. This structure assigns leadership of each functional cluster to a specific UN agency or designated organization, with the expectation of technical expertise, coordination capacity, and strategic vision. Such allocation enables clear responsibility chains while reducing fragmentation in multi-actor environments (UNHCR, 2025b, 2025e).

Each cluster's lead agency is tasked with convening partners, setting standards, managing inter-agency coordination, and ensuring operational delivery. For instance, UNHCR leads the Protection Cluster, focusing on displaced populations' safety and rights (UNHCR, 2025e). The CCCM Cluster is co-led by UNHCR and IOM, addressing site management and coordination across displacement settings (UNHCR, 2025a). UNICEF leads several clusters, including WASH, while WFP often leads Food Security. This division reflects both organizational mandates and technical specialization, contributing to institutional legitimacy and operational clarity.

Sector leadership plays a crucial role in facilitating cross-sector coordination, especially at national and sub-national levels. In decentralized or localized settings, such leadership helps navigate power dynamics and align humanitarian efforts with governmental actors. Evidence from localized cluster implementations in the Philippines illustrates how lead agencies act as brokers between local governance and global coordination mechanisms, enabling contextual adaptation without compromising system integrity (Abaya et al., 2020). Additionally, sector leads often provide the administrative infrastructure for inter-agency needs assessments, funding appeals, and performance monitoring (UNHCR, 2025d).

Despite its benefits, the sectoral division of labor also faces structural and political challenges. Field assessments in South Sudan and the Central African Republic have shown that resource asymmetries between clusters can undermine collaboration and exacerbate coordination gaps, especially when less-resourced clusters lack the convening power of larger ones (Grigoli et al., 2024). In contexts of overlapping mandates—such as between Food Security and Nutrition Clusters—inter-agency tension may emerge unless managed through formalized coordination forums.

Moreover, during food security crises, response quality often depends on the lead agency's ability to orchestrate diverse actors under time pressure, revealing the critical role of leadership in ensuring accountability and coherence (Maxwell & Parker, 2012). Similarly, coordination breakdowns can occur when cluster leads are insufficiently empowered or when leadership is contested at field level (Clarke & Campbell, 2018).

3.1.3 Professionalization and Operational Roles

The institutionalization of the Cluster System has led to the emergence of distinct professional roles that are central to the governance of humanitarian coordination. Among these, the Cluster Coordinator and the Information Management (IM) Officer have become standard positions within cluster-led operations, reflecting a shift toward the professionalization of humanitarian coordination as a specialized function rather than an ad hoc managerial task (UNHCR, 2024).

These roles are formalized within cluster terms of reference, with clearly defined responsibilities, reporting lines, and competency frameworks. For example, the Cluster Coordinator is tasked with facilitating strategic planning, partner engagement, resource mobilization, and monitoring mechanisms. Their effectiveness hinges not only on technical expertise but also on soft skills such as negotiation, consensus-building, and multi-level communication. In parallel, IM Officers are responsible for data collection, needs analysis, response mapping, and information dissemination—a role increasingly critical in ensuring evidence-based coordination and accountability (UNHCR, 2025c).

Empirical findings from health cluster operations in Afghanistan highlight how under-resourced or poorly trained coordination staff can significantly undermine response efficiency. Inadequate IM staffing, in particular, has been associated with data gaps, inconsistent reporting, and reduced service coverage in high-need areas (Mirzazada et al., 2020). Similarly, experience from reproductive health coordination platforms shows that reliable frontline data systems depend on the presence of skilled information officers who understand both technical data tools and field-level realities (Kobeissi et al., 2021).

The consolidation of coordination into a recognized profession also brings new governance expectations. Coordination roles now often require cross-agency experience and familiarity with humanitarian architecture, enabling actors to mediate between UN mandates, donor expectations, and local partners. Conceptual work on coordination dynamics argues that such roles are best understood within self-managed, adaptive organizational models, where responsibility is shared across agencies and leadership is relational rather than hierarchical (Dupont & Skjold, 2022).

At the structural level, the creation and funding of these professional roles signal a deepening institutional embedding of the Cluster System. Their presence supports not only internal functionality but also external legitimacy. Field-level stakeholders, including governments and NGOs, often perceive the visibility and professionalism of cluster leads as proxies for operational credibility and coordination efficacy (Abaya et al., 2020; UNHCR, 2024).

3.1.4 Accountability and Ethical Mandates

Accountability within the Cluster System is not solely a managerial concern but a foundational ethical principle embedded in its institutional architecture. Central to this framework is the "Provider of Last Resort" commitment, which obliges Cluster Lead Agencies to step in and ensure service provision when no other actor is able or willing to do so (UNHCR, 2025b). This mechanism reflects a system-wide commitment to safeguarding humanitarian response, particularly in high-risk or under-resourced settings where coordination failures may directly translate into human suffering.

This responsibility is not symbolic. In practice, lead agencies are expected to mobilize internal capacities or leverage partner networks to fill operational gaps. For example, within the Food Security Cluster, this principle has reinforced the accountability of lead agencies in protracted crises, compelling them to act decisively even when operational mandates are stretched or funding is uncertain (Maxwell & Parker, 2012). The ethical weight of this provision becomes particularly salient in volatile contexts such as Syria, where health cluster actors faced sustained attacks on healthcare facilities. In those cases, the commitment to "last resort" delivery placed moral and operational pressure on the Health Cluster leadership to maintain services under extreme duress (Elamein et al., 2017).

Despite its normative clarity, accountability in the field often encounters structural and political ambiguity. Field observations have shown that multiple clusters operating in the same context may blur lines of responsibility, especially when sectoral mandates overlap or when national authorities assert divergent priorities (Clarke & Campbell, 2018). Moreover, efforts to localize coordination raise new accountability dilemmas, particularly when international lead agencies are expected to devolve authority without compromising performance. Evidence from localized cluster settings in the Philippines suggests that accountability becomes diffused when institutional authority is not equally shared or clearly delineated among national and local actors (Abaya et al., 2020).

Institutional mechanisms such as Humanitarian Programme Cycle monitoring and Inter-Agency Standing Committee (IASC) guidance attempt to operationalize accountability through planning, reporting, and performance benchmarks (UNHCR, 2018). However, these mechanisms depend heavily on the leadership strength and ethical stance of individual coordinators, reinforcing the idea that ethical accountability in humanitarian coordination is both systemic and agent-driven.

Ultimately, the Cluster System's accountability logic rests on a balance between institutional mandate and moral obligation. Ensuring that ethical commitments like the "Provider of Last Resort" are consistently upheld requires not only technical competence and formalized procedures but also normative clarity, resource parity across clusters, and context-sensitive leadership strategies.

3.2 Platform-Based Coordination and Information Mechanisms: The Logic of Dynamic Collaboration

As humanitarian emergencies grow increasingly complex, the Cluster System has evolved from a static coordination structure into a dynamic, platform-based architecture. This transformation enables timely activation, cross-sectoral synchronization, and community-integrated feedback. By deploying modular tools such as the IASC Scale-Up Protocol, MIRA assessments, CCCM data platforms, and CwC loops, the system facilitates flexible coordination among diverse actors. These infrastructures operationalize real-time decision-making, ensure needs-based targeting, and embed accountability mechanisms. This section analyzes how technical platforms, information systems, and feedback circuits underpin the adaptive functionality of the Cluster System in volatile humanitarian environments.

3.2.1 Emergency Activation and Coordination Triggers: From Static Mandates to Dynamic Platforms

The capacity of the Cluster System to respond effectively to large-scale humanitarian emergencies depends heavily on its ability to transition from static coordination mandates to dynamic, context-responsive platforms. The Inter-Agency Standing Committee (IASC) Scale-Up Activation Protocol offers a structured yet flexible mechanism to mobilize inter-agency coordination in situations of system-wide emergency, enabling clusters to be rapidly scaled according to need and context (UNHCR, 2018).

The activation process begins when the scale, complexity, or urgency of a crisis exceeds the capacity of the country-level Humanitarian Coordinator and existing cluster arrangements. Upon this trigger, an IASC Principals meeting is convened to decide on a system-wide Scale-Up Activation. This activation includes not only the deployment of surge personnel and funding instruments but also the reconfiguration of cluster coordination structures to match operational demands. Clusters are not merely activated as discrete units, but are reassembled into an interoperable coordination platform—capable of absorbing new partners, adapting sectoral leadership, and enabling cross-sector responses (UNHCR, 2018).

What distinguishes this mechanism is its platform-based coordination logic. Unlike traditional command-and-control approaches, activation under the Scale-Up Protocol allows clusters to function as modular, interoperable components within a shared governance framework. This form of coordination emphasizes agility, distributed leadership, and functional integration across clusters, which is particularly vital in high-tempo emergencies such as sudden-onset disasters or large-scale displacements (Clarke & Campbell, 2018).

Furthermore, platform governance within the Cluster System reflects characteristics of self-managed organizations, where adaptation emerges through shared standards, digital interfaces, and negotiated roles rather than centralized control. This governance model facilitates multi-actor synchronization and horizontal accountability, making the system more resilient to breakdowns in any single cluster or institutional node (Dupont & Skjold, 2022).

Importantly, the activation mechanism is not limited to initial deployment but also enables dynamic reconfiguration over time, allowing clusters to scale down, transition to national ownership, or integrate new coordination layers as needs evolve. Such flexibility ensures that the Cluster System operates not as a static framework but as a responsive platform architecture, capable of both rapid mobilization and iterative adaptation to complex humanitarian environments (UNHCR, 2018).

3.2.2 *Multi-cluster Needs Assessment and Shared Information Systems*

Effective humanitarian coordination relies not only on organizational alignment but also on the production, circulation, and synchronization of shared information. The Cluster System addresses this through structured assessment mechanisms and interoperable data platforms that support evidence-based decision-making across clusters and agencies.

The Multi-Cluster/Sector Initial Rapid Assessment (MIRA) exemplifies this information-centric approach. Developed as a standardized methodology under the IASC framework, MIRA facilitates joint needs assessments across clusters, typically within the first two weeks of a sudden-onset emergency. It combines secondary data analysis with field-level assessments to produce a consolidated overview of humanitarian needs, priority populations, and recommended response strategies (UNHCR, 2025d). This process enables humanitarian actors to align their interventions around a common situational picture, minimizing duplication and ensuring a coordinated, needs-based response.

Shared information systems are not limited to assessment phases. Tools such as 3W/4W matrices (Who is doing What, Where, and When), Flash Appeals, and Humanitarian Response Plans (HRPs) allow clusters to monitor partner presence, identify gaps, and allocate resources dynamically. These data mechanisms operate as coordination infrastructures, providing platforms for transparency, joint planning, and strategic funding requests. However, their efficacy depends on the timely and accurate input of disaggregated data from field actors, something often constrained by capacity or access limitations (Maxwell & Parker, 2012; UNHCR, 2025d).

The breakdown of such information systems can significantly undermine service delivery. In conflict settings like Afghanistan, limited information flow within the Health Cluster has been directly linked to uneven service coverage and unmet maternal health needs. Data gaps—particularly around local health infrastructure and mobility patterns—hampered cross-cluster coordination and hindered outreach planning (Mirzazada et al., 2020). These findings reinforce the idea that information is a critical public good within humanitarian governance—its absence is as consequential as a resource shortfall.

Despite advances in tool standardization, the system still faces challenges in data harmonization across clusters, especially in complex emergencies involving multiple actors and overlapping mandates. Interoperability and shared definitions remain fragile, and the process of aggregating localized data into global platforms often leads to loss of contextual nuance (Maxwell & Parker, 2012).

3.2.3 Technical Infrastructure in Sector Platforms: The Case of CCCM

Within the Cluster System, the Camp Coordination and Camp Management (CCCM) Cluster offers a concrete illustration of how technical infrastructure underpins daily coordination practices. The CCCM platform not only facilitates service delivery in displacement sites but also functions as a data-driven coordination interface, integrating real-time information tools, standardized reporting protocols, and geospatial visualizations to ensure coherent action among partners (UNHCR, 2025a).

Central to this infrastructure is the deployment of Information Management (IM) Officers, who operate at national and sub-national levels. These roles are tasked with managing tools such as Kobo Toolbox, ActivityInfo, and Google Data Studio dashboards, all of which enable partners to track service coverage, population movement, and site-level gaps with regular frequency. For example, CCCM dashboards typically consolidate data on WASH, shelter, and protection services in over 300 displacement sites, updated bi-weekly to support coordinated decision-making (UNHCR, 2025a). This routine institutionalization of data flows exemplifies how sector platforms transition from reactive mechanisms to embedded systems of governance.

These infrastructures also facilitate visual analytics and GIS mapping, allowing cluster partners to layer population data with access constraints and operational timelines. By combining standardized reporting templates (e.g., weekly 3Ws) with geographic overlays, the CCCM Cluster can guide both site-level interventions and strategic planning at national humanitarian response levels. The daily operationalization of such systems enables feedback loops between field realities and decision-making bodies, anchoring coordination in both data and context.

The significance of such systems extends beyond CCCM. In public health emergencies, similar tools have proven vital in enabling front-line actors to respond in real time to service disruptions or population movements (Kobeissi et al., 2021). Furthermore, simulation-based studies in humanitarian operations show that timely and distributed information diffusion enhances system adaptiveness, particularly in highly complex and uncertain environments (Altay & Pal, 2014).

3.2.4 Communication with Communities: Feedback as a Platform Loop

Within the architecture of platform-based humanitarian coordination, Communication with Communities (CwC) functions as a crucial feedback loop, transforming affected populations from passive recipients into active nodes in the coordination ecosystem. The CwC framework provides a dual interface: it allows communities to input concerns, perceptions, and complaints through mechanisms such as hotlines, mobile surveys, community consultations, and feedback boxes; simultaneously, it supports the output of standardized information from humanitarian actors via community radio, posters, SMS alerts, and direct engagement (UNHCR, 2025c).

CwC mechanisms are increasingly embedded in sectoral and inter-cluster coordination platforms, often through technical working groups and shared accountability dashboards. For instance, in displacement settings managed under the CCCM cluster, real-time community feedback is captured through the Site Monitoring Tool, feeding directly into multi-sectoral planning (UNHCR, 2025a). The strategic use of tools like Rapid Feedback Forms and "4Ws + Accountability" matrices helps streamline community insights into operational decisions, particularly in camp planning, service prioritization, and protection case referrals (UNHCR, 2025c).

More than a communication strategy, CwC acts as an operational interface for institutional accountability. Its embeddedness in humanitarian coordination enables localized decision-making and offers an early warning system for systemic service breakdowns. In conflict settings such as Syria, for example, community feedback mechanisms have become key to identifying service disruptions caused by attacks on health infrastructure, especially where formal monitoring systems were compromised (Elamein et al., 2017). Similarly, in localized cluster coordination models, the legitimacy of sector lead agencies is often tied to their ability to translate feedback into responsive programming, particularly in culturally or politically sensitive contexts (Abaya et al., 2020).

However, challenges persist. The fragmentation of data platforms and lack of standardized indicators across clusters have limited the interoperability of CwC systems, while protection risks in insecure settings can deter open feedback (UNHCR, 2025c). Despite these limitations, CwC continues to evolve as a participatory layer of the platform logic, shaping how humanitarian systems understand, anticipate, and adjust to evolving community needs.

3.3 Accountability, Feedback, and System Performance: Governance Foundations of Coordination Mechanisms

The governance legitimacy of the Cluster System rests on its capacity to institutionalize accountability, evaluate coordination performance, and integrate community feedback into operational decision-making. This section examines how formal accountability structures link Cluster Leads to humanitarian leadership hierarchies, how performance is systematically monitored through tools like CCPM and HNO, and how downward accountability—via AAP and participatory interfaces—anchors coordination in affected populations' realities. These mechanisms form a multi-directional governance framework, balancing strategic oversight with localized responsiveness and reinforcing the ethical foundations of humanitarian coordination.

3.3.1 Formal Accountability Architecture Within the Cluster System

The Cluster System embeds a multi-tiered accountability structure that anchors operational leadership within an overarching humanitarian governance framework. At the core of this structure lies a vertical chain of accountability extending from individual Cluster Lead Agencies to the Humanitarian Coordinator (HC), and ultimately to the Emergency Relief Coordinator (ERC) through the Office for the Coordination of Humanitarian Affairs (OCHA) and the Inter-Agency Standing Committee (IASC) (UNHCR, 2025b, 2025e). This formal architecture aims to clarify decision-making authority, reinforce coordination oversight, and ensure system-wide responsibility for humanitarian outcomes.

Within this configuration, Cluster Leads are not merely coordinators but designated duty-bearers. They are expected to guide strategic direction, maintain minimum standards, ensure partner engagement, and report regularly on performance to HCs and sectoral governance mechanisms. The designation of "Provider of Last Resort" further obliges lead agencies to deliver or guarantee services when other actors are unable or unwilling to act, underscoring their non-delegable responsibilities in critical settings (UNHCR, 2025b). This provision institutionalizes ethical commitment into operational mandate, particularly in sectors such as Protection or Food Security where service gaps can have life-threatening implications.

However, formal accountability mechanisms often encounter tension in field operations. The intersection between cluster mandates, national sovereignty, and operational fluidity frequently produces blurred lines of responsibility. In decentralized contexts, humanitarian actors have reported "accountability vacuums" when cluster leadership at subnational levels is either contested or unsupported by clear delegation from national counterparts (Clarke & Campbell, 2018). Similarly, attempts to localize coordination—such as in the Philippines and Ethiopia—have highlighted the fragility of vertical accountability chains when transnational agencies are expected to share or transfer leadership to local governance systems without robust co-management structures in place (Abaya et al., 2020).

Despite these challenges, the formal accountability architecture remains a central pillar of the Cluster System's legitimacy. The HC, supported by OCHA, plays a key role in enforcing coordination standards, resolving disputes, and ensuring cross-cluster coherence. Tools such as Humanitarian Response Plans (HRPs), inter-agency evaluations, and IASC oversight mechanisms serve to translate institutional expectations into actionable benchmarks (UNHCR, 2018, 2025e). These instruments, when effectively applied, transform hierarchical oversight into a system of distributed governance capable of maintaining both operational effectiveness and ethical responsibility in complex humanitarian settings.

3.3.2 Performance Evaluation and Inter-cluster Benchmarking

Ensuring the functional integrity and adaptive quality of humanitarian coordination requires more than structural design—it demands robust and institutionalized performance evaluation. Within the Cluster System, several tools have been developed to monitor coordination effectiveness, benchmark inter-cluster performance, and link evaluative processes to both internal learning and external accountability. Chief among these instruments is the Cluster Coordination Performance Monitoring (CCPM), which enables clusters to assess their own functionality through standardized indicators, periodic self-assessment, and partner feedback (UNHCR, 2025e).

The CCPM framework encompasses key coordination domains, including information sharing, inclusivity, leadership, and strategic planning. Typically conducted annually or biannually, the CCPM process requires each cluster to gather structured feedback from partners and synthesize findings into actionable improvement plans. This model of evaluative governance—combining internal reflection with peer inputs—has contributed to greater transparency and facilitated cross-context learning within the humanitarian system (UNHCR, 2025e). Implementation remains uneven. In resource-constrained or insecure environments, clusters often struggle to operationalize CCPM cycles due to staff turnover, low partner participation, or limited analytical capacity (Grigoli et al., 2024).

Beyond CCPM, tools such as the Humanitarian Needs Overview (HNO) and Protection Analysis Updates support inter-cluster performance by aligning assessments, response planning, and advocacy. These mechanisms offer a system-wide diagnostic of humanitarian needs and response gaps, enabling clusters to coordinate not just vertically with lead agencies but horizontally across sectors. The integration of inter-sectoral benchmarks in HNO processes has helped to identify duplications, misalignments, and emerging risks, allowing clusters to recalibrate strategic priorities and adjust resource allocations accordingly (UNHCR, 2025d, 2025e).

However, performance monitoring is not merely a technical exercise—it also interfaces with donor logics and external accountability pressures. Evidence from multi-donor pooled fund settings indicates that evaluative tools increasingly serve as instruments for strategic funding decisions. Clusters that demonstrate higher performance on CCPM or contribute effectively to HNOs are often better positioned to access financing, creating both incentives and risks of bureaucratization (Pizzi et al., 2020). This entanglement between performance and resource flow underscores the need for balance: performance systems must remain credible to external actors without compromising their utility as tools for internal learning and adaptive coordination.

3.3.3 Downward Accountability and Accountability to Affected Populations (AAP)

Beyond upward accountability to donors and coordination hierarchies, the Cluster System has increasingly institutionalized Accountability to Affected Populations (AAP) as a governance imperative. AAP reframes the coordination logic by recognizing crisis-affected individuals not merely as beneficiaries but as rights-holding participants whose voices, perceptions, and agency must inform humanitarian decision-making. This shift reflects a normative evolution in humanitarian coordination—from managerial control to participatory legitimacy (UNHCR, 2025c).

The AAP framework is operationalized through a combination of communication, participation, and feedback mechanisms. These include community help desks, complaints and feedback boxes, hotlines, focus group discussions (FGDs), and community liaison staff. Technical tools such as community perception surveys and accountability dashboards are increasingly embedded within clusters—particularly those operating in protection, health, and CCCM contexts—to ensure that population feedback informs both strategy and service delivery (UNHCR, 2025c). For instance, in CCCM-managed displacement sites, community feedback loops have been integrated into site monitoring frameworks, enabling real-time adjustment of camp services and incident response (UNHCR, 2025a).

More than a technical layer, AAP functions as a moral and operational check within volatile settings. In conflict-affected areas like Syria, formal monitoring systems are often disrupted or politically constrained. There, community-based feedback mechanisms have served as surrogate accountability channels—reporting service interruptions, unmet needs, and abuse patterns that would otherwise go undetected (Elamein et al., 2017). Similarly, in public health coordination platforms, transparent communication and participatory assessments have enhanced trust between service providers and affected populations, especially in reproductive and maternal health responses (Kobeissi et al., 2021).

However, the implementation of AAP remains uneven across clusters and contexts. Field assessments show that while the principle of downward accountability is widely accepted, its practical realization is often constrained by security risks, data fragmentation, and capacity limitations. In highly localized coordination environments, accountability mechanisms may be co-opted by local elites or undermined by unclear communication strategies (Abaya et al., 2020).

3.3.4 Community-Based Feedback and Embedded Participation Tools

Within the governance framework of the Cluster System, community-based feedback mechanisms are increasingly institutionalized as integral components of adaptive coordination. These mechanisms extend beyond ad hoc consultation, embedding structured participation tools into the information management (IM) architecture and humanitarian programming cycles. The aim is to translate localized insights into actionable data, ensuring that community voices inform both strategic planning and operational adjustments (UNHCR, 2025a, 2025c).

Key participatory instruments include focus group discussions (FGDs), help desks, complaint and feedback boxes, and real-time digital reporting tools. These are deployed across displacement sites and sector-specific service areas—often embedded within CCCM and Protection Clusters—to gather structured feedback on access, service gaps, and perceived quality. For example, the CCCM Cluster has developed templates for site-level FGDs, which are synchronized with broader multi-cluster needs assessments and performance monitoring dashboards (UNHCR, 2025a). Through this integration, participation becomes not a peripheral activity but a recurring input within governance loops.

The conceptual underpinning of such tools aligns with non-hierarchical governance models, wherein decision-making emerges through iterative negotiation, rather than centralized authority. Feedback is not only transmitted upward but translated horizontally into cross-cluster dialogue, enabling peer-based correction and adaptive coordination (Dupont & Skjold, 2022). As coordination systems evolve toward platform-based models, such embedded tools reinforce the system's responsiveness and legitimacy.

In practice, the presence of community participation interfaces also functions as an early warning system. For instance, anonymous complaints submitted through feedback boxes have alerted camp authorities to protection risks or gender-based violence incidents that formal monitoring failed to detect. Similarly, mobile-based quick polls and SMS surveys have helped identify gaps in shelter or WASH coverage, especially in remote or rapidly evolving contexts (UNHCR, 2025c).

Simulation studies support these real-world findings: coordination systems that include structured community input are more capable of self-correction and performance stabilization during crises. Particularly under uncertainty or resource constraints, these inputs improve targeting accuracy and reduce coordination lags (Altay & Pal, 2014).

However, the effectiveness of these mechanisms depends on their institutional anchoring. Without designated accountability pathways and response protocols, feedback risks becoming symbolic. Embedding participation within both technical platforms and governance cycles is therefore essential for transforming fragmented insights into collective intelligence and adaptive humanitarian response (UNHCR, 2025a, 2025c).

3.4 From International Leadership to Local Embedding

The Cluster System, originally designed as a temporary coordination mechanism, is undergoing a strategic transformation toward localized ownership and institutional sustainability. This transition entails more than operational scaling-down; it involves a reconfiguration of governance, capacity, and legitimacy. This section explores four interlinked dimensions of the transition process: the logic and practice of responsible disengagement, the institutionalization of co-leadership with national actors, the role of capacity building in embedding coordination functions, and the Protection Cluster's unique positioning within the peace–development–humanitarian nexus. These dynamics illustrate how clusters evolve from international instruments into locally anchored governance platforms.

3.4.1 Cluster Exit and "Responsible Disengagement"

The Cluster System was never intended as a permanent governance architecture, but rather as a transitional coordination mechanism activated in response to major humanitarian emergencies. From its inception, the system has included an explicit exit logic, guided by the principle of "responsible disengagement"—a strategic transition in which clusters reduce or relinquish their direct coordination role while ensuring continuity of essential services through national or local actors (UNHCR, 2024, 2025a). This departure is not synonymous with withdrawal, but with a calibrated shift from operational control to technical advisory and capacity-strengthening roles.

Exit planning typically hinges on a triad of conditions: (1) assessed national capacity to assume coordination functions, (2) stable governance environments where risks of humanitarian collapse are minimized, and (3) demonstrable redundancy of cluster structures relative to existing systems. For example, in Shelter and CCCM clusters, field guidance outlines phased disengagement strategies—ranging from co-coordination with government authorities to full handover to national disaster management agencies (UNHCR, 2025a). These transitions are embedded in broader Humanitarian Programme Cycle milestones, including response reviews and inter-agency planning updates (UNHCR, 2025d).

Despite formal frameworks, responsible disengagement in practice faces multiple obstacles. Field experiences reveal that prolonged cluster activation—particularly in protracted crises like South Sudan or the Democratic Republic of Congo—can unintentionally entrench international coordination, marginalize national institutions, and create dependency dynamics (Clarke & Campbell, 2018). Without clear disengagement strategies, the platform risks becoming institutionalized as a parallel system, undermining state legitimacy and duplicating governance functions.

From an organizational design perspective, the Cluster System operates as a modular and platform-based coordination model capable of scaling both up and down depending on context (Dupont & Skjold, 2022; UNHCR, 2018). This architecture supports flexible transitions: clusters may shift from direct operational leadership to support functions such as technical training, information management, or cross-sectoral integration. In CCCM contexts, for instance, post-disengagement engagement often continues through localized capacity-building initiatives and shared monitoring tools (UNHCR, 2025a).

Ultimately, successful disengagement depends not only on internal readiness but also on the ability to embed coordination capacities within national systems. Responsible transition demands more than exit plans—it requires institutional anchoring, political commitment, and sustainable knowledge transfer, ensuring that the departure of cluster mechanisms does not precipitate the collapse of humanitarian functions, but rather signals a shift toward nationally led governance models capable of absorbing coordination responsibilities over time.

3.4.2 Localization and Co-leadership Mechanisms

The transition from internationally led coordination to nationally embedded systems necessitates not only technical handovers but a transformation in governance structures. One of the most salient institutional innovations supporting this shift is the co-leadership model, which integrates national NGOs (NNGOs) or government agencies as equal partners in cluster coordination. This model reconfigures the traditional power asymmetry between UN agencies and local actors, offering a pathway for shared authority, contextual legitimacy, and sustained local ownership (UNHCR, 2025e).

Within the Protection Cluster, for instance, recent policy guidance promotes the structured inclusion of NNGOs as co-leads at subnational levels, aiming to foster equitable decision-making and joint accountability in protection analysis, advocacy, and service delivery (UNHCR, 2025e). Co-leadership arrangements have also emerged in CCCM and Shelter clusters, often linked to capacity-building programs and co-financing mechanisms that support local institutional development (UNHCR, 2024, 2025a).

This shift is not merely procedural—it reshapes the coordination ecosystem. When NNGOs move from consultative roles to formal co-leadership positions, they gain access to operational planning, funding negotiations, and inter-agency strategy setting. However, empirical evidence suggests that the depth of participation varies significantly. In localized responses such as the Philippines, co-leadership has increased visibility and contextual sensitivity but has not always translated into real decision-making power, due to resource imbalances, language barriers, or donor-driven agenda setting (Abaya et al., 2020). Similarly, in contexts like the Central African Republic and South Sudan, national actors report symbolic inclusion without full integration into technical or strategic coordination processes (Grigoli et al., 2024).

The risk of "tokenized co-leadership"—where local actors are formally present but functionally peripheral—raises critical questions about the sustainability and legitimacy of localization efforts. Formal co-lead structures must therefore be accompanied by redistribution of coordination resources, including budget allocations, IM systems access, and leadership training. When these conditions are met, co-leadership can serve as a critical transition mechanism—enabling gradual transfer of governance authority and strengthening national response systems without compromising humanitarian standards (UNHCR, 2025b, 2025d).

3.4.3 Capacity Building as Institutional Embedding

Within the logic of sustainable humanitarian coordination, capacity building is not a secondary activity but a structural prerequisite for responsible disengagement. The Cluster System's transition toward national ownership hinges on the depth and durability of institutional capacities developed across technical, financial, and policy domains (UNHCR, 2024, 2025a).

Effective capacity building enables the progressive replacement of international coordination structures with embedded national mechanisms, transforming the cluster from a service provider to a system enabler.

Technically, capacity development focuses on information management (IM), needs assessments, contingency planning, and response monitoring. In the Shelter and CCCM Clusters, structured capacity-strengthening plans have been integrated into national response frameworks, including joint trainings, standardized toolkits, and national IM dashboards. These instruments are designed to ensure continuity and interoperability between international and local actors, even after cluster deactivation (UNHCR, 2024). However, standardized tools are not always context-sensitive, and tensions persist between global benchmarks and local implementation capacities (Diaz-Sarachaga et al., 2018).

Financially, the ability of local actors to access and manage resources—particularly pooled funds and bilateral grants—represents a key threshold for coordination autonomy. While donors increasingly emphasize localization targets, field-level evidence suggests that many NNGOs lack the administrative infrastructure or audit capabilities to directly manage large-scale humanitarian financing (Pizzi et al., 2020). In response, some clusters have piloted financial mentoring systems, pairing international agencies with local partners to support grant management and compliance processes (UNHCR, 2025a).

At the policy level, capacity building entails enabling national authorities and civil society to lead or co-lead coordination efforts, interface with other ministries, and align cluster priorities with national development strategies. This requires not only technical knowledge but recognition of the political economy of coordination, including inter-ministerial negotiation and crisis governance authority (UNHCR, 2025d). In practice, the absence of clear national mandates often hinders this transition, and capacity-building initiatives risk becoming projectized rather than institutionally embedded (Abaya et al., 2020; Grigoli et al., 2024).

3.4.4 Protection Cluster and the Nexus Logic

Among all clusters within the humanitarian coordination architecture, the Protection Cluster most clearly embodies the shift from emergency response toward governance-oriented institutional embedding. Its strategic positioning aligns directly with the peace–development–humanitarian (PDH) nexus, making it a critical interface for inter-sectoral collaboration and long-term systems resilience (UNHCR, 2025e). While clusters are generally designed for temporary coordination, the Protection Cluster operates with a broader mandate—encompassing legal frameworks, access to justice, community-based protection, and transitional justice—that inherently requires multi-year planning and cross-pillar coherence.

Institutionally, the Protection Cluster is situated to bridge humanitarian action with national governance systems. It does so by aligning emergency response with development programming through legal reform, community empowerment, and integration of protection standards into national policies (UNHCR, 2025c, 2025e). In contexts such as South Sudan and Iraq, cluster activities have supported the development of local protection strategies, influencing national legislation on civil documentation, return processes, and child protection (UNHCR, 2025e). This embedded role contrasts with more technical clusters and underscores the cluster's dual identity—as both emergency coordinator and normative actor.

Operationally, protection actors are increasingly involved in nexus-based coordination platforms that bring together humanitarian, development, and peacebuilding stakeholders. These platforms aim to ensure coherence between immediate protection needs (e.g., SGBV response, IDP registration), structural drivers of risk (e.g., legal exclusion, displacement-linked land disputes), and long-term institutional strengthening (Varela et al., 2025). At the field level, inter-agency mechanisms such as Protection Analysis Updates and Area-Based Coordination initiatives allow for shared situational awareness and strategic alignment between clusters and non-humanitarian actors (UNHCR, 2025d).

However, the shift toward nexus logic introduces governance challenges, particularly regarding mandate clarity, funding alignment, and institutional leadership. Fragmented coordination across humanitarian and development actors has, in several cases, led to overlapping interventions or policy incoherence—especially in fragile governance contexts (Raman et al., 2025). To address this, the Protection Cluster increasingly employs area-based and community-centered approaches that localize coordination and support multi-sectoral integration at subnational levels (Abaya et al., 2020; UNHCR, 2024).

3.5 Conclusion

This chapter has examined the UN Cluster System as a modular, platform-based coordination architecture that bridges emergency response with institutional governance. Structurally, the system's functional modularity, sectoral leadership, and professionalized roles enable operational coherence across diverse humanitarian settings. Technically, dynamic information platforms, assessment tools, and feedback loops enhance responsiveness, transparency, and cross-sectoral alignment. Governance mechanisms such as formal accountability chains, performance benchmarking, and community-based feedback tools ensure ethical legitimacy and adaptive learning. Critically, the system is not static: it is evolving toward localization and institutional embedding through responsible disengagement, co-leadership, and capacity transfer. The Protection Cluster illustrates this transformation most clearly, linking humanitarian response with legal, developmental, and peacebuilding agendas. These dynamics position the Cluster System not merely as a coordination tool, but as a governance platform capable of facilitating ethical, adaptive, and locally anchored humanitarian action across protracted and complex crises.

References

Abaya, M. R. T., Le Dé, L., & Lopez, Y. (2020). Localising the UN cluster approach: The Philippines as a case study. *Environmental Hazards, 19*(4), 360–374. https://doi.org/10.1080/17477891.2019.1677209

Altay, N., & Pal, R. (2014). Information diffusion among agents: Implications for humanitarian operations. *Production and Operations Management, 23*(6), 1015–1027. https://doi.org/10.1111/poms.12102

Clarke, P. K., & Campbell, L. (2018). Coordination in theory, coordination in practice: The case of the clusters. *Disasters, 42*(4), 655–673. https://doi.org/10.1111/disa.12282

Diaz-Sarachaga, J. M., Jato-Espino, D., & Castro-Fresno, D. (2018). Is the sustainable development goals (SDG) index an adequate framework to measure the progress of the 2030 agenda? *Sustainable Development, 26*(6), 663–671. https://doi.org/10.1002/sd.1735

Dupont, C., & Skjold, A. (2022). Coordination conundrum in the United Nations development system: Solutions from self-managed organizations. *Global Perspectives, 3*(1), 57083. https://doi.org/10.1525/gp.2022.57083

Elamein, M., Bower, H., Valderrama, C., Zedan, D., Rihawi, H., Almilaji, K., Abdelhafeez, M., Tabbal, N., Almhawish, N., Maes, S., & AbouZeid, A. (2017). Attacks against health care in Syria, 2015–16: Results from a real-time reporting tool. *The Lancet, 390*(10109), 2278–2286. https://doi.org/10.1016/S0140-6736(17)31328-4

Grigoli, G. D. A., Silva Júnior, M. F. D., & Pedra, D. P. (2024). Challenges and perspectives for humanitarian logistics: A comparative study between the Democratic Republic of Congo, the Central African Republic and the Republic of South Sudan. *Journal of Humanitarian Logistics and Supply Chain Management, 14*(4), 384–398. https://doi.org/10.1108/JHLSCM-07-2022-0087

Kobeissi, L. H., Ashna, M., Messier, K., Moran, A. C., Say, L., Strong, K. L., & Foster, A. (2021). Exploring the feasibility of establishing a core set of sexual, reproductive, maternal, newborn, child and adolescent health indicators in humanitarian settings: A multimethods, multicountry qualitative study protocol. *British Medical Journal Open, 11*(12), e041270. https://doi.org/10.1136/bmjopen-2020-041270

Maxwell, D., & Parker, J. (2012). Coordination in food security crises: A stakeholder analysis of the challenges facing the global food security cluster. *Food Security, 4*(1), 25–40. https://doi.org/10.1007/s12571-012-0166-3

Mirzazada, S., Padhani, Z. A., Jabeen, S., Fatima, M., Rizvi, A., Ansari, U., Das, J. K., & Bhutta, Z. A. (2020). Impact of conflict on maternal and child health service delivery: A country case study of Afghanistan. *Conflict and Health, 14*(1), 38. https://doi.org/10.1186/s13031-020-00285-x

Pizzi, S., Caputo, A., Corvino, A., & Venturelli, A. (2020). Management research and the UN sustainable development goals (SDGs): A bibliometric investigation and systematic review. *Journal of Cleaner Production, 276*, 124033. https://doi.org/10.1016/j.jclepro.2020.124033

Raman, R., Sreenivasan, A., Kulkarni, N. V., Suresh, M., & Nedungadi, P. (2025). Analyzing the contributions of biofuels, biomass, and bioenergy to sustainable development goals. *iScience, 28*(4), 112157. https://doi.org/10.1016/j.isci.2025.112157

UNHCR. (2018). *IASC scale up activation*. UNHCR.

UNHCR. (2024). *Shelter cluster*. UNHCR.

UNHCR. (2025a). *Camp coordination and camp management CCCM cluster IASC*. UNHCR.

UNHCR. (2025b). *Cluster approach*. UNHCR.

UNHCR. (2025c). *Coordination and communication*. UNHCR.

UNHCR. (2025d). *Multi-cluster/sector initial rapid needs assessment (MIRA)*. UNHCR.

UNHCR. (2025e). *Protection cluster*. UNHCR.

Varela, E., Jay, M., Flinzberger, L., Mobarak, C., & Plieninger, T. (2025). A review of high nature value farming systems in Europe: Biodiversity, ecosystem services, drivers, innovations and future prospects. *People and Nature*. https://doi.org/10.1002/pan3.70048

Chapter 4
The Sphere Project: Global Standards and Local Adaptation in Humanitarian Response

Abstract The Sphere Project has grown beyond a technical handbook to become a dynamic platform for shaping humanitarian action through shared values, operational standards, and collaborative governance. This chapter begins by unpacking Sphere's normative architecture, showing how ethical commitments and legal principles are translated into adaptable, measurable standards. It then explores the pathways through which these standards diffuse globally and are adapted locally, revealing both structural challenges and innovative strategies in diverse contexts. The analysis further highlights how social organizations reinterpret, implement, and institutionalize Sphere's guidance, not as passive users but as active contributors. Finally, the chapter reflects on Sphere's evolving governance model—one that increasingly embraces flexibility, co-ownership, and pluralistic negotiation across the humanitarian field.

Keywords Normative negotiation · Contextual adaptation · Networked co-governance · Ethical foundations · Standard circulation

4.1 Normative Architecture: From Ethical Commitments to Operational Standards

The Sphere Project constructs a unique humanitarian norm system by integrating ethical imperatives, technical methodologies, and institutional governance. Anchored in rights-based commitments, it reframes humanitarian aid as an obligation grounded in law and dignity rather than discretionary charity. Through a structured methodology of minimum standards, key actions, and indicators, it operationalizes these principles while allowing contextual flexibility. This normative-technocratic framework is sustained by an evolving governance model that formalizes decision-making, diversifies global membership, and promotes peer-driven adaptation. These elements illustrate Sphere's transformation from a standard-setting initiative into a dynamic, ethically grounded, and institutionally embedded humanitarian platform.

Q. Du and R. Shaw, *Social Organizations in Disaster Response and Recovery*, Disaster Risk Reduction, https://doi.org/10.1007/978-981-92-2846-1_4

4.1.1 *The Genesis of Humanitarian Norms: Ethical and Legal Foundations*

The Sphere Project establishes its standards not merely as technical recommendations, but as normative claims grounded in ethics and international law. At its core is the Humanitarian Charter, which articulates three foundational commitments: the right to life with dignity, the right to receive humanitarian assistance, and the right to protection and security (Sphere Association, 2018). These principles assert that affected populations are not passive recipients of aid but holders of rights, whose dignity and well-being must be preserved regardless of political, geographical, or organizational boundaries.

What distinguishes Sphere is its integration of these ethical commitments into the architecture of operational standards. Rather than treating standards as flexible guidelines or procedural benchmarks, Sphere presents them as obligations derived from legal and moral sources. This rights-based orientation reframes humanitarian action from charitable intervention to institutional responsibility (Darcy, 2004; Young et al., 2004). The standards do not exist in a normative vacuum but are directly linked to international humanitarian law, human rights law, and refugee law, as explicitly cited in the Handbook (Sphere Association, 2018).

To operationalize these values, the Sphere Handbook embeds the Protection Principles and the Core Humanitarian Standard (CHS). These frameworks guide organizations in ensuring non-discrimination, participation, transparency, and accountability throughout the humanitarian response cycle (Sphere Association, 2018). They serve as a practical translation of ethical commitments into governance mechanisms, helping organizations align everyday actions with broader normative goals.

However, the universalizing ambition of these standards has not been without critique. Questions have been raised about the feasibility of applying rights-based norms in complex, resource-constrained environments. For instance, challenges arise when local actors are expected to uphold universal standards without corresponding capacity or political support (Dufour et al., 2004). This tension reveals a structural paradox: while the Charter aspires to universality, implementation is inevitably shaped by context-specific constraints, including cultural interpretation, legal pluralism, and humanitarian access.

Despite these tensions, the rights-based foundation remains a defining characteristic of Sphere. By embedding ethical and legal imperatives within technical standards, the project aims to shift the humanitarian paradigm—from discretionary action to accountable obligation, from “best practice” to “minimum rights.”

4.1.2 Evolution of Technical Structure and Methodological Design

The technical architecture of the Sphere Standards reflects a deliberate attempt to translate humanitarian values into operational tools. This structure is organized around four key domains—water, sanitation and hygiene (WASH); food security and nutrition; shelter and settlement; and health. Each domain includes a coherent triad: minimum standards, key actions, and key indicators, forming a normative yet adaptable framework that guides humanitarian responses without imposing rigid compliance (Sphere Association, 2018).

The emphasis on "minimum standards" reflects Sphere's ambition to define a universal threshold of humanitarian quality, but the operational logic remains intentionally flexible. This is reinforced in the Handbook's preamble, which positions the standards as adaptable tools that should be "contextualized" in line with local needs, capacities, and risks (Sphere Association, 2018). Technical adjustments are explicitly encouraged in settings with limited resources, fragile infrastructure, or cultural specificities.

Multiple empirical studies have shown that this "soft regulation" approach allows for technical sophistication while avoiding prescriptive rigidity. In the nutrition sector, for instance, indicators have evolved to include refined parameters such as caloric adequacy, micronutrient content, and anthropometric benchmarks (O'Donnell et al., 2002; Seal & Thurstans, 2013). This evolution also introduces complexity. Practitioners often struggle with interpreting and applying these standards in unstable or data-poor environments.

One persistent challenge is the measurement and verification of compliance. According to a review of public health applications, less than 15% of Sphere indicators are quantifiable in a way that allows consistent tracking or benchmarking (Shalash et al., 2022). This raises concerns about the practicality of applying Sphere's technical standards for real-time decision-making or outcome evaluation. For instance, in conflict-affected contexts, the disjunction between normative standards and operational capacity can render the indicators symbolically powerful but functionally limited (Fliss-Isakov et al., 2025).

Despite these critiques, the methodological structure of Sphere serves as a crucial reference point in global humanitarian coordination. Its value lies not in enforcing uniform outputs but in offering a shared vocabulary and set of expectations across diverse actors. By linking ethics with metrics, and flexibility with structure, Sphere provides a normative compass for humanitarian professionalism while acknowledging the limits of universal technocratic design.

4.1.3 Institutional Structure and Governance: From Advocacy to Organizational Formalization

The institutional consolidation of the Sphere Project marks a significant transition from normative advocacy to a structured governance platform. As defined in its legal statutes, Sphere is constituted as an international non-profit association registered in Switzerland, equipped with a defined legal personality and internal decision-making mechanisms (Sphere Association, 2021a). The governance architecture includes a governing Board, a General Assembly of members, and a Secretariat, ensuring both strategic oversight and operational coordination.

This formalized structure enables Sphere to operate not only as a standards-setting body but also as a legitimate actor in the global humanitarian regime. The governance design facilitates participatory processes among diverse stakeholders, ranging from international NGOs to local organizations. According to its membership registry, Sphere had, by 2025, incorporated over 60 full members from diverse geographic regions including South Asia, East Africa, and Latin America—illustrating an expanding pluralistic base of ownership (Sphere Association, 2025). Such a governance model reinforces its legitimacy and enhances the mutual accountability between the central organization and its constituency.

Strategically, Sphere has shifted from a static custodian of standards to a dynamic facilitator of localized implementation. The "Strategy 2021–2025" outlines a paradigm change from merely developing technical standards to enabling their contextualized application through capacity development, peer learning, and decentralized partnerships (Sphere Association, 2021b). This reorientation is operationalized through a growing emphasis on knowledge sharing platforms, co-creation of training tools, and engagement with national coordination systems.

Internal review mechanisms also support this shift. The "Mid-Term Review" published in 2024 identifies key performance gaps—notably, insufficient progress in localization and limited uptake of standards in complex emergencies—and calls for a deeper investment in member support systems and regional hubs (Sphere Association, 2024). Rather than relying solely on top-down dissemination, the institutional strategy promotes a "community of practice" logic, facilitating horizontal exchange and long-term engagement among actors.

By formalizing its governance, expanding its membership base, and embedding mechanisms of learning and reflexivity, Sphere has evolved into an institutional ecosystem rather than a one-dimensional standard-setter. This transformation reflects its commitment not only to normative leadership but also to inclusive, adaptable, and network-based humanitarian governance.

4.2 Pathways of Global Diffusion and Local Adaptation: Toward a Multi-scalar Humanitarian Standard

Rather than unfolding through linear dissemination, the expansion of the Sphere Project reflects a dynamic interplay between global norm-making and local adaptation. Over the past decade, its strategic orientation has shifted from centralized standard-setting to a facilitative model that emphasizes capacity-building, ownership, and accessibility. This transformation faces significant friction—from institutional bottlenecks and cultural mismatches to uneven data environments. At the same time, diverse pathways have emerged across the Global South, where local actors engage not only in implementation but also in reinterpretation and selective institutionalization. This section examines these overlapping trajectories and the tensions they reveal within a plural humanitarian landscape.

4.2.1 From Standard-Setting to Enabling Strategies: The Globalization Vision of Sphere

The Sphere Project's current strategic vision reflects a decisive departure from traditional top-down norm diffusion toward a more facilitative and participatory model. Rather than remaining a centralized standard-setter, Sphere positions itself as an "enabler, facilitator, and convenor," a role articulated clearly in its Strategy 2021–2025 (Sphere Association, 2021b). This strategic repositioning emphasizes supporting localized implementation, strengthening national actors, and promoting inclusive engagement across the humanitarian ecosystem.

Central to this transformation are three interconnected pillars: localization, ownership, and accessibility. Localization prioritizes the integration of Sphere standards into national and community-based systems, seeking to reduce dependency on external actors. Ownership involves empowering local stakeholders—not only to apply standards but to participate in their adaptation and revision. Accessibility ensures that standards and guidance are available in multiple formats, languages, and modalities tailored to diverse operational environments (Sphere Association, 2021b).

To operationalize this vision, Sphere employs a variety of diffusion mechanisms. These include national-level focal points, Sphere-certified trainers, regional champions, and institutional partnerships. Standard materials are disseminated through localized translations, digital platforms, and training-of-trainers programs. Community workshops and contextualized toolkits have also been promoted to bridge knowledge gaps across literacy and language divides.

In practice, these strategies have yielded promising, albeit uneven, results. For instance, the integration of Sphere principles into national systems has been demonstrated in the public health domain, where indicators from Sphere's handbook were proposed for routine health information systems in Southeast Asia (Aung &

Whittaker, 2013). In settings with low literacy, visual and oral communication methods have been developed to translate complex standards into accessible community messages (Jagoe et al., 2023). These approaches reflect a growing recognition that normative legitimacy must be accompanied by epistemic and operational adaptability.

Nonetheless, the Strategy itself acknowledges that the pathway from global standard to local adoption remains incomplete. Structural asymmetries, limited local capacity, and contextual complexity continue to challenge the universal applicability of Sphere standards. The shift in institutional self-conception—from authoritative publisher to collaborative enabler—signals a long-term commitment to normative pluralism and distributed humanitarian governance.

4.2.2 Tensions and Institutional Bottlenecks in Localization: Challenges in Standard Adaptation

While the Sphere Project has made significant strides in establishing globally recognized humanitarian standards, its localization process remains fraught with implementation challenges. The Mid-Term Review of Sphere's Strategy 2021–2025 reveals a critical self-assessment: "localisation is not progressing as expected" (Sphere Association, 2024). Despite the formal diffusion of standards, meaningful integration into diverse sociopolitical and cultural contexts has proven uneven. Structural asymmetries, contextual complexity, and limited institutional capacity are repeatedly cited as key bottlenecks.

In conflict-affected and linguistically fragmented regions, standards often fail to take root. For instance, in the context of protracted crisis in Gaza, caloric food assistance fell significantly below the Sphere-recommended threshold of 2100 kilocalories per person per day, reflecting a substantial gap between normative targets and operational realities (Fliss-Isakov et al., 2025). Similar mismatches are reported in psychosocial and mental health interventions. Standardized approaches have shown limited cultural sensitivity, particularly in trauma-affected populations such as children in post-disaster contexts (Henderson et al., 2016; Morris et al., 2007).

Beyond cultural and environmental barriers, technical constraints also hamper effective use. Quantitative evaluation of public health programs has revealed a significant limitation: fewer than 15% of Sphere's indicators are currently usable for data-driven assessments in humanitarian settings, raising concerns about the framework's operational measurability (Shalash et al., 2022). This undermines both accountability mechanisms and evidence-informed implementation.

Crucially, power asymmetries embedded in the humanitarian system exacerbate these difficulties. While Sphere's standards are often imposed on local organizations, the financial and logistical control remains with international donors. This has created situations where standards function more as compliance checklists for funding eligibility than as tools for contextualized program design (Darcy, 2004; Dufour et al.,

2004). Such instrumentalization risks hollowing out the normative and participatory aspirations that underlie the Sphere model.

Altogether, these dynamics suggest that localization is not a linear technical task, but a deeply political and relational process. Effective adaptation requires more than translation—it necessitates negotiation between global norms and local priorities, with sustained investment in local capacity, cultural sensitivity, and power-sharing mechanisms.

4.2.3 Diversified Pathways and Localized Mechanisms: Institutionalizing Sphere Standards in the Global South

The institutional reach of the Sphere Project increasingly depends not only on its normative coherence but on its ability to embed standards into the infrastructures and practices of humanitarian actors in the Global South. As of 2025, the project's membership structure reflects significant global diversity, incorporating organizations from regions such as South Asia, East Africa, and Latin America—including national platforms from Yemen, the Democratic Republic of Congo, and Pakistan (Sphere Association, 2025). This widespread organizational uptake signals a degree of formal legitimacy and contextual relevance in historically underrepresented humanitarian geographies.

Dissemination of Sphere standards occurs through a multilayered architecture of outreach and localization. One entry point is community-based communication: adapting Sphere's normative language into accessible and visual formats has enabled rural populations to better understand entitlements such as food rations, reinforcing inclusiveness and awareness at the last mile (Jagoe et al., 2023). In parallel, the integration of Sphere content into formal education and training represents another dissemination channel. In emergency medicine and disaster management curricula, Sphere has become a foundational component—contributing to professionalization and capacity-building among frontline responders (Williams & Downes, 2017).

Institutional embedding at the national level also plays a critical role. A notable example comes from Myanmar, where the proposal to incorporate Sphere indicators into national health information systems represents a deliberate strategy to align humanitarian assessment with domestic governance infrastructure (Aung & Whittaker, 2013). Such alignment not only improves coordination but reinforces the institutional legitimacy of Sphere-based monitoring frameworks.

Localization does not always imply full translation of standards into practice. In many cases, Sphere serves as a foundational reference rather than an operational blueprint. Field evidence suggests that local NGOs often treat Sphere as a "starting point" from which to adapt, negotiate, or selectively implement standards based on available capacity and political constraints (Gostelow, 1999; Young et al., 2004). This reflects an embedded tension between normative standardization and practical adaptability.

Thus, the diffusion of Sphere in the Global South is best understood as a hybrid process: it combines top-down institutional adoption with bottom-up reinterpretation and selective use. While formal alignment with Sphere has grown, deeper transformation depends on local agency, iterative adaptation, and sustained investment in contextual knowledge.

4.3 Practical Translation and Institutional Embedding by Social Organizations

As Sphere standards move from global texts into lived humanitarian practice, their significance is increasingly shaped by how social organizations interpret, internalize, and institutionalize them. This section shifts the analytical focus from diffusion to practice, examining how standards reconfigure organizational behavior, enable local actors to act as translators and co-producers of knowledge, and interact with national governance systems. These dynamics reveal Sphere not as a static rulebook, but as a set of norms continually reshaped through organizational learning, contextual negotiation, and institutional alignment across multiple levels of humanitarian action.

4.3.1 Normative Reconfiguration and Professionalization of Organizational Behavior

The institutionalization of the Sphere standards has catalyzed a fundamental shift in how humanitarian organizations conceptualize responsibility, project governance, and operational accountability. Rather than functioning as a purely technical toolkit, the Sphere Handbook embeds humanitarian action within an ethical regime defined by accountability, quality assurance, and a rights-based approach to service delivery (Sphere Association, 2018). These normative foundations have reoriented many NGOs and community-based actors toward more formalized, structured systems of "project governance," whereby actions are no longer guided solely by ad hoc field knowledge but by codified obligations rooted in human dignity and protection rights.

At the level of project design and implementation, Sphere has influenced the recalibration of the humanitarian project cycle. Training modules built around Sphere principles—especially in health, WASH, and shelter—now serve as foundational content in disaster response education and humanitarian certification programs. As noted in training-based evaluations, exposure to Sphere fosters procedural thinking around needs assessment, baseline setting, and results-based monitoring, reinforcing a more metrics-oriented and systematic logic among field actors (Williams & Downes, 2017). This indicates a shift from reactive action toward anticipatory planning, guided by standard-aligned benchmarks.

The ethical commitments codified in Sphere have also influenced organizational identity and internal governance. By positioning humanitarian actors as accountable not just to donors but also to affected populations, the standard creates a moral imperative that extends beyond compliance. In this sense, Sphere does not merely provide "technical guidance," but seeks to redefine the very identity of humanitarian organizations as ethically anchored, rights-obligated entities (Gostelow, 1999). This identity shift is further underlined by the Handbook's emphasis on participatory assessment, transparency, and non-discrimination as default operational norms (Sphere Association, 2018).

However, embedding ethical accountability into operational structures remains uneven. Studies have identified persistent gaps in risk assessment systems and feedback loops, especially in contexts where organizational incentives remain donor-driven. Sphere's framework explicitly seeks to fill this accountability void by establishing minimum standards that center the rights and dignity of beneficiaries as the primary metric of success, thereby rebalancing the accountability dynamic from upward (donor-oriented) to downward (population-oriented) responsibility (Darcy, 2004).

In sum, the uptake of Sphere standards has contributed not only to the professionalization of humanitarian practice, but also to its ethical re-legitimation. This dual transformation—technical and normative—marks a significant departure from earlier, voluntarist models of humanitarian action.

4.3.2 Translational Capacity and Knowledge Co-production by Local Actors

The effectiveness of the Sphere standards increasingly hinges on the capacity of local organizations to not merely implement, but actively reinterpret and co-develop the standards within their operational and cultural contexts. Recent strategic directions emphasize that Sphere should be understood not only as a normative framework but also as a "community of purpose and practice" in which local actors are empowered as co-authors of humanitarian standards rather than passive recipients (Sphere Association, 2021b). This approach reframes standard diffusion as a process of iterative translation—linguistic, institutional, and cultural—within locally governed systems.

Translational practices begin with the adaptation of communication modalities. In rural areas with limited literacy, for example, information about food assistance standards has been transformed into visual guides and oral formats that make key principles accessible to marginalized populations (Jagoe et al., 2023). This visual-symbolic translation process expands the inclusivity of humanitarian norms, allowing populations historically excluded from written technical materials to engage with and shape their meaning.

Institutional translation further occurs when humanitarian indicators are embedded into state-run systems. In Myanmar, the integration of Sphere indicators into national health information systems was not a simple act of adoption, but involved negotiated simplification, redefinition of data categories, and adaptation to local bureaucratic capacities (Aung & Whittaker, 2013). This points to the co-evolution of global norms and national infrastructure—a shift from "diffusion" to "alignment."

The cultural and psychological dimensions of translation are equally critical. In post-conflict Sri Lanka, mental health guidelines based on Sphere were reinterpreted by community social workers to reflect local religious idioms and culturally resonant practices (Henderson et al., 2016). Here, the standard was not diluted but made more contextually legitimate through culturally embedded delivery mechanisms.

Horizontal learning mechanisms also play a vital role in reinforcing local ownership. Through Sphere's membership system, actors in countries such as Yemen and Pakistan have developed peer learning structures that enable shared adaptation strategies across Global South contexts (Sphere Association, 2025). These exchanges facilitate what may be termed "South–South standardization," a model in which technical norms are co-produced through cross-contextual negotiation and mutual learning.

Ultimately, the construction of humanitarian standards such as those on food security has itself emerged as a site of epistemic negotiation, where global expertise and local knowledge are co-mingled, negotiated, and reconstituted (Young et al., 2004). Translation, in this sense, is not merely linguistic but ontological—shaping what is considered valid knowledge and acceptable practice in humanitarianism.

4.3.3 Institutional Integration and Alignment with National Systems

The long-term viability of Sphere standards increasingly depends on their ability to transition from humanitarian tools into pre-positioned institutional frameworks embedded within national systems. Rather than treating standards as reactive instruments used only during emergencies, recent efforts have promoted their incorporation into state planning and preparedness frameworks. This shift repositions Sphere as a "preconfigured capacity model," intended for integration before crises occur (Aung & Whittaker, 2013).

However, this institutionalization trajectory is fraught with challenges. One critical issue lies in the operational disconnect between Sphere's normative ambition and the structural readiness of many national actors. During the 1998 Bangladesh floods, Sphere was introduced as a guideline for nutritional response, yet the majority of responding agencies lacked the training, tools, or baseline data to operationalize its indicators effectively. As a result, standards functioned more as symbolic reference points than actionable protocols (O'Donnell et al., 2002). This highlights the danger of over-relying on technical norms without concurrent investments in capacity development and system alignment.

Further, the political sensitivity of Sphere's rights-based language has presented barriers to its institutional adoption. In contexts where humanitarian aid is perceived as infringing upon state sovereignty or where governance structures are highly centralized, Sphere's emphasis on accountability to affected populations is often muted or resisted. Critics have noted that the assertion of universal humanitarian rights can provoke defensiveness or dilution when translated into national policies, especially where political legitimacy is contested (Darcy, 2004; Dufour et al., 2004). This suggests the need for flexible integration strategies that recognize institutional path dependencies and political constraints.

Moreover, the perceived donor-driven nature of Sphere has raised concerns about external imposition. When standards are tied to funding conditionalities rather than domestically driven demand, they risk being institutionalized as compliance checklists rather than tools for genuine system improvement. The Mid-Term Strategic Review similarly noted persistent asymmetries in ownership, with many national actors excluded from the early design or prioritization of standards, undermining local legitimacy (Sphere Association, 2024).

Successful integration, therefore, depends on soft alignment mechanisms—such as joint planning, indicator co-development, and context-sensitive language adaptation—rather than rigid standard transfer. Embedding Sphere meaningfully into national systems requires not only technical adaptation but political trust-building, institutional compatibility, and the cultivation of co-ownership across multiple governance levels.

4.4 Rebalancing Global Normativity and Local Practice in the Sphere Framework

As the Sphere Project evolves, a central institutional challenge has emerged: how to balance its global normative ambition with the diversity and complexity of local humanitarian contexts. This section examines how Sphere functions not merely as a fixed standard system, but as an institutional arena of ethical negotiation, how its model of "flexibility with integrity" enables adaptive yet principled application on the ground, and how its governance has shifted toward decentralized, networked co-production. These dynamics illustrate a paradigmatic shift—from normative diffusion to pluralistic negotiation, and from centralized authority to shared humanitarian stewardship.

4.4.1 *The Multifaceted Identity of Sphere: Standard System or Institutional Arena?*

The Sphere Project is often perceived as a compendium of humanitarian standards, yet its function extends far beyond static technical guidelines. It operates as a normative arena—one that is shaped by contested values, evolving practices, and the complex interplay of ethics, politics, and professionalism. The Sphere Handbook is not merely a codified manual, but rather a space of normative negotiation where diverse actors engage in shaping humanitarian principles and translating them into operational commitments (Sphere Association, 2018).

This dynamic quality positions Sphere less as a prescriptive tool and more as a platform for institutional deliberation. Rather than providing definitive solutions, it enables consensus-building and ethical coordination among actors with differing mandates and cultural contexts. This was one of the core arguments behind its creation: to act as a catalyst for dialogue and convergence, especially in situations where humanitarian norms remain contested or ambiguous (Dufour et al., 2004). In this sense, Sphere serves not only to standardize but to mediate—offering a shared language through which humanitarian actors can negotiate their roles and responsibilities.

At the heart of this mediation lies the tension between the universalistic aspirations of rights-based humanitarianism and the particularities of operational realities. Rather than resolving this tension through rigid enforcement, Sphere facilitates a "middle ground" between the moral imperatives of rights and the constraints of context. This conceptual space—between what should be done and what can be done—is where many of its most valuable negotiations take place (Young et al., 2004).

Furthermore, the Handbook's value lies in its ability to accommodate ambiguity without undermining normative coherence. By embracing the notion of "standards with flexibility," Sphere permits contextual adjustments while insisting on minimum thresholds of dignity and protection. Its evolution reflects this institutional complexity—where the production of humanitarian norms is not linear or technocratic, but iterative, political, and situated within broader struggles over authority and legitimacy (Darcy, 2004; Gostelow, 1999).

Sphere, therefore, must be understood as more than a framework of metrics—it is a normative infrastructure in which power, ethics, and technical rationality are continuously renegotiated. Its strength lies not in resolving tensions, but in holding them productively open for collective deliberation.

4.4.2 *Normative Flexibility with Ethical Integrity: A Practice-Oriented Logic*

The Sphere Project's attempt to balance global normative coherence with local relevance is best captured in its evolving principle of "flexibility with integrity." Rather than envisioning standardization and contextualization as a binary, Sphere promotes a co-constructive model in which ethical commitments are upheld even as technical guidance is adapted to local constraints and cultural contexts. The Sphere Handbook explicitly frames its standards as adaptable: not rigid metrics, but context-sensitive benchmarks meant to be interpreted and applied with judgment in relation to available resources, risks, and socio-political realities (Sphere Association, 2018).

This flexibility, however, introduces inherent tensions. In low-literacy environments, local actors often face challenges in accessing and applying textual standards. To address this, localization efforts have translated abstract technical norms into visual tools and culturally intelligible media, preserving the core message of dignity while increasing operational accessibility (Jagoe et al., 2023). This demonstrates that adaptation does not necessarily dilute normative intent—it can enhance inclusivity and community ownership.

In contexts of extreme resource scarcity, such as armed conflict zones, the literal application of Sphere's standards may become not only impractical but ethically counterproductive. A study of humanitarian nutrition in Gaza revealed significant gaps between Sphere's recommended caloric minimums and actual provision levels due to siege conditions and political blockades, prompting questions about the limits of normative insistence without corresponding resource guarantees (Fliss-Isakov et al., 2025). In these cases, rigid adherence risks converting rights-based principles into unattainable aspirations, undermining their credibility.

Similarly, technical standards that depend on robust data infrastructures—such as public health surveillance indicators—can become inapplicable when such systems are absent. Without localized data collection capacity, some of Sphere's indicators may be seen less as accountability tools and more as administrative burdens (Shalash et al., 2022). These instances point to the importance of defining "integrity" not in terms of technical uniformity but as fidelity to the underlying ethical commitments: to protect life, dignity, and agency (Darcy, 2004).

In this framework, Sphere's integrity lies in its refusal to impose uniformity at the expense of relevance. Flexibility is not a compromise but a condition of meaningful application. Upholding universal humanitarian values requires the capacity to negotiate their expression in particular and often precarious local realities.

4.4.3 *Emerging Governance Models: From Global Standardization to Networked Co-governance*

The institutional trajectory of the Sphere Project increasingly signals a shift from centralized standard dissemination toward a distributed, network-based model of humanitarian governance. Rather than positioning itself as the sole arbiter of best practice, Sphere now articulates a facilitative role—acting as enabler, convenor, and collaborative platform to support diverse actors in co-constructing context-responsive humanitarian standards (Sphere Association, 2021b). This shift reflects broader recognition that legitimacy in norm-making increasingly depends on inclusive participation, horizontal accountability, and adaptive integration across organizational scales.

The formal expansion of Sphere's membership architecture illustrates this transition. Once dominated by large international NGOs, the network now includes national and regional humanitarian organizations, research institutes, and civil society coalitions from over 60 countries, comprising a "community of purpose and practice" that sustains norm diffusion through peer learning, shared implementation challenges, and iterative revisions (Sphere Association, 2025). This shift from central production to distributed stewardship transforms Sphere into a governance node rather than a hierarchical rule-setter.

Such distributed co-governance is particularly visible in localized standard adaptation and knowledge production. Empirical studies show that national and subnational actors are not merely applying global guidelines, but actively participating in their translation, reinterpretation, and institutionalization. For example, health and disaster management agencies in Myanmar adapted Sphere indicators through national coordination platforms, creating hybrid frameworks aligned with both global commitments and local data infrastructures (Aung & Whittaker, 2013). Likewise, training processes among local humanitarian workers in East Africa have moved from prescriptive dissemination to dialogic formats, fostering inter-organizational consensus and mutual validation (Jagoe et al., 2023).

This evolution indicates that Sphere is no longer functioning as a closed normative regime, but as a dynamic governance platform where multiple actors engage in standard negotiation, contextualization, and co-ownership. Its legitimacy, therefore, stems not from universality alone, but from its ability to accommodate normative pluralism while preserving foundational commitments to human dignity and accountability.

4.5 Conclusion

The evolution of the Sphere Project reveals a profound transformation from a static compendium of humanitarian standards into a dynamic arena of ethical negotiation, local translation, and shared governance. While rooted in universal rights-based commitments, Sphere increasingly embraces contextual flexibility—allowing standards to be reinterpreted through visual, institutional, and cultural lenses. This balance between normative coherence and practical adaptability is sustained through its principle of "flexibility with integrity," which preserves ethical foundations while enabling diverse implementation. As governance shifts from centralized authorship to networked co-ownership, Sphere now operates less as a rule-setter and more as a platform for collective norm-making. However, persistent asymmetries in capacity, politics, and data infrastructure continue to shape how standards are localized and legitimized. Ultimately, the strength of the Sphere framework lies not in rigid enforcement, but in its ability to hold space for pluralism—enabling humanitarian actors across scales to co-produce a shared, yet adaptable, vision of accountability and dignity.

References

Aung, E., & Whittaker, M. (2013). Preparing routine health information systems for immediate health responses to disasters. *Health Policy and Planning, 28*(5), 495–507. https://doi.org/10.1093/heapol/czs081

Darcy, J. (2004). Locating responsibility: The Sphere humanitarian charter and its rationale. *Disasters, 28*(2), 112–123. https://doi.org/10.1111/j.0361-3666.2004.00247.x

Dufour, C., De Geoffroy, V., Maury, H., & Grünewald, F. (2004). Rights, standards and quality in a complex humanitarian space: Is Sphere the right tool? *Disasters, 28*(2), 124–141. https://doi.org/10.1111/j.0361-3666.2004.00248.x

Fliss-Isakov, N., Nitzan, D., Magnazi, M. B., Mendlovic, J., Preis, S. A., Twig, G., Troen, A. M., & Endevelt, R. (2025). Food supplied to Gaza during seven months of the Hamas-Israel war. *Israel Journal of Health Policy Research, 14*(1), 8. https://doi.org/10.1186/s13584-025-00668-6

Gostelow, L. (1999). The Sphere Project: The implications of making humanitarian principles and codes work. *Disasters, 23*(4), 316–325. https://doi.org/10.1111/1467-7717.00121

Henderson, S. E. K., Elsass, P., & Berliner, P. (2016). Mental and social health in disasters: The Sphere standards and post-tsunami psychosocial interventions in Asia. *Disasters, 40*(3), 432–451. https://doi.org/10.1111/disa.12159

Jagoe, C., O'Reilly, C. F., Gunnell, H., Tirzi, K., Lancaster, C., & Brahmbhatt, K. (2023). Communicating accessible messages for food insecure communities in Northern Mozambique: Supporting sustainable development goal 2. *International Journal of Speech-Language Pathology, 25*(1), 62–67. https://doi.org/10.1080/17549507.2022.2140829

Morris, J., Van Ommeren, M., Belfer, M., Saxena, S., & Saraceno, B. (2007). Children and the Sphere standard on mental and social aspects of health. *Disasters, 31*(1), 71–90. https://doi.org/10.1111/j.1467-7717.2007.00341.x

O'Donnell, M. R., Bacos, D., & Bennish, M. L. (2002). Nutritional response to the 1998 Bangladesh flood disaster: Sphere minimum standards in disaster response. *Disasters, 26*(3), 229–241. https://doi.org/10.1111/1467-7717.00202

Seal, A., & Thurstans, S. (2013). Derivation of nutrient requirements for disaster-affected populations: Sphere Project 2011. *Food and Nutrition Bulletin, 34*(1), 45–51. https://doi.org/10.1177/156482651303400106

Shalash, A., Abu-Rmeileh, N. M. E., Kelly, D., & Elmusharaf, K. (2022). The need for standardised methods of data collection, sharing of data and agency coordination in humanitarian settings. *BMJ Global Health, 7*(Suppl 8), e007249. https://doi.org/10.1136/bmjgh-2021-007249

Sphere Association. (2018). *The Sphere handbook: Humanitarian charter and minimum standards in humanitarian response* (4th ed.). https://spherestandards.org/handbook-2018/

Sphere Association. (2021a). *Sphere Association statutes and bylaws (amended May 27, 2021).* https://spherestandards.org/about/

Sphere Association. (2021b). *Sphere strategy 2021–2025: Global humanitarian standards for local action and preparedness.* https://spherestandards.org/about/

Sphere Association. (2024). *Sphere mid-term strategic review (2021–2025).* https://spherestandards.org/about/

Sphere Association. (2025). *Sphere full members list (as of May 30, 2025).* https://spherestandards.org/about/

Williams, H., & Downes, E. (2017). Development of a course on complex humanitarian emergencies: Preparation for the impact of climate change. *Journal of Nursing Scholarship, 49*(6), 661–669. https://doi.org/10.1111/jnu.12339

Young, H., Taylor, A., Way, S., & Leaning, J. (2004). Linking rights and standards: The process of developing 'rights-based' minimum standards on food security, nutrition and food aid. *Disasters, 28*(2), 142–159. https://doi.org/10.1111/j.0361-3666.2004.00249.x

Chapter 5
ADRRN: Strengthening Asian NGOs Through Regional Networks and Capacity-Building

Abstract Through the lens of regional solidarity and local innovation, ADRRN emerges as a distinctive actor reshaping disaster governance across Asia. Its lightweight structure, anchored in thematic Hubs, enables decentralized coordination, peer learning, and the amplification of grassroots knowledge. Rather than replicating top-down models, ADRRN fosters trust-based collaboration, supports locally defined capacity-building, and builds legitimacy through moral capital and embedded practice. From field-level experimentation to global policy engagement, the network demonstrates how southern-led alliances can produce not only effective responses, but also new norms for equitable and locally grounded humanitarian action.

Keywords Networked governance · Embedded innovation · Hybrid capacity-building · Normative representation · Epistemic agency

5.1 Institutional Architecture and Regional Positioning: ADRRN as a Networked Platform

As a regional civil society platform, the Asian Disaster Reduction and Response Network (ADRRN) exemplifies a distinctive model of governance that integrates decentralized coordination, embedded innovation, and global–local interface. This section examines how ADRRN's institutional architecture—anchored in a lightweight governance structure and thematic Hubs—enables adaptive, member-driven collaboration. It further explores how the network fosters inclusivity through a southern-oriented ethos, cultivates institutional entrepreneurship via regionally grounded innovations, and engages with global disaster governance while preserving local epistemic agency. These dimensions illustrate how ADRRN redefines humanitarian coordination from the Global South, not as a recipient, but as a norm-setting actor.

Q. Du and R. Shaw, *Social Organizations in Disaster Response and Recovery*, Disaster Risk Reduction, https://doi.org/10.1007/978-981-92-2846-1_5

5.1.1 The Formation and Governance Architecture of ADRRN: A Lightweight Regional Platform

Founded in 2002 with support from the United Nations Office for the Coordination of Humanitarian Affairs (UNOCHA), the Asian Disaster Reduction and Response Network was created to enhance the disaster preparedness and response capabilities of local and national NGOs in Asia. Unlike centralized, bureaucratic systems of humanitarian governance, ADRRN was deliberately designed as a flexible, member-driven regional platform. Its governance model is anchored in a tripartite structure: an Executive Committee responsible for strategic leadership, a Secretariat located in Tokyo managing daily coordination, and an Advisory Board that provides technical guidance and external linkages (ADRRN, 2025a).

This lightweight architecture allows for efficient, decentralized management while preserving cohesion through shared principles and mutual accountability. Rather than enforcing top-down control, ADRRN facilitates peer-led collaboration through what it terms "light-touch governance." Thematic Hubs form the core of this structure—specialized, member-led sub-networks that enable horizontal coordination across issues such as innovation, accountability, localization, and digital storytelling (ADRRN, 2025a). As stated in network documents, "ADRRN does not operate through a centralized command-and-control structure but facilitates member-led collaboration through light-touch governance" (ADRRN, 2025a).

Each Hub is hosted by a member organization with relevant expertise. For instance, the Asian Technical Innovation Hub (ATIH), led by CWS Japan, supports the institutionalization of community-based innovations, helping to embed them into disaster planning processes. Its aim is to "institutionalize creative solutions that are grounded in the local context," ensuring that innovation emerges from, rather than is imposed upon, affected communities (ADRRN, 2025a). Similarly, the Localization Hub coordinated by SEEDS India promotes community leadership in defining needs and shaping responses, reinforcing a bottom-up governance logic.

This governance model exemplifies the adaptive, self-organizing logic that characterizes effective networked civil society platforms in the Global South. Studies of other transnational NGO networks in Asia confirm that trust-based, decentralized, and thematic coordination structures tend to foster greater local legitimacy and responsiveness (Galkina & Yang, 2020; Izumi & Shaw, 2012). ADRRN's architecture reflects not only institutional entrepreneurship, but also a deliberate departure from donor-driven models, grounding its authority in the priorities of local actors embedded in diverse socio-political contexts.

5.1.2 A Southern-Oriented, Inclusive, and Open Network Structure

The organizational structure of ADRRN reflects a conscious attempt to move away from hierarchical and donor-centric paradigms of humanitarian coordination. Instead, it adopts a southern-oriented, inclusive, and open architecture that prioritizes equity, participation, and locally rooted knowledge systems. With over 50 member organizations spanning South Asia, Southeast Asia, and the Pacific, ADRRN positions itself not only as a regional coordination platform but also as a community of practice where diverse actors can co-create knowledge and strategies (ADRRN, 2025a).

Membership in ADRRN is deliberately non-exclusive. It welcomes both large international NGOs and small grassroots organizations, enabling representation from across different operational scales and thematic areas. This openness ensures that voices from underrepresented regions and marginalized communities are not just included but actively shape network agendas. The operational logic is not anchored in a vertical chain of command, but in principles of horizontal learning and South-South solidarity. This approach contrasts sharply with conventional donor-driven models, where knowledge and funding typically flow unidirectionally from global North to South (Izumi & Shaw, 2012).

The Localization Hub, coordinated by SEEDS India, exemplifies this ethos. It promotes a model where "local actors lead, fund, and shape programs based on their real needs" (ADRRN, 2025a). Rather than treating localization as a technical compliance issue, the Hub frames it as a structural reordering of authority within disaster governance. Initiatives like locally led adaptation (LLA) grants and community-to-community exchange mechanisms are operationalized through this hub, reinforcing ADRRN's commitment to subsidiarity and participatory practice (ADRRN, 2025c).

Such an inclusive design helps challenge dominant narratives of vulnerability and passivity often ascribed to southern actors in global frameworks. Instead, ADRRN offers a counter-logic where regional NGOs become agenda-setters, drawing from situated experiences and cultural specificities. This is consistent with broader observations that regional networks in Asia have the potential to reshape the moral and operational terrain of humanitarianism by centering local knowledge and rejecting externally imposed templates (Shipper, 2012; Williams, 2018).

In sum, ADRRN's inclusive and southern-led architecture demonstrates how civil society networks can reconfigure governance from below. Through its open membership, thematic Hubs, and emphasis on local leadership, it enables a more just and effective distribution of voice and agency in regional disaster response and risk reduction.

5.1.3 Institutional Entrepreneurship: ADRRN as a Regional Innovation Actor

Rather than functioning merely as a coordination platform, ADRRN has evolved into a site of institutional entrepreneurship that actively shapes regional disaster governance through embedded innovation. This role is not externally imposed but generated through iterative, localized practices and adaptive experimentation. Unlike traditional NGO networks that often replicate Northern governance templates, ADRRN facilitates context-sensitive institutional development grounded in regional relationships and cultural legitimacy (Galkina & Yang, 2020).

A core mechanism of this innovation dynamic lies in ADRRN's thematic Hub system, which decentralizes initiative and expertise to network members. The Asia Technical Innovation Hub (ATIH), led by CWS Japan, exemplifies how creative, practice-based solutions—such as inclusive early warning systems or low-tech adaptive infrastructure—are institutionalized within daily operations. As described, "ATIH helps institutionalize creative solutions that are grounded in the local context" (ADRRN, 2025a). These Hubs function not just as knowledge repositories but as catalysts for scaling informal practices into structured programs.

The LEAPS Hub demonstrates another form of institutional entrepreneurship by mobilizing a regional human resource pool that can be deployed for rapid post-disaster response. This mechanism reduces dependency on international surge teams and builds intra-regional readiness through shared protocols and culturally attuned actors (ADRRN, 2025a). In doing so, ADRRN asserts a "region-from-within" logic—grounded in capacity already embedded in member organizations rather than externally sourced expertise.

Crucially, ADRRN also formalizes unconventional tools of practice such as digital storytelling, community dialogues, and oral history curation. The Digital Storytelling Hub enables local actors to document and circulate lived experiences from disaster-affected communities, thereby transforming narrative practices into institutional memory and advocacy tools (ADRRN, 2025a). This is not merely communication work—it constitutes a form of "epistemic innovation," whereby marginalized knowledge is encoded into governance frameworks (Dressler, 2017).

Institutional entrepreneurship within ADRRN is therefore neither abstract nor technocratic. It reflects a form of governance anchored in adaptive agency, peer legitimacy, and practice-led standard-setting. As such, the network embodies what might be termed embedded innovation: innovations that grow organically from local contexts but gain durability and legitimacy through regional institutional scaffolding. This mode of networked governance reaffirms ADRRN's distinct position as both a learning community and a norm entrepreneur in Asia's disaster governance landscape.

5.1.4 Global Embedding with a Local Mandate: Balancing Regional Autonomy and Global Integration

ADRRN's evolution reflects a dual commitment to regional autonomy and meaningful engagement in global disaster governance. Rather than functioning as a passive recipient of global agendas, the network actively intervenes in shaping transnational frameworks through grounded, regionally anchored contributions. Since the 2015 Sendai Framework for Disaster Risk Reduction (SFDRR), ADRRN has become an established voice within global platforms such as APMCDRR and UNDRR regional consultations, often providing direct input into thematic priorities such as anticipatory action and localization (ADRRN, 2022, 2024).

This global positioning is not driven by elite diplomacy, but by a distributed ecosystem of technical Hubs that channel regional expertise into international forums. For instance, the Quality and Accountability Hub has worked to reinterpret global humanitarian standards through an Asian lens, proposing adaptive frameworks more attuned to the socio-political contexts of South and Southeast Asia (ADRRN, 2025a). Such efforts signal a shift from compliance with imported norms toward the co-production of standards that emerge from practice-led learning within the Global South.

While embedded in global processes, ADRRN maintains a firm commitment to what it defines as a "South-first knowledge flow." The emphasis is on transforming regionally situated innovations—such as the LLA microgrant mechanism or community storytelling methodologies—into frameworks that inform broader policy dialogue (ADRRN, 2025c). This orientation counters the traditional top-down logic of Northern-led capacity-building, and instead centers epistemic agency among Southern actors.

This approach echoes dynamics in other sectors of transnational governance, where Southern NGO networks increasingly assert agenda-setting roles. The case of South Asian health organizations that influenced global eye health policy illustrates how embedded regional partnerships can invert knowledge hierarchies and challenge normative dependencies (Williams, 2018). ADRRN similarly leverages its institutional architecture to not only implement, but shape, what constitutes resilience and accountability within global DRR discourses.

Ultimately, ADRRN's model offers an alternative to binary framings of local versus global. Its structure allows for a porous, multi-scalar integration, where global embedding is not a dilution of local mandates but an amplification of them. By translating situated knowledge into transnational norms, the network repositions Asia not as a recipient but as a co-author of global disaster governance.

5.2 Coordination and Collaboration: Mechanisms of Trust, Horizontal Governance, and Local Embeddedness

ADRRN's coordination model redefines disaster response through decentralized, trust-based, and context-sensitive mechanisms. This section explores four interlinked dimensions: the modular hub-based system that enables flexible and rapid deployment across diverse crisis contexts; the cultivation of trust through iterative, emotionally grounded collaboration among members; the emergence of horizontal governance structures that support peer-based decision-making in complex emergencies; and the central role of locally embedded actors as moral brokers and knowledge intermediaries. These mechanisms illustrate how ADRRN operationalizes an adaptive, socially rooted approach to humanitarian governance grounded in regional legitimacy and relational accountability.

5.2.1 Adaptive Coordination Models: Flexible, Decentralized, and Action-Oriented

ADRRN's coordination approach is structured around a modular and adaptive model designed to respond swiftly to complex, dynamic crisis environments across Asia. This approach relies on a decentralized "Hub-based" structure that enables member organizations to specialize, innovate, and coordinate across thematic and geographic priorities. Instead of central command systems, ADRRN fosters collaborative platforms that can rapidly adapt through localized leadership and technical integration (ADRRN, 2025a).

A central mechanism of this flexibility lies in the Hotspot-based coordination model, which identifies high-risk or underserved regions for proactive engagement. By mapping vulnerability hotspots and pre-positioning capacity, ADRRN facilitates anticipatory governance rather than reactive response. This approach is particularly critical in contexts where national-level systems are overstretched or underresourced. The Hotspot model enabled local organizations to act as first responders within their own contexts, rather than waiting for externally coordinated interventions (ADRRN, 2024).

The LEAPS Hub, led by RedR India, exemplifies how modular coordination is operationalized through regionally embedded mechanisms. It maintains a certified roster of trained responders—technical and non-technical—from member organizations across Asia, offering tailored training, regional deployment capacity, and rapid response tools (ADRRN, 2025a). This system strengthens intra-network surge capability and reduces dependency on international actors. As described, the Hub aims to "ensure rapid, high-quality humanitarian action across the Asia–Pacific" by mobilizing expertise already situated in local contexts.

The Drone Hub, coordinated by Humanitarian Aid International, further expands ADRRN's action-oriented capacity by integrating real-time data collection into emergency response. Leveraging unmanned aerial systems and local partnerships, this Hub supports damage assessment, situational mapping, and early warning dissemination in remote or inaccessible areas. Importantly, it empowers local NGOs to directly access and use spatial intelligence without waiting for externally driven assessments. As noted in ADRRN documents, "The Drone Hub enables rapid response, real-time data collection, and informed decision-making during emergencies" (ADRRN, 2025c).

Collectively, these mechanisms reflect a strategic shift from rigid command hierarchies toward a distributed, interoperable ecosystem. Thematic Hubs act not as silos but as dynamic platforms for peer learning, innovation, and cross-sector alignment. This coordination logic, rooted in flexibility, specialization, and local agency, offers a compelling model for disaster response networks navigating increasingly complex risk landscapes in the Global South (Dodd et al., 2023; Partelow et al., 2023).

5.2.2 Trust-Based Collaboration: Social Embeddedness and Everyday Ties

In ADRRN's networked governance structure, collaboration is not enforced through formal contracts or top-down directives, but emerges through trust-based relationships cultivated over time. This form of "socially embedded collaboration" relies on repeated interaction, emotional empathy, and shared practical experience across diverse organizational contexts. Trust, in this sense, functions as both a moral resource and an operational mechanism, enabling coordination under conditions of uncertainty and institutional diversity (Lai & Fu, 2021; Ng, 2024).

Rather than treating trust as an abstract cultural value, ADRRN operationalizes it through structured yet informal practices. The CBDRRM Hub, coordinated by the Center for Disaster Preparedness in the Philippines, provides a compelling example. Through facilitated community dialogues, story-sharing circles, and participatory workshops, the Hub fosters a common language of risk rooted in lived experience. These engagements promote what ADRRN identifies as "peer-based recognition" and help build consensus across socio-cultural and linguistic divides (ADRRN, 2025a). The trust generated through such grassroots mechanisms often proves more resilient and actionable than formal agreements.

Horizontal modes of interaction, such as inter-Hub peer visits and co-hosted field exercises, further deepen relational trust across the network. These micro-level engagements function as iterative sites of learning and boundary negotiation, where members test cooperation, exchange insights, and reaffirm shared values. This aligns with broader findings that organizational trust in Asian civil society is often constructed through incremental and empathetic practices rather than bureaucratic standardization (Ng, 2024; Shipper, 2012).

Importantly, these trust dynamics are not incidental but are actively cultivated as part of ADRRN's governance design. Trust is institutionalized through what may be called "emotional infrastructures"—ritualized encounters, collaborative routines, and shared storytelling platforms that bind the network together over time. The Digital Storytelling Hub contributes to this process by equipping local actors to narrate and circulate their disaster experiences, thereby reinforcing mutual recognition and affirming community voice within the network (ADRRN, 2025a).

In effect, ADRRN's trust-based collaboration model functions as a distributed capacity: a relational web that enables coordination without centralized enforcement. By privileging empathy, local anchoring, and peer legitimacy, the network demonstrates how social trust—when intentionally embedded—can underpin flexible and durable disaster governance from below.

5.2.3 Horizontal Governance Under Stress: Navigating Complexity Through Peer-Based Coordination

In the face of intersecting crises—such as climatic shocks, protracted displacement, and sudden-onset disasters—ADRRN has advanced a mode of horizontal governance that privileges distributed coordination over centralized command. Instead of relying on hierarchical authority or formal chains of command, the network sustains operational coherence through decentralized leadership, mutual accountability, and knowledge circulation among peers (ADRRN, 2025a; Galkina & Yang, 2020).

The Quality and Accountability (Q&A) Hub, facilitated by Community World Service Asia, exemplifies this logic by providing a pool of trained technical personnel who can be rapidly deployed across the region. These individuals are equipped not only with standards-based knowledge but also with the relational capacity to work across organizational and cultural boundaries. As noted in network records, "the Q&A Hub supports contextually informed decision-making by deploying trained personnel during emergencies" (ADRRN, 2025a). This approach reinforces shared accountability without introducing hierarchical control, relying instead on trust-based standards and peer recognition.

In situations marked by overlapping vulnerabilities, rigid institutional hierarchies often struggle to adapt to rapid shifts. By contrast, ADRRN's governance strategy emphasizes informal yet reliable channels of communication, drawing upon the strength of cross-hub familiarity and shared practices (Partelow et al., 2023). This includes flexible task-sharing, co-authorship of protocols, and consensus-based problem-solving—forms of horizontal coordination that allow for agile responses without requiring centralized approval. Case reviews suggest that such NGO-led coordination often outperforms bureaucratic responses in terms of timeliness, contextual fit, and local legitimacy (Dodd et al., 2023).

What enables this model to function effectively under compound stress is the embedded culture of mutual aid and equality among members. Hubs are not technical silos but relational infrastructures, where response capacity is intertwined with everyday peer engagement. Even in the absence of formal mandates, ADRRN actors exhibit high alignment due to a strong normative culture and embedded routines of interaction (Ng, 2024). As observed, "informal but effective coordination mechanisms demonstrate a shared ethos of mutual support, rather than control" (ADRRN, 2025a).

Ultimately, ADRRN's experience challenges the assumption that effective crisis response requires command hierarchies. Instead, it demonstrates how peer legitimacy, horizontal trust, and flexible coordination—when institutionalized through hubs—can sustain collaborative governance even in fragmented, multi-crisis contexts.

5.2.4 Embeddedness and Local Leadership: Trust Actors, Moral Brokers, and Contextual Interpreters

The effectiveness of decentralized disaster governance hinges not only on structural coordination mechanisms but also on the presence of locally embedded actors who possess social legitimacy and contextual fluency. Within ADRRN, collaboration is often mediated by individuals and organizations that operate as moral brokers and cultural interpreters—figures who hold trust within their communities and serve as conduits for both mobilization and sense-making during crisis (Carter-Tod et al., 2025; Dressler, 2017). These trust actors, such as religious leaders, local volunteers, and long-standing community organizers, play an outsized role in translating abstract frameworks into culturally resonant practices.

Rather than imposing external leadership models, ADRRN's Localization Hub—hosted by SEEDS India—prioritizes shared leadership through co-design, co-implementation, and co-ownership of response strategies. This is grounded in the principle that "local actors lead, fund, and shape programs based on their real needs" (ADRRN, 2025a). Such practices move beyond tokenistic participation to foster deep collaboration between networks and embedded local authorities, where credibility is derived not from formal credentials but from long-standing relational presence within communities (Izumi & Shaw, 2012; Shipper, 2012).

The Digital Storytelling Hub further amplifies the role of local actors by enabling communities to narrate their own disaster experiences, thereby shifting representational power from external intermediaries to those directly affected. Through facilitated storytelling, documentation, and peer exchange, the Hub nurtures what has been described as "epistemic voice"—a form of knowledge authority that is rooted in lived experience rather than technocratic abstraction (ADRRN, 2025a; Dressler, 2017). These stories are not merely symbolic; they feed into regional learning processes, advocacy campaigns, and standard-setting within the network.

Local legitimacy is also relational, built through consistent presence and iterative engagement. Initiatives such as inter-community visits, peer shadowing, and shared narrative spaces serve as mechanisms for reinforcing bonds of trust across different socio-cultural contexts (Ng, 2024). In this regard, ADRRN's approach reflects a broader reconfiguration of disaster governance: where horizontal trust, moral proximity, and culturally attuned leadership supplant formal authority as the primary enablers of effective coordination.

5.3 Capacity-Building and Knowledge Flows: Empowering Local Actors Through Regional Learning

ADRRN's approach to capacity-building foregrounds locally embedded, network-enabled learning that strengthens the agency of Southern actors across Asia. This section explores how the network localizes training and funding mechanisms, facilitates peer-to-peer knowledge transfer, integrates digital platforms with trust-based relationships, and catalyzes innovation diffusion from the grassroots. Moving beyond standardized models, ADRRN fosters a decentralized learning ecology where community-based organizations co-produce, adapt, and circulate knowledge. By aligning learning with local priorities and institutional realities, the network not only enhances technical competencies but also advances epistemic justice and resilience through regional solidarity.

5.3.1 Localizing Capacity-Building: Training, Microgrants, and Community Anchoring

Capacity-building within ADRRN is deliberately framed as a locally anchored and demand-driven process rather than a standardized transfer of technical skills. The network's approach prioritizes strengthening the agency of local and national NGOs by embedding learning within community realities and organizational practice. This orientation is reflected in a diverse set of mechanisms, including national-level workshops, community-led training initiatives, and small-scale Locally Led Adaptation (LLA) grants implemented through the SIDRRA program (ADRRN, 2025b, 2025c).

A defining feature of ADRRN's capacity-building strategy is the flexibility granted to member organizations in setting priorities and designing activities. Rather than prescribing thematic agendas, ADRRN enables organizations to identify training needs based on locally perceived risks, institutional gaps, and contextual constraints. This "needs-first" model allows learning processes to remain responsive to evolving vulnerabilities and capacities, reinforcing local ownership over both content and outcomes. As documented in program reviews, "The SIDRRA LLA scheme promoted adaptive learning by giving local organizations the flexibility to define their own priority areas" (ADRRN, 2025c).

Training under this model is closely tied to practice. National and sub-national workshops are often organized in collaboration with community-based organizations and are structured around applied learning, scenario-based exercises, and reflection on lived experience. Within the Hub framework, the Community-Based Disaster Risk Reduction and Management (CBDRRM) Hub, coordinated by the Center for Disaster Preparedness in the Philippines, plays a central role in operationalizing this approach. Its activities integrate community knowledge, participatory risk analysis, and leadership development, emphasizing the inseparability of knowledge and action. As stated in network documentation, the CBDRRM Hub "promotes DRR knowledge and practice rooted in communities, training local leaders as catalysts for sustained action" (ADRRN, 2025a).

Beyond skill acquisition, ADRRN's capacity-building efforts seek to activate self-organized learning at the community level. Micrograants under the LLA framework support experimentation, peer learning, and iterative adaptation, allowing local actors to test solutions, reflect on outcomes, and recalibrate strategies. This process strengthens not only technical competencies but also organizational confidence and collective problem-solving capacity. Such locally embedded learning dynamics resonate with broader findings that durable capacity emerges when training is integrated into everyday institutional practice rather than delivered as one-off interventions (Izumi & Shaw, 2012; Lai & Fu, 2021).

Through these mechanisms, ADRRN advances a model of capacity-building that is less about disseminating predefined standards and more about cultivating locally grounded capabilities. By anchoring learning in community contexts and reinforcing local decision-making authority, the network contributes to more sustainable and context-sensitive disaster risk reduction across Asia.

5.3.2 Peer-to-Peer Learning: Horizontal Knowledge Transfer and South-South Solidarity

ADRRN's capacity-strengthening strategy is rooted in a commitment to peer-based knowledge circulation and South-South solidarity. Rejecting extractive or donor-driven paradigms, the network foregrounds horizontal learning processes that valorize context-specific expertise among its members. Knowledge here is not passively transmitted from a center to the periphery but actively co-produced by frontline actors across diverse geographies (Izumi & Shaw, 2012; Williams, 2018).

This commitment materializes through structured peer mentoring, cross-country learning visits, joint simulation exercises, and shared documentation of practice. These modalities enable organizations to exchange operational insights in ways that are culturally resonant and technically grounded. Instead of relying on external trainers, learning was driven by peers who shared contextualized practices across borders (Izumi & Shaw, 2012). Such approaches ensure that knowledge flows are embedded in lived realities, facilitating adaptive uptake rather than mechanical replication.

The LEAPS Hub, led by RedR India, exemplifies how peer learning is institutionalized through mentorship systems and surge rosters embedded within the network. It offers practice-based certification, technical guidance, and regionally coordinated deployments to facilitate the transfer of experience from seasoned practitioners to emerging responders (ADRRN, 2025a). This peer-to-peer model also enhances the responsiveness of the network, ensuring that interventions are not only timely but informed by local knowledge embedded within the region. As noted, "The LEAPS Hub provides technical mentoring, peer learning, and a regional surge roster to ensure rapid, high-quality action" (ADRRN, 2025a).

Crucially, these exchanges are not purely technical but also epistemic and political. Hubs such as the Q&A Hub and CBDRRM Hub support co-authored protocols and guidelines, reflecting a broader shift toward distributed authority in knowledge production. Rather than privileging universalist standards, these platforms foster what (Lai & Fu, 2021) term "co-produced expertise," grounded in collective experience and peer recognition.

In effect, ADRRN's peer learning mechanisms embody a reconfiguration of power within regional humanitarian governance. Through dialogic exchange and mutual validation, the network positions southern actors as knowledge generators, challenging hierarchies of expertise and advancing a pluralistic model of disaster risk reduction rooted in solidarity, relevance, and shared capacity.

5.3.3 Hybrid Knowledge Platforms: Blending Digital Tools with Relational Infrastructure

ADRRN's knowledge-sharing architecture reflects a deliberate blending of technological infrastructure and social trust, creating hybrid platforms that optimize both reach and relational depth. This approach challenges the notion that digital innovation alone can facilitate effective capacity-building. Instead, it recognizes that knowledge flows are most impactful when mediated by both technological access and interpersonal trust, a dynamic particularly evident in Asia's civil society networks (Lai & Fu, 2021; Ng, 2024).

Across ADRRN's thematic Hubs, digital platforms serve as foundational tools for coordination, training dissemination, and peer exchange. The Innovation Hub, for instance, curates context-sensitive technologies and hosts an open-access digital library, enabling member organizations to adapt tools and approaches across varied risk contexts (ADRRN, 2025a). Meanwhile, the Q&A Hub provides standardized guidance on accountability frameworks, hosting virtual clinics and a database of trained personnel, accessible across countries and sectors (ADRRN, 2025a).

These digital interfaces facilitate rapid diffusion of standards, methodologies, and innovations across the network.

However, ADRRN has consistently avoided overreliance on virtual platforms. Hubs such as CBDRRM and LEAPS deliberately pair digital tools with in-person learning formats—community workshops, mentoring visits, and simulation exercises—ensuring that abstract knowledge is absorbed through situated interaction (ADRRN, 2025a, 2025c). This strategic duality aligns with the observation that "digital tools increased access, but it was long-term trust and in-person exchanges that ensured actual learning took place" (Lai & Fu, 2021). Knowledge is not only transmitted but translated, negotiated, and contextualized through face-to-face dialogue and peer validation.

Relational infrastructures also play a central role in validating knowledge and ensuring its legitimacy. Trust-based practices such as joint protocol development, shared storytelling platforms, and inter-hub field visits function as "social glue," supporting knowledge absorption and reducing asymmetries between organizations with uneven technical capacity (Galkina & Yang, 2020; Ng, 2024). The success of ADRRN's hybrid platforms lies not in technological sophistication alone, but in their embeddedness within a peer-driven, trust-rich network ecology.

5.3.4 *Enabling Local Innovation: Regional Learning as a Catalyst for Practice Transfer*

Beyond conventional capacity-building, ADRRN foregrounds the importance of enabling local actors not only to adopt external solutions but to generate, institutionalize, and disseminate their own innovations. This reflects a shift from a top-down model of knowledge transfer toward a regional learning ecology where local practices, embedded knowledge, and lived experiences are recognized as valid and transferable sources of resilience (ADRRN, 2025a; Dressler, 2017).

A central mechanism for this is the Digital Storytelling Hub, coordinated by SEEDS India. This Hub supports communities in documenting and sharing their own disaster narratives and adaptive strategies through accessible digital media. Rather than being passive recipients of best practices, community members are positioned as knowledge producers, whose insights inform the broader network. According to ADRRN documentation, "The Digital Storytelling Hub empowers local actors to document and disseminate contextual innovations across the network" (ADRRN, 2025a). These narratives are not merely symbolic; they feed into cross-regional learning forums, peer-to-peer trainings, and advocacy platforms.

Such practices reinforce the role of moral and narrative capital in shaping institutional change. As discussed in the literature, innovations that are embedded in local norms, values, and histories often carry greater legitimacy and uptake potential than externally imposed standards (Dressler, 2017). The network's emphasis on storytelling as both a pedagogical and political act allows local actors to reclaim representational authority while also contributing to collective problem-solving.

The Gili Trawangan case illustrates how locally derived practices can be scaled and reframed as regional assets. Originally developed as an ecological tourism recovery strategy, the model was later integrated into disaster recovery frameworks through networked knowledge-sharing platforms. As observed, "Knowledge-sharing platforms helped reframe small island tourism not as vulnerability, but as resilience capacity to be shared" (Partelow et al., 2023). This case underscores ADRRN's role as a conduit through which local practices are transformed into regional learning assets, rather than being treated as isolated exceptions.

Institutional structures within ADRRN, such as the Innovation Hub and peer-led forums, further support the horizontal diffusion of localized innovation, enabling mutual adaptation without homogenization. In this way, ADRRN operationalizes regional learning not simply as knowledge exchange, but as a political project of epistemic justice, amplifying the legitimacy of Southern-led responses in the global humanitarian discourse (ADRRN, 2025a; Williams, 2018).

5.4 Legitimacy, Influence, and the Politics of Representation in Regional Disaster Governance

Building legitimacy in regional disaster governance is not merely about occupying institutional seats, but about how representation, authority, and knowledge are constructed from the ground up. This section traces ADRRN's evolving role as both a policy interlocutor and a locally rooted actor. It begins with how the network speaks for the region in global frameworks, then turns to the normative authority it derives from community embeddedness. It further examines internal tensions around measurement and coalition dynamics, before highlighting how ADRRN reshapes knowledge hierarchies by institutionalizing practice-based innovations across governance scales.

5.4.1 Speaking for the Region: ADRRN's Representation in Global Disaster Frameworks

ADRRN plays a strategic intermediary role in bridging grassroots disaster risk concerns with global policy processes. Its engagement with platforms such as the Asia–Pacific Ministerial Conference on Disaster Risk Reduction (APMCDRR), the Sendai Framework for Disaster Risk Reduction (SFDRR), and UNDRR consultations has positioned the network as a critical voice articulating the needs and perspectives of local and national NGOs across Asia (ADRRN, 2022, 2024). These engagements are not merely symbolic but serve to open policy windows for often-excluded constituencies, enabling bottom-up agendas to be embedded into formal governance mechanisms. As stated in the network's internal reflection, "ADRRN's presence in APMCDRR was critical in articulating localized adaptation priorities often excluded from national agendas" (ADRRN, 2024).

This representative function is not carried out solely through participation in high-level meetings but is institutionalized through the network's Hub architecture. The LEAPS Hub, for example, works in close collaboration with UN agencies and national disaster management authorities to enhance regional surge readiness and professional deployment coordination (ADRRN, 2025a). By creating shared operational rosters and capacity certification mechanisms, it strengthens the interface between regional civil society responders and multilateral institutions, demonstrating ADRRN's capacity to align grassroots expertise with formal institutional mandates.

Similarly, the Q&A Hub serves as a bridge between local actors and policy institutions through its work on accountability and standard-setting. It facilitates the co-development of guidance documents and training curricula in partnership with governments, academic bodies, and civil society organizations, reinforcing ADRRN's role in shaping normative frameworks within and beyond the region (ADRRN, 2025a). These efforts extend the network's influence from implementation arenas into policy formulation, allowing ADRRN to act as both a field-level convener and a governance interlocutor.

Crucially, ADRRN's representative legitimacy derives not from top-down mandates but from its ability to connect diverse actors across scales while maintaining alignment with local realities. Its decentralized governance model enables frontline organizations from smaller or underrepresented countries to contribute meaningfully to transnational policy dialogues under a shared regional identity (Williams, 2018). In this way, ADRRN's representational strategy avoids the pitfalls of tokenistic participation by embedding legitimacy in both its institutional infrastructure and its long-standing commitment to locally grounded advocacy.

5.4.2 *Moral Capital and Normative Authority: How Legitimacy Is Built from Below*

Legitimacy within ADRRN is not derived from formal mandates or technical accreditation, but from embeddedness in local communities and the network's sustained presence in humanitarian response and recovery. This moral capital—accumulated through action, proximity, and accountability—is central to ADRRN's ability to claim normative authority in regional governance. As observed in disaster research, "legitimacy came not from formal mandates, but from being the first to act, and the last to leave" (Dressler, 2017). Such grounded authority is especially critical in Asia, where civil society organizations often operate in hybrid political systems and where official recognition does not always equate to social trust (Shipper, 2012).

Multiple Hubs within ADRRN operationalize this grassroots legitimacy. The Community-Based Disaster Risk Reduction and Management (CBDRRM) Hub, coordinated by the Center for Disaster Preparedness in the Philippines, embeds disaster risk reduction in participatory community processes. Through localized risk mapping, scenario building, and leadership development, the Hub cultivates trust-based governance practices that resonate with community values and lived realities (ADRRN, 2025a). These embedded practices allow ADRRN to anchor its advocacy in experiential knowledge rather than abstract metrics, reinforcing its credibility with both local constituencies and international actors.

The Localization Hub further institutionalizes this moral authority by supporting community-driven mechanisms for humanitarian coordination. Rather than positioning local actors as beneficiaries, the Hub reconfigures them as co-producers of response strategies, governance structures, and accountability frameworks. These efforts disrupt the prevailing humanitarian hierarchies that often marginalize community knowledge and challenge the technocratic bias of international aid systems (Galkina & Yang, 2020; Ng, 2024).

Meanwhile, the Digital Storytelling Hub amplifies this bottom-up legitimacy by facilitating community-authored narratives. Through training in oral history, digital media, and participatory communication, the Hub enables marginalized actors to document and disseminate their perspectives on risk, vulnerability, and resilience. This not only expands the epistemic space of disaster discourse but also builds narrative legitimacy—what Dressler describes as the power of "moral and cultural capital" in shaping institutional response (Dressler, 2017). As noted, "the Digital Storytelling Hub empowers local actors to document and disseminate contextual innovations across the network" (ADRRN, 2025a).

5.4.3 *Challenges in Measurement and Coalition Politics: Navigating Fragmented Governance*

While ADRRN has emerged as a significant regional voice in disaster governance, its organizational coherence and accountability mechanisms remain shaped by the tensions inherent in transnational NGO networks. As a decentralized coalition of diverse actors—ranging from large international NGOs to grassroots organizations—ADRRN must navigate complex coalition politics and persistent challenges in impact measurement and internal coordination (Izumi & Shaw, 2012; Ng, 2024).

A central issue concerns the lack of systematic and network-wide evaluation tools to assess the effectiveness of ADRRN's interventions. Although individual Hubs have made progress in professionalization—for example, the LEAPS Hub has introduced a regional roster and practitioner certification schemes to standardize deployment readiness (ADRRN, 2025a)—these efforts remain unevenly institutionalized. ADRRN lacks a unified framework to measure outcomes, track learning across Hubs, or assess policy influence. This reflects broader structural asymmetries in the network, where resource-rich members are better equipped to adopt evaluation tools, while smaller organizations struggle to align with technical standards (Galkina & Yang, 2020).

Such fragmentation is also evident in coalition governance. While ADRRN promotes a shared commitment to localization and equity, the diversity of its member organizations—both in terms of political space and operational capacity—renders internal consensus-building a fragile exercise. Tensions between central coordination and local autonomy often emerge during strategic decision-making or the rollout of joint initiatives (ADRRN, 2024; Izumi & Shaw, 2012). The Q&A Hub, for instance, has made strides in developing accountability standards in consultation with state and academic actors, yet the adoption of these standards across the network remains inconsistent (ADRRN, 2025a). Without enforceable governance instruments, the principles of transparency and mutual accountability risk becoming symbolic rather than actionable.

Furthermore, ADRRN's collective voice is sometimes diluted by disparities in representation. Larger NGOs with greater institutional capacity may dominate policy dialogues, leaving smaller or more regionally confined actors underrepresented in high-level forums. This creates legitimacy dilemmas around "who speaks for whom"—a recurring concern in transnational advocacy networks (Lai & Fu, 2021; Shipper, 2012).

To address these challenges, ADRRN has endorsed voluntary alignment tools such as the CSO Commitment Declaration and has promoted inter-Hub coordination platforms (ADRRN, 2024). However, in the absence of binding mechanisms for policy integration or mutual accountability, governance within the network remains adaptive but uneven—mirroring the fragmented landscape of disaster governance in Asia more broadly.

5.4.4 From Local Experience to Global Norms: Reversing the Knowledge Hierarchy

ADRRN's strategic emphasis on regional learning not only enhances localized response but reconfigures how knowledge is produced, validated, and institutionalized in disaster governance. Rather than positioning local actors as passive recipients of external expertise, the network foregrounds them as epistemic agents whose situated practices inform regional—and increasingly global—policy frameworks (Dressler, 2017; Williams, 2018). This shift disrupts the conventional top-down flow of norms and reflects ADRRN's broader commitment to knowledge sovereignty.

Several thematic Hubs operationalize this inversion of knowledge hierarchy. The Innovation Hub curates grassroots innovations that emerge from frontline response contexts and embeds them into formal institutional routines. These innovations are not abstract prototypes but derived from lived experience—ranging from community-led early warning mechanisms to context-specific recovery models—subsequently scaled through inter-hub sharing and regional forums (ADRRN, 2025a). The Hub functions as a laboratory for adaptive governance, facilitating policy feedback loops between field-level experimentation and strategic coordination.

The Digital Storytelling Hub complements this by amplifying vernacular knowledge through narrative platforms. It empowers communities to document their own adaptive strategies, risk perceptions, and histories of resilience using accessible digital tools. These narratives are not merely symbolic; they serve as evidence in cross-regional learning exchanges and advocacy efforts. In doing so, the Hub democratizes representation in regional forums and challenges dominant discourses that privilege technocratic expertise (ADRRN, 2025a; Carter-Tod et al., 2025).

Further, Hubs such as CBDRRM and the Drone Hub demonstrate how local practices evolve through technical co-production. CBDRRM initiatives foreground participatory risk mapping and leadership training, while the Drone Hub enables grassroots actors to integrate geospatial tools into disaster risk reduction planning. This intersection of community engagement and technological innovation exemplifies the "embedded innovation" model emphasized in regional learning (ADRRN, 2025a; Lai & Fu, 2021).

These efforts have tangible normative impact. Elements of ADRRN's localized adaptation models—such as the Locally Led Adaptation (LLA) framework—have been referenced in global discussions on equitable climate governance, and community-led recovery strategies piloted through ADRRN have informed UNDRR regional consultations (ADRRN, 2022, 2024). Such instances affirm that knowledge produced at the margins can shape policy centers, provided there are institutional pathways for recognition and dissemination.

5.5 Conclusion

ADRRN exemplifies how regional civil society networks can reconfigure disaster governance from below—through decentralized coordination, moral legitimacy, and the elevation of local knowledge into global norms. Its hub-based architecture fosters adaptive collaboration, while trust-based governance and peer learning reinforce cohesion across diverse political and institutional contexts. Rather than passively aligning with international frameworks, ADRRN asserts representational agency in global forums, grounded in experiential authority and community embeddedness. Challenges in evaluation, internal asymmetries, and legitimacy politics remain. Even so, the network's ability to transform localized innovations into normative frameworks signals a shift in humanitarian power dynamics—one where Southern actors do not merely implement global agendas but actively shape them. ADRRN thus represents both a practical model and a political project of regional self-determination.

References

ADRRN. (2022). *Strategy 2030: A resilience movement for Asia.* https://www.adrrn.net/wp-content/uploads/2025/05/ADRRN-Strategy-2021-2030-F.pdf

ADRRN. (2024). *2024 CSO commitment to action: Addressing hotspots of vulnerabilities to advance SFDRR implementation.* https://www.adrrn.net/wp-content/uploads/2025/05/2024-CSO-Commitment-to-Action_ADRRN.pdf

ADRRN. (2025a). *ADRRN's HUBS.* https://www.adrrn.net/adrrns-hubs/

ADRRN. (2025b). *SIDRRA-supported locally led adaptations (LLA) grant.* https://www.adrrn.net/sidrra-project/

ADRRN. (2025c). *The SIDRRA Project—Strengthening inclusive disaster risk governance for climate resilience in Asia.* https://www.adrrn.net/wp-content/uploads/2025/05/SIDRRA-Project-Brief-2.pdf

Carter-Tod, F., Ford, J. V., & Weissman, J. L. (2025). "You start with the community": The value of community-based approaches to COVID-19 in Sri Lanka. *Community Health Equity Research & Policy.* https://doi.org/10.1177/2752535X251337660

Dodd, W., Brubacher, L. J., Speers, S., Servano, D., Go, D. J., & Lau, L. L. (2023). The contributions of religious leaders in addressing food insecurity during the COVID-19 pandemic in the Philippines: A realist evaluation of the Rapid Emergencies and Disasters Intervention (REDI). *International Journal of Disaster Risk Reduction, 86*, 103545. https://doi.org/10.1016/j.ijdrr.2023.103545

Dressler, W. H. (2017). Contesting moral capital in the economy of expectations of an extractive frontier. *Annals of the American Association of Geographers, 107*(3), 647–665. https://doi.org/10.1080/24694452.2016.1261684

Galkina, T., & Yang, M. (2020). Bringing Nordic slush to Asia: Entrepreneurial internationalization of an NGO as a social movement. *International Business Review, 29*(6), 101749. https://doi.org/10.1016/j.ibusrev.2020.101749

Izumi, T., & Shaw, R. (2012). Effectiveness and challenges of an Asian NGO network for disaster reduction and response. *Risk, Hazards & Crisis in Public Policy, 3*(2), 1–15. https://doi.org/10.1515/1944-4079.1106

Lai, C.-H., & Fu, J. S. (2021). Exploring the linkage between offline collaboration networks and online representational network diversity on social media. *Communication Monographs, 88*(1), 88–110. https://doi.org/10.1080/03637751.2020.1869797

Ng, I. (2024). Just around the corner: How cross-ethnic solidarity is established between locals and the asylum-seekers and refugees in a local NGO. *Journal of International Migration and Integration.* https://doi.org/10.1007/s12134-024-01166-5

Partelow, S., Fujitani, M., Williams, S., Robbe, D., & Saputra, R. A. (2023). Disaster impacts, resilience, and sustainability opportunities for Gili Trawangan, Indonesia: Transdisciplinary reflections following COVID-19. *Disasters, 47*(2), 499–518. https://doi.org/10.1111/disa.12554

Shipper, A. W. (2012). Influence of the weak: The role of foreigners, activism, and NGO networks in democratizing Northeast Asia. *International Studies Quarterly, 56*(4), 689–703. https://doi.org/10.1111/j.1468-2478.2012.00742.x

Williams, L. D. A. (2018). The South Asian origins of the global network to eradicate blindness: WHO, NGOs, and decentralization. *Endeavour, 42*(1), 27–41. https://doi.org/10.1016/j.endeavour.2018.03.001

Chapter 6
U.S. VOAD: Institutional Embedding and Coordination Through the 4Cs Principle

Abstract The evolution of the U.S. VOAD system reflects a broader shift from informal volunteer networks to institutionally embedded governance actors. This chapter examines how VOADs become integrated into multi-level disaster governance through formal structures, procedural tools, and the operationalization of the 4Cs—Communication, Coordination, Collaboration, and Cooperation. First, it analyzes the institutional embedding of VOADs across national, state, and local levels, highlighting persistent gaps and emerging reform strategies. Second, it explores how the 4Cs have evolved from normative principles into procedural mechanisms that govern inter-organizational action. Third, it assesses how professionalization and standardization enable VOADs to function as credible, risk-aware partners. Finally, it examines how volunteerism, when institutionally structured, becomes a vehicle for equity and community-based long-term recovery.

Keywords Institutional embedding · 4Cs governance · Institutional professionalization · Procedural standardization · Equity and community recovery

6.1 Embedding VOAD into Multi-level Governance Systems

From a loosely connected volunteer coalition to a more structured governance actor, the U.S. Voluntary Organizations Active in Disaster (VOAD) system illustrates an ongoing process of institutional embedding across national, state, and local levels. Effective disaster coordination depends not only on shared values but also on formalized structures, clearly defined roles, and sustainable mechanisms. By tracing the evolution of VOAD's multi-level architecture, this chapter highlights both the strengths of localized engagement and the persistent gaps in vertical integration. Through standardization efforts, procedural linkages, and digital coordination strategies, VOADs are progressively formalizing their place within the broader ecosystem of disaster governance—though structural imbalances and under-institutionalization remain key challenges.

Q. Du and R. Shaw, *Social Organizations in Disaster Response and Recovery*, Disaster Risk Reduction, https://doi.org/10.1007/978-981-92-2846-1_6

6.1.1 Why Institutional Embedding Matters

Institutional embedding is essential for transforming loosely organized volunteer coalitions into durable, legitimate actors within formal governance systems. For voluntary organizations active in disaster response, embedding refers not only to their physical presence or frequency of participation in disaster events but also to their structural integration into multi-level emergency management systems through formal roles, protocols, and accountability frameworks. In the context of the USA, the National Voluntary Organizations Active in Disaster (NVOAD) network has gradually evolved from an informal convening space into an increasingly structured coordination platform. However, its degree of institutional embedding remains uneven across governance layers, particularly at the state and local levels (Jensen, 2024; NVOAD, 2023).

Despite NVOAD's official recognition as a key coordinating body in the U.S. National Response Framework (NVOAD, 2024c), many of its affiliated State VOADs (SVOADs) lack legal incorporation, permanent staff, or formalized partnerships with state emergency management agencies. In a cross-state comparative study, 60% of SVOADs reported having no permanent funding mechanism, and nearly half lacked a memorandum of understanding with their respective state governments (Jensen, 2024). These gaps severely limit the capacity of NVOAD structures to function as embedded governance actors during response and recovery operations.

The consequences of weak institutional embedding become more visible in times of crisis. During the COVID-19 pandemic, digital communication networks on platforms like Twitter showed that while many national VOAD member organizations occupied central roles in disseminating information, their actual authority in decision-making or resource coordination was limited, especially when compared with public agencies or large corporate foundations (Gong et al., 2023). Similarly, multi-level network analysis of resource flows found that national VOAD actors were often "central in name, peripheral in exchange" when it came to material support, underscoring the structural thinness of vertical coordination (Clay Mathews et al., 2024).

While the NVOAD system promotes normative principles such as the 4Cs—Communication, Coordination, Collaboration, and Cooperation—these ideals alone cannot substitute for institutionalized mechanisms that guarantee interoperability, shared authority, and sustainability. Without formal structures that bind NVOADs to public systems, participation remains contingent, coordination ad hoc, and accountability diffuse (Jensen, 2024; NVOAD, 2023). Embedding NVOADs into disaster governance, therefore, is not simply a matter of inclusion—it is foundational for enabling them to function as reliable, predictable, and professionally recognized actors across the full disaster lifecycle.

6.1.2 Structural Analysis: The Multi-level Organizational System of VOAD

Understanding the structural configuration of the VOAD system is critical to evaluating its capacity for institutional embedding and disaster governance. The architecture of the National Voluntary Organizations Active in Disaster is organized across three interrelated tiers: national, state, and community. While this multi-level model aims to foster cross-scalar coordination, its institutionalization remains highly uneven, with gaps in authority, integration, and resource flow among the different levels.

At the national level, VOAD functions as a convening and standard-setting platform for more than 70 member organizations involved in disaster response, including major nonprofits, faith-based groups, and philanthropic institutions (NVOAD, 2023). It plays a critical role in advancing shared principles—most notably the 4Cs framework (Communication, Coordination, Collaboration, and Cooperation)—and in producing operational guidance such as Points of Consensus, training curricula, and technical standards (NVOAD, 2024a, 2024c). While its national coordination role is symbolically strong, it does not directly manage field operations, and its influence over state and local VOADs is largely normative rather than directive.

At the state level, the structure is notably more fragmented. State VOADs (SVOADs) are generally underfunded, inconsistently staffed, and lack formal mandates within emergency management frameworks. Fewer than half of U.S. SVOADs have legal incorporation or defined governance protocols, and many operate without sustainable funding (Jensen, 2024). Their authority is largely relational, often limited to facilitating inter-organizational dialogue rather than executing or coordinating service delivery. This limited institutionalization makes them vulnerable to role ambiguity, especially during large-scale disaster events that require rapid, cross-sector coordination.

Community or local VOADs (often referred to as COADs) tend to exhibit higher levels of operational density. Their proximity to affected populations enables more agile response and recovery interventions, particularly through Long-Term Recovery Groups and volunteer mobilization networks (NVOAD, 2024b, 2025a). However, local COADs are heavily reliant on informal leadership, volunteer labor, and inconsistent funding, which raises concerns about sustainability and accountability.

The contrast between strong horizontal coordination at the local level and weak vertical integration across governance tiers reveals a structural tension within VOAD-based disaster governance. Multi-level network analyses indicate that local VOAD networks are characterized by dense horizontal ties, whereas connections linking local, state, and national levels remain comparatively thin (Clay Mathews et al., 2024). This vertical fragmentation limits cross-level information transmission, weakens strategic–operational alignment, and reduces the system's capacity for rapid, coordinated resource mobilization under crisis conditions.

A truly embedded VOAD system must therefore address these vertical gaps by institutionalizing formal ties between NVOAD, SVOADs, and COADs, ensuring not only alignment of principles but also coordination of authority, funding, and accountability mechanisms.

6.1.3 *Mechanisms of Embedding: How VOAD Advances Institutionalized Governance*

Institutional embedding of the VOAD system depends not only on its structural expansion but also on the formalization of its tools, processes, and coordination roles within the broader emergency management ecosystem. Over the past decade, VOAD has developed a variety of instruments and integration strategies to support this transition from informal coordination to procedural governance.

One key mechanism lies in the creation and diffusion of standardized tools and normative frameworks. National-level consensus documents such as the Volunteer Management Points of Consensus, Cleanup Definitions, and Long-Term Recovery Group Committee Guidelines serve to align expectations across member organizations, clarify responsibilities, and professionalize disaster services (NVOAD, 2024a, 2024b, 2024d). For example, the Volunteer Management Points of Consensus outlines protocols for liability, cultural sensitivity, and spontaneous volunteer intake—components that help institutionalize accountability and interoperability among VOAD actors at different levels (NVOAD, 2024d). These documents function not merely as ethical guidelines but as quasi-regulatory templates embedded into recovery operations and state-level planning tools (NVOAD, 2025a).

In addition to normative alignment, VOAD organizations have increasingly pursued procedural integration into existing public emergency management systems, particularly through the Incident Command System (ICS). As noted in early consensus-based guidelines, mental health responders affiliated with VOADs are expected to be ICS-trained in order to operate within FEMA-aligned Multi-Agency Coordination Systems (Everly et al., 2008). This move reflects a shift from autonomous response to synchronized participation, effectively embedding VOAD capacities into incident-based command hierarchies.

A third and more recent embedding strategy concerns network-centered role consolidation. During the COVID-19 pandemic, nonprofit organizations within the VOAD network emerged as communication and information hubs across social media ecosystems. Network analysis shows that many VOAD members occupied high-centrality positions in disaster-related Twitter networks, effectively shaping real-time information flows and facilitating organizational coordination (Gong et al., 2023). These findings indicate that VOAD is evolving into a de facto platform for connective governance, one that transcends sectoral boundaries while reinforcing operational legitimacy.

Together, these mechanisms demonstrate how VOAD and its affiliates are institutionalizing their roles through standardization, procedural coupling, and digital centrality. This embedding process allows VOADs to shift from being auxiliary actors to becoming embedded nodes within the broader architecture of disaster governance—capable of sustained coordination, professional accountability, and strategic integration.

6.1.4 Structural Gaps and Reform Opportunities

While the U.S. VOAD network aspires to function as a vertically integrated system of disaster coordination, significant asymmetries persist across its national, state, and local layers. These structural imbalances constrain the institutional embedding of VOADs, particularly in their ability to function as formal governance actors beyond the voluntary domain.

A major source of fragility stems from the limited institutional centrality of the national VOAD body. Despite being nominally positioned at the top of the coordination hierarchy, NVOAD often struggles to translate its legitimacy into operational leverage. Empirical findings show that while it serves as a symbolic center for convening and advocacy, its actual role in resource allocation and decision-making remains peripheral. As noted in a multi-level network study, "NVOAD is central in name, but peripheral in exchange," reflecting its constrained authority within disaster resource ecosystems (Clay Mathews et al., 2024). This disconnect weakens the capacity of the national level to serve as a robust integration node, thereby fragmenting the vertical coherence of the VOAD system.

The challenge is further compounded at the state VOAD (SVOAD) level, where structural underdevelopment remains a persistent barrier. Analysis of multiple state contexts reveals that SVOADs frequently operate without formal mandates, dedicated funding, or embedded liaison roles within emergency management systems (Jensen, 2024). These organizations are simultaneously under-defined, under-resourced, yet over-expected to fulfill coordination roles during disasters. Their "soft authority" often fails to translate into durable partnerships or cross-sector integration, leading to siloed responses and minimal vertical alignment with COADs and NVOAD (Jensen, 2024).

Despite these weaknesses, several reform strategies are emerging. NVOAD's recent push to institutionalize "Points of Consensus," standard operating protocols, and long-term recovery guidance has started to address normative fragmentation across layers (NVOAD, 2024b, 2025a). However, deeper integration may require establishing embedded liaison positions that link SVOADs to both emergency management agencies and COADs on a continuous basis (Clay Mathews et al., 2024). Furthermore, improved digital infrastructure and interoperable data-sharing platforms could serve to reduce coordination latency and enhance vertical situational awareness (Gong et al., 2023).

Addressing these structural imbalances is not merely a matter of administrative efficiency—it is essential for ensuring that VOADs can function as institutionally embedded nodes in the broader architecture of disaster governance. Without deliberate reforms to clarify roles, assign responsibilities, and align incentives across layers, the transformative potential of the VOAD model will remain unrealized.

6.2 Operationalizing the 4Cs: From Norms to Governance Mechanisms

The institutionalization of the 4Cs—Communication, Coordination, Collaboration, and Cooperation—represents a foundational shift in VOAD's governance trajectory. Initially conceived as normative ideals, the 4Cs have evolved into actionable governance principles embedded in protocols, procedures, and sectoral infrastructures. This section traces their transformation from symbolic values to operational frameworks through four analytical lenses: the historical evolution of the 4Cs as institutional logics; the standardization of language, roles, and agreements; their adaptability in digital and specialized governance contexts; and their role in repositioning VOADs from voluntary networks to structured nodes of collaborative authority.

6.2.1 From Normative Principle to Institutional Practice: The Evolution of the 4Cs

Originally introduced by NVOAD as a guiding framework for voluntary collaboration, the 4Cs—Communication, Coordination, Collaboration, and Cooperation—were intended to establish a shared behavioral ethos among diverse actors responding to disasters (NVOAD, 2024b, 2025c). At inception, the 4Cs served primarily as a normative compass, offering a vocabulary of partnership and mutual respect. However, as the operational demands of disaster response intensified, the 4Cs have undergone a functional evolution, shifting from rhetorical ideals to embedded governance principles that underpin protocols, role definitions, and inter-organizational accountability.

This evolution reflects a deliberate institutional turn. Rather than remaining a set of aspirational values, the 4Cs have been systematically incorporated into key programmatic tools and regulatory templates. The goal is to ensure that multi-organizational participation during crises is not merely ad hoc but governed by predictable procedures and distributed responsibilities. For example, the Volunteer Management Points of Consensus document (NVOAD, 2024d) translates the abstract tenets of "cooperation" and "coordination" into specific operational standards—covering areas such as data sharing, liability coverage, spontaneous volunteer intake, and cross-cultural sensitivity. These standards ensure that collaboration becomes a mechanism for distributed governance rather than discretionary action.

The transformation is also evident in the production of field-specific technical norms. The Cleanup Definitions (NVOAD, 2024a) distinguish between activities like "debris removal" and "demolition," enabling organizations to communicate clearly within multi-agency environments. In this case, "communication" is no longer an abstract moral good, but a regulatory condition for workflow synchronization. Similarly, the Long-Term Recovery Group Committee Guidelines institutionalize "collaboration" by assigning functional roles to subcommittees such as finance, case management, and spiritual care (NVOAD, 2025a). This reflects a movement away from volunteerism as flexible aid toward embedded governance via procedural clarity.

The 4Cs framework has thus been repurposed from a symbolic platform for unity into a structural mechanism for collaborative governance. Through formalization across documents, procedures, and training curricula, VOAD organizations have internalized the 4Cs not only as ethical commitments but as rule-based operational logic. This transformation is uneven: as recent analysis shows, many State VOADs still struggle to institutionalize these norms due to unclear mandates and fragile resourcing (Jensen, 2024). Nonetheless, the 4Cs remain the conceptual backbone of VOAD's governance identity, enabling coordination to shift from voluntarism to structured interdependence.

6.2.2 Institutional Tools and the Standardization of Terms, Procedures, and Agreements

The operationalization of the 4Cs framework—Communication, Coordination, Collaboration, and Cooperation—has been made possible through the institutional embedding of standardized tools that translate abstract values into actionable mechanisms. At the heart of this transformation are NVOAD's Points of Consensus documents, which function as de facto regulatory instruments, aligning diverse organizations through shared terminology, predefined workflows, and clearly distributed responsibilities (NVOAD, 2024d, 2025a).

A core entry point for standardization lies in the realm of communication, particularly the effort to develop shared operational language. The Cleanup Definitions document, for instance, delineates distinctions between terms such as "debris removal," "muck and gut," and "demolition," ensuring consistency in service expectations, eligibility criteria, and liability management during recovery phases (NVOAD, 2024a). Such semantic precision facilitates cross-organizational coordination, minimizing misunderstandings, and reducing redundancy during field operations.

For coordination, the Volunteer Management Points of Consensus provides detailed procedural guidance for organizations working with unaffiliated volunteers. This includes protocols for background checks, cultural and linguistic matching, liability insurance, and data protection practices (NVOAD, 2024d). Notably, the document specifies responsibilities around volunteer registration and intake coordination, which are critical for organizations operating across state or federal lines. These standards not only align expectations but also serve as formalized risk management strategies within multi-agency environments.

Institutionalized collaboration and cooperation are especially evident in the Long-Term Recovery Group operational guidelines. These frameworks formalize organizational roles via designated subcommittees—such as "case management," "construction," or "spiritual and emotional care"—each with specific scopes of work, reporting relationships, and accountability mechanisms (NVOAD, 2025a). These substructures effectively transform cooperation into an institutionalized architecture of interdependence, embedded in both planning templates and disaster recovery workflows.

Taken together, these tools do more than enable smoother interactions; they constitute the procedural infrastructure of collaborative governance. Rather than relying on goodwill or informal networks, VOADs now increasingly operate through codified agreements and standardized operating procedures that are interoperable with public systems. However, persistent asymmetries—particularly in how State VOADs adopt and enforce these tools—underscore the continued need for institutional investment and cross-level alignment (Jensen, 2024).

6.2.3 Expanding the Governance Scenarios: Digital Communication and Sectoral Adaptation

The institutionalization of the 4Cs framework extends beyond static procedural standardization; its adaptability in dynamic and specialized governance scenarios—such as digital communication environments and sector-specific interventions—demonstrates the evolving complexity of VOAD's collaborative infrastructure.

In the digital sphere, disaster governance increasingly hinges on the rapid exchange of information across organizational boundaries. Analysis of Twitter networks during the COVID-19 pandemic reveals that VOAD-affiliated nonprofits consistently occupied high-centrality positions in digital communication flows (Gong et al., 2023). Rather than merely disseminating public messaging, these organizations facilitated real-time coordination of resources, volunteer mobilization, and cross-sector response. The centrality of VOAD actors within these networks supports the view that the 4Cs—particularly communication and coordination—are no longer confined to analog interaction but have been embedded into emerging digital governance architectures.

This digital shift is further supported by formal guidance from NVOAD documents that recommend integrating situational awareness tools and centralized information dashboards in disaster response operations (NVOAD, 2024c). These mechanisms allow members to exchange needs assessments, logistics data, and deployment schedules in formats that support shared situational understanding, especially under the time constraints typical of large-scale emergencies.

Beyond communication technologies, the institutional logic of the 4Cs has been effectively transplanted into sectoral subsystems, notably mental health and psychosocial support. In post-disaster psychological interventions, multi-agency teams have adopted standardized training models, interoperable terminology, and role-specific protocols to ensure clarity in emotional and spiritual care delivery (Everly et al., 2008; NVOAD, 2025a). In this context, the principle of collaboration is operationalized through predefined committee structures, such as the "emotional and spiritual care subcommittee," which formalizes inter-organizational cooperation within the Long-Term Recovery Group framework (NVOAD, 2025a).

6.2.4 From Normative Vision to Procedural Governance: VOAD's Institutional Turn

The transformation of VOAD from a network premised on voluntary participation to a structured coordination platform marks a significant shift in the institutional logic of disaster governance. At the heart of this evolution is the proceduralization of the 4Cs—Communication, Coordination, Collaboration, and Cooperation—not simply as aspirational norms, but as embedded mechanisms that shape inter-organizational behavior and accountability.

This proceduralization is evident in the proliferation of codified tools such as Memoranda of Understanding (MOUs), joint operating procedures, and standard Points of Consensus. For instance, formal agreements now routinely define role expectations between national and state VOADs, set criteria for activation, and stipulate protocols for cross-organizational data sharing and liability management (NVOAD, 2024d, 2025a). In the context of Long-Term Recovery Groups, meeting structures and decision-making processes are explicitly designed around collaborative protocols, ensuring that coordination is not left to informal goodwill but follows operational logics (NVOAD, 2025b).

Organizationally, VOADs have undergone a governance shift from consensus-driven collectives to multi-actor institutional nodes with defined leadership roles, committee systems, and embedded coordination authorities. This is reflected in the increasing formalization of task-specific structures—such as volunteer management subcommittees or spiritual care units—whose responsibilities are encoded in procedural manuals and capacity-building workshops (NVOAD, 2024a, 2025a). Such arrangements enable not only horizontal cooperation but also vertical accountability across national, state, and local levels.

Despite these advances, structural tensions persist—particularly at the state VOAD level—where coordination authority remains uneven and implementation of national protocols is often hindered by fragmented funding streams and unclear mandates (Jensen, 2024). Inconsistent access to training, institutional knowledge, and legal frameworks further exacerbates the gap between normative aspirations and operational execution.

These institutional frictions underscore the dual challenge VOAD faces: scaling procedural governance while retaining the flexible ethos of volunteerism. The 4Cs framework has proven essential in anchoring this transition, serving as a bridge between collaborative intent and institutional accountability. However, its full institutionalization depends on continuous investment in legal clarity, resource parity, and adaptive governance infrastructures that can sustain multi-level coordination during both acute crises and long-term recovery.

6.3 Professionalization and Standardization in VOAD Practice

The professionalization of VOAD practice reflects a fundamental shift in the role of voluntary organizations—from morally driven actors to system-integrated providers of institutional services. This chapter examines how VOADs enhance their legitimacy and coordination capacity through standardization. It begins by explaining why professionalism is essential in increasingly complex disaster environments. It then analyzes how consensus-based tools translate volunteer actions into procedural norms, and how capacity-building efforts ensure sustainable replication. Finally, it explores the institutional implications of these transformations, arguing that professionalism enables VOADs to gain trust, ensure accountability, and operate as credible partners in disaster governance.

6.3.1 Why Professionalization Matters: From Moral Commitment to Systemic Trust

As disasters have grown in scale, frequency, and systemic complexity, the informal ethos of volunteerism—once a source of moral legitimacy—has increasingly come under pressure. In large-scale or multi-jurisdictional emergencies, the absence of shared standards and procedural expectations has produced coordination gaps, service duplication, and, in some cases, inadvertent harm. While VOAD organizations historically operated as value-based communities premised on shared goodwill, contemporary disaster governance demands their transition into professional practice communities capable of operating within formal institutional settings.

In this context, professionalization has emerged not as a retreat from the spirit of volunteerism, but as its necessary evolution. The legitimacy of VOADs in multi-organizational disaster response now rests less on the intensity of civic mobilization and more on their ability to deliver consistent, predictable, and interoperable services. Recent guidance documents emphasize "safety," "accountability," and "operational consistency" as non-negotiable dimensions of volunteer engagement—particularly in high-risk environments such as post-disaster debris removal, spiritual care, or shelter management (NVOAD, 2024a, 2024d). These values reflect a shift from moral voluntarism to procedural reliability.

This transformation is also mirrored in the expectations of governmental and institutional partners. In assessing the performance and trustworthiness of nonprofit actors, public agencies increasingly prioritize organizational clarity and standardized practices over mere capacity for mobilization. As highlighted in recent empirical work, trust in voluntary organizations is tightly coupled with institutional embeddedness—i.e., the extent to which roles, responsibilities, and interfaces are formally defined and operationally intelligible (Özkan & Uzun, 2024). For VOADs, failing to professionalize is not only a coordination risk, but a legitimacy deficit in the eyes of state and philanthropic actors.

Moreover, the risks of unprofessionalized engagement are not theoretical. Instances of untrained or unaffiliated volunteers jeopardizing safety, mishandling sensitive information, or undermining local recovery plans have led to growing demands for credentialing, liability protocols, and formal coordination mechanisms (NVOAD, 2024d, 2025a). In this light, professionalization becomes a prerequisite for trust-based integration into broader disaster governance systems.

Ultimately, the push toward professionalization reflects a deeper institutional shift: VOADs are no longer peripheral actors supplementing official response but are now expected to function as accountable, system-compatible agents of recovery. Making this possible requires codifying expectations, building replicable capacities, and translating moral commitment into institutional form.

6.3.2 *Core Mechanisms: Translating Volunteerism into Institutional Language Through Standardized Tools*

The institutional turn of VOADs toward professionalized disaster governance hinges on their capacity to convert moral commitment into operational standards. At the center of this transformation lies a suite of standardized instruments—particularly the Points of Consensus—which serve as soft-law governance tools, articulating a shared vocabulary, role boundaries, and baseline performance expectations across member organizations (NVOAD, 2024a, 2024d). These documents do not function as top-down mandates but as negotiated benchmarks, acceptable across religious, cultural, and organizational diversity, thereby enabling interoperability in complex multi-agency environments.

One illustrative example is the Cleanup Points of Consensus, which translates what was once a largely intuitive practice—post-disaster cleanup—into a standardized typology. By differentiating between "debris removal," "muck and gut," and "demolition," the document ensures consistency in eligibility assessments, service planning, and legal liability (NVOAD, 2024a). This semantic precision is not a bureaucratic detail but a precondition for safe, coordinated field operations involving multiple actors with varied capacities.

Similarly, the Volunteer Management Points of Consensus establishes protocols that reframe volunteers not as ad hoc labor but as managed actors within a governed system. It sets out expectations around registration, background screening, cross-cultural placement, insurance, and data privacy (NVOAD, 2024d). In high-pressure disaster contexts, these guidelines enable VOADs to reduce risks, streamline intake, and align with public sector accountability standards, thus enhancing their credibility as institutional partners.

Sector-specific standards also reflect this professionalizing impulse. The Long-Term Recovery Group Guidelines offer a modular template that specifies committee structures for case management, construction, and emotional and spiritual care (NVOAD, 2025a). Within these, standards of care are articulated for each service stream, transforming compassion-based work into procedural routines with defined responsibilities and reporting lines. For instance, the Emotional and Spiritual Care (ESC) subcommittee operates under a formal mandate, with ethical guidelines, referral protocols, and documentation requirements to ensure that care delivery aligns with both community needs and professional expectations (NVOAD, 2025a).

6.3.3 Capacity Building and Organizational Reproduction: Sustaining Professionalization in VOAD Practice

The professionalization of VOADs cannot be sustained through documentation alone; it requires deliberate mechanisms for replication, adaptation, and continuity. Standardized procedures are only as effective as the organizational capacity to internalize and reproduce them. Recognizing this, NVOAD and its member organizations have invested in a layered infrastructure of training workshops, modular curricula, and technical assistance platforms aimed at institutionalizing competence across multiple operational domains (NVOAD, 2015, 2025a).

One of the most significant tools for professional learning is the Recovery Tools Workshop, a training initiative designed to equip Long-Term Recovery Groups and local VOADs with the practical knowledge required for case management, construction coordination, volunteer supervision, and financial oversight. These workshops are not ad hoc learning sessions but structured pedagogical environments where Points of Consensus documents are unpacked, applied, and contextualized for local needs (NVOAD, 2025a). This localized transfer of expertise ensures that national standards are not merely disseminated but adapted through situated practices.

Modular training programs serve a complementary function by enabling targeted capacity development within specialized subfields. For example, Emotional and Spiritual Care (ESC) providers receive training on boundaries of care, documentation protocols, and interfaith coordination (NVOAD, 2025a), while Disaster

Case Managers (DCMs) are instructed in data security, intake prioritization, and case closure procedures. These modules support the operationalization of standards through role-specific learning, creating a culture of competence that is both horizontally scalable and vertically accountable.

The emphasis on reproducibility is particularly salient in VOAD's international engagements. Drawing on the lessons from global humanitarian coordination, the 2015 guidelines on International VOAD Coordination underscore the need for rapid deployment teams to align with pre-established templates and shared ethical codes (NVOAD, 2015). This reflects a broader institutional logic: professionalization must not only address the present workforce but also prepare future personnel through transferable and interoperable systems.

In this regard, VOADs function not merely as conveners of organizations but as "capacity infrastructures" that enable the systemic reproduction of disaster governance competencies. The capacity-building mechanisms embed professional norms within organizational routines, thereby ensuring that the legitimacy of VOAD practice is sustained across turnover, geographic variation, and crisis types. Professionalization thus becomes not a finite intervention but a cyclical process of learning, adaptation, and institutional memory.

6.3.4 Institutional Implications: From Professionalization to Legitimacy and Trust

The institutionalization of professional standards within VOAD operations is not merely an exercise in technical refinement. It serves as a foundational strategy for building legitimacy and trust—both within the disaster response ecosystem and in the eyes of external stakeholders such as public agencies, funders, and affected communities. In complex, multi-organizational environments, predictability of behavior is a prerequisite for collaboration. By codifying procedures and role expectations, VOADs reduce uncertainty and render their actions auditable, thus reinforcing their status as trustworthy governance actors (NVOAD, 2024d; Özkan & Uzun, 2024).

At the operational level, professionalization enables risk-sensitive and rights-based coordination. For example, the Volunteer Management Points of Consensus outlines procedures for background checks, liability coverage, and data protection, offering public authorities assurance that volunteers are not only well-intentioned but also properly vetted and managed (NVOAD, 2024d). In disaster contexts where state actors delegate responsibility to non-governmental partners, such standardization becomes essential for legitimizing third-party action. Similarly, the Emotional and Spiritual Care Guidelines define ethical boundaries and referral protocols, ensuring that spiritual support does not violate professional norms or individual rights (NVOAD, 2025a). These practices convert moral commitment into regulatory reliability.

Legitimacy also depends on visibility and evaluability. The increasing use of shared data dashboards, situational awareness platforms, and standardized needs assessments facilitates a transparent operational environment in which partner organizations, including government agencies, can monitor contributions and adjust support accordingly (Gong et al., 2023; NVOAD, 2024c). This alignment between operational data and governance oversight strengthens institutional clarity and enhances governability in cross-sector coordination, which constitutes a key enabling condition for sustained NGO–state cooperation (Özkan & Uzun, 2024).

Importantly, trust is not generated through compliance alone, but through institutional complementarity. VOADs gain credibility when they are seen not as autonomous moral actors but as embedded components of the disaster governance architecture—actors who fill institutional gaps with competence rather than contest authority (Jensen, 2024). By adopting standardized procedures and training systems, VOADs make themselves legible to bureaucratic partners, and thereby indispensable in policy implementation, resource targeting, and recovery planning.

In this sense, professionalization allows VOADs to transcend their roots in informal volunteerism. It transforms them into dependable governance actors—governed themselves, and capable of governing others through structured, ethical, and accountable engagement. This institutional turn is not the end of volunteerism but its reinvention as a legitimate, scalable mode of public service.

6.4 Volunteerism, Equity, and Community-Based Long-Term Recovery

Volunteerism plays a vital yet double-edged role in disaster recovery: while it mobilizes civic commitment and flexible labor, its effectiveness hinges on institutional design. This section examines how VOAD structures transform spontaneous volunteerism into an engine for equitable, community-led, and sustainable recovery. It first outlines the limits of unregulated volunteer action and the need for governance-based safeguards. It then analyzes how Long-Term Recovery Groups embed equity through local coordination, and how VOADs, as multi-level intermediaries, address systemic disparities by aligning data, resources, and accountability. Across all levels, institutionalized volunteerism emerges as the foundation for just and resilient recovery.

6.4.1 Rethinking Volunteerism: Institutional Preconditions for Equitable and Sustainable Recovery

In the aftermath of disasters, volunteer labor often serves as the most immediate and visible form of response, offering emotional support, physical reconstruction, and symbolic solidarity. As recovery processes extend beyond initial relief into longer-term rebuilding, the limits of ad hoc volunteer mobilization become increasingly evident. Disaster recovery is not merely about restoring infrastructure; it is fundamentally about reconstructing social systems, restoring community trust, and addressing pre-existing vulnerabilities. In this context, volunteerism—if left unstructured—risks exacerbating, rather than alleviating, post-disaster inequalities.

Unregulated volunteer engagement can result in resource duplication, unmet local needs, and inadvertent harm. For example, the influx of unaffiliated volunteers has, in past responses, overwhelmed local capacity, misaligned with community priorities, and even caused "second-order damage" through culturally inappropriate or technically unqualified interventions (NVOAD, 2007). Without institutional mechanisms for coordination, volunteers may become disconnected from the very recovery processes they intend to support. This underscores a critical design problem: equitable recovery requires not just moral commitment, but institutional scaffolding that embeds volunteer action within local governance structures.

Recent frameworks from NVOAD have increasingly emphasized this shift. Volunteer engagement must be made "safe, consistent, and locally responsive" through clearly defined roles, liability coverage, training protocols, and mechanisms for cultural competence (NVOAD, 2024d). Such principles reflect a move from goodwill-based mobilization to system-integrated volunteer governance. More broadly, disaster recovery systems must pivot from episodic mobilization to sustained, community-led processes—a logic embodied in the Long-Term Recovery Group model (LTRG). Rather than treating recovery as an external intervention, the LTRG approach institutionalizes volunteer support within community-driven planning structures, allowing for context-sensitive allocation of time, labor, and resources (NVOAD, 2025a).

Importantly, this institutional orientation is not a bureaucratic constraint on volunteerism but a precondition for its legitimacy and equity. As seen in recent evaluations, volunteer efforts embedded in structured frameworks such as LTRGs are more likely to align with local priorities, reach marginalized populations, and support holistic recovery goals (Gong et al., 2023). In this sense, the question is not whether volunteers can contribute to equitable recovery, but under what conditions their efforts become vehicles for inclusive and durable change.

6.4.2 Volunteer Governance: Embedding Ethical and Operational Safeguards in Volunteer Engagement

Volunteer engagement in disaster recovery cannot be treated as inherently benign or effective. While often driven by altruism, volunteers may inadvertently disrupt local recovery processes if their participation lacks operational oversight, cultural awareness, or legal accountability. Effective disaster recovery demands a shift from informal volunteerism to structured volunteer governance—an approach that defines the rights, responsibilities, and relational boundaries of volunteers as institutional actors.

NVOAD's Volunteer Management Points of Consensus offers a foundational model for this transition. It articulates operational protocols for role clarity, background checks, liability insurance, supervision, and emotional support (NVOAD, 2024d). These provisions move beyond procedural control to reflect a broader philosophy of "responsibility-based volunteerism," in which volunteers are not merely well-meaning helpers but accountable participants embedded in a governed system. Particular emphasis is placed on cultural and linguistic inclusion, trauma-informed care, and the need to avoid retraumatization or "second harm" during engagements with survivors (NVOAD, 2024d). This reflects earlier lessons captured in the 2007 Guide to Volunteer Management, which warned that well-intentioned but uncoordinated interventions could undermine community trust, displace local actors, and generate conflict over resource allocation (NVOAD, 2007).

Rather than constraining volunteerism, such institutional design mechanisms expand its legitimacy and impact. For example, by ensuring that volunteers are assigned roles appropriate to their skill level and contextual understanding, VOAD member organizations reduce the risk of cultural misinterpretation, operational inefficiency, and liability exposure. In post-disaster contexts—especially those involving historically marginalized or linguistically diverse populations—such safeguards are not bureaucratic formality, but prerequisites for ethical and equitable recovery.

Furthermore, institutionalized volunteer governance enables coordination with public sector actors who may otherwise hesitate to delegate tasks to unaffiliated or unvetted individuals. Standardized expectations regarding data privacy, risk mitigation, and interagency communication make volunteers "legible" to the broader governance system, ensuring their efforts are not isolated but integrated into collective recovery planning (NVOAD, 2024d; Özkan & Uzun, 2024).

6.4.3 Community-Based Long-Term Recovery: Embedding Equity Through Localized Governance Structures

Long-term recovery cannot be sustained through external interventions alone; it requires embedded, community-driven mechanisms capable of integrating diverse local needs, actors, and recovery trajectories. Within the NVOAD framework, this imperative has been institutionalized through the development of Long-Term Recovery Groups—multi-organizational platforms designed to facilitate equitable, locally anchored recovery governance. Rather than treating recovery as a technical process of rebuilding infrastructure, the LTRG model reframes it as a socio-institutional process of rebuilding trust, capacities, and distributive fairness.

Operational guidance from Recovery Tools Workshop defines the LTRG as a collaborative structure initiated within the first 90 days following a disaster, comprising workgroups on case management, construction, financial administration, and emotional and spiritual care (NVOAD, 2025a). These groups are intentionally composed of local faith-based organizations, nonprofits, VOAD affiliates, and community leaders, ensuring that decision-making is rooted in situated knowledge and responsive to context-specific vulnerabilities. Unlike externally managed recovery programs, LTRGs are not "implementers" of pre-designed interventions but platforms for co-producing recovery priorities, timelines, and resource allocation strategies.

The governance logic of the LTRG model hinges on two principles: "community-driven" and "needs-based" recovery (NVOAD, 2024b, 2025b). Case management protocols, for example, begin with locally conducted unmet needs assessments, which are used to prioritize assistance based on household vulnerability, not political visibility or administrative ease. In this framework, volunteers are deployed not on a first-come, first-served basis, but in alignment with equity-sensitive prioritization systems facilitated by LTRG committees. Emotional and spiritual care services, likewise, are guided by culturally attuned engagement protocols that reflect local faith traditions and community sensibilities (NVOAD, 2025a).

Crucially, LTRGs are not merely vehicles for delivering services; they are mechanisms for building recovery capacity. Their structure allows for the onboarding of local actors into formal governance roles, transforming community members from passive aid recipients into active recovery agents. This embeddedness enhances both social legitimacy and institutional resilience, particularly in under-resourced or historically marginalized areas (Clay Mathews et al., 2024).

By nesting disaster recovery within locally governed LTRGs, VOAD creates a recovery architecture that is not only more equitable, but also more durable. In doing so, it moves beyond episodic disaster response and toward a model of participatory governance—where equity, capacity, and accountability are built into the recovery process itself.

6.4.4 Addressing Structural Inequities: Building Institutional Resilience Through VOAD Platforms

Equity in disaster recovery is often undermined not by the absence of resources but by asymmetries in their distribution, visibility, and institutional flow. As pointed out in recent multi-level analyses of voluntary aid networks, disaster response frequently suffers from a mismatch between policy directives at higher governance levels and actionable data or feedback from community-based actors—a condition summarized as "policy at the top, data at the bottom" (Clay Mathews et al., 2024). In this context, VOAD platforms are increasingly expected not only to coordinate service delivery but to function as resilience-enabling intermediaries that bridge these systemic gaps.

Addressing such structural inequities requires a shift from reactive coordination to anticipatory infrastructure. One institutional strategy promoted by NVOAD is the formalization of feedback loops connecting local, regional, and state/national levels. These loops are sustained through regular reporting, standardized data sharing, and interoperable communication protocols, many of which are embedded in the Recovery Tools Workshop and affiliated documentation (NVOAD, 2025a). For instance, VOADs are encouraged to utilize unified needs assessment templates and deploy data dashboards that facilitate visibility across jurisdictional boundaries (Gong et al., 2023; NVOAD, 2024c). Such mechanisms allow resource requests and response priorities to be coordinated in real time, mitigating the marginalization of underserved communities.

To further institutionalize equity-sensitive recovery, NVOAD guidance urges local COADs to adopt metrics that track inclusivity across vulnerable groups—such as women, older adults, people with disabilities, and racial or ethnic minorities—thereby embedding distributive fairness into performance evaluation (NVOAD, 2025a). These "equity-sensitive recovery indicators" also inform volunteer deployment strategies, ensuring that aid does not disproportionately favor the most accessible or most visible populations (NVOAD, 2007). Rather than relying on volunteer self-selection or ad hoc referrals, NVOADs promote the use of centralized volunteer registration and credentialing systems to match skillsets with community-defined needs (NVOAD, 2024d).

Ultimately, VOADs' contribution to structural resilience lies in their ability to act as information integrators and normative translators—entities that align spontaneous civic energy with the logic of procedural fairness. By standardizing information flows, codifying inter-organizational interfaces, and promoting equity benchmarks, VOADs help build a governance ecology where responsiveness is not only rapid but just. In doing so, they do not merely mitigate inequity but institutionalize inclusion as a core dimension of recovery governance.

6.5 Conclusion

The institutional evolution of VOAD reflects a deliberate shift from informal volunteer coordination to structured, multi-level disaster governance. Through mechanisms such as the 4Cs framework, Points of Consensus, and standardized recovery protocols, VOADs have become integral to disaster response systems—functioning not only as service providers but as governance actors. However, challenges remain: structural gaps across national, state, and local levels persist, and equitable outcomes depend on institutional safeguards that embed volunteerism within community-led frameworks. By operationalizing professional standards, enhancing vertical coordination, and embedding equity into recovery architectures, VOADs demonstrate how civic capacity can be transformed into institutional resilience. This chapter underscores that sustainable disaster recovery requires not just mobilization, but accountable, inclusive, and procedurally embedded volunteer governance.

References

Clay Mathews, M., Vickery, J., & Peek, L. (2024). Resource exchange patterns between Voluntary Organizations Active in Disaster (VOADs): A multilevel network assessment to improve disaster response capacity. *International Journal of Disaster Risk Reduction, 108*, 104455. https://doi.org/10.1016/j.ijdrr.2024.104455

Everly, G. S., Hamilton, S. E., Tyiska, C. G., & Ellers, K. (2008). Mental health response to disaster: Consensus recommendations: Early Psychological Intervention Subcommittee (EPI), National Volunteer Organizations Active in Disaster (NVOAD). *Aggression and Violent Behavior, 13*(6), 407–412. https://doi.org/10.1016/j.avb.2008.05.004

Gong, X., Peng, S., Lu, Y., Wang, S., Huang, X., & Ye, X. (2023). Social network analysis of nonprofits in disaster response: The case of Twitter during the COVID-19 pandemic in the United States. *Social Science Computer Review, 41*(6), 2029–2054. https://doi.org/10.1177/08944393221130674

Jensen, J. (2024). Conceptualizing State Voluntary Aid Organizations in Disaster (SVOAD) effectiveness. *Risk, Hazards & Crisis in Public Policy, 15*(1), 25–46. https://doi.org/10.1002/rhc3.12268

NVOAD. (2007). *When disaster strikes... how to donate or volunteer successfully!* https://www.nvoad.org/wp-content/uploads/when_disaster_strikes_-2.pdf

NVOAD. (2015). *Guide to developing a VOAD in your country.* https://www.nvoad.org/wp-content/uploads/guidetodevelopingavoad.pdf

NVOAD. (2023). *2023 triennial review.* https://www.nvoad.org/wp-content/uploads/Triennial-Review-2023.pdf

NVOAD. (2024a). *Cleanup definitions and practices.* https://www.nvoad.org/wp-content/uploads/Cleanup-Definitions-and-Practices-2024.pdf

NVOAD. (2024b). *Long-term recovery group committee points of consensus.* https://www.nvoad.org/wp-content/uploads/LTRG-Points-of-Consensus.pdf

NVOAD. (2024c). *Strategic plan 2024–2026.* https://www.nvoad.org/wp-content/uploads/Strategic-Plan-2024-2026.pdf

NVOAD. (2024d). *Volunteer management points of consensus.* https://www.nvoad.org/wp-content/uploads/NVOAD-Volunteer-Management-POC_Final-Ratified-May-2024.pdf

NVOAD. (2025a). *Long term recovery orientation—Recovery tools workshop.* https://www.nvoad.org/wp-content/uploads/Long-Term-Recovery-Orientation-Recovery-Tools-Workshop-One-Pager.pdf

NVOAD. (2025b). *National VOAD guidance for disaster response and recovery for renters and rental properties.* https://www.nvoad.org/wp-content/uploads/Rental-Guidance-Document-2025.pdf

NVOAD. (2025c). *State of disaster recovery: Expending assistance.* https://www.nvoad.org/wp-content/uploads/State-of-Disaster-Recovery-Expediting-Assistance-Report.pdf

Özkan, O., & Uzun, A. (2024). The democratic touch of disasters: The perception of non-governmental organizations in public institutions in the context of disaster governance. *SSRN.* https://doi.org/10.2139/ssrn.4989771

Chapter 7
JVOAD: Civil Society Leadership in Coordinated Disaster Response in Japan

Abstract Japan Voluntary Organizations Active in Disaster (JVOAD) represents a rare case of civil society leadership that is both operationally effective and structurally embedded within Japan's disaster governance system. Its emergence was driven by repeated coordination failures in major disasters from 2011 to 2016, prompting the development of a formally institutionalized platform grounded in legal status, procedural access, and financial stability. Rather than relying on top-down authority, JVOAD organizes disaster response through decentralized networks of NGOs, faith-based groups, and public agencies—facilitating horizontal coordination built on trust and shared protocols. This governance model was put to the test during the 2024 Noto Peninsula Earthquake, where JVOAD served as a connective hub, enabling multi-sector collaboration without commanding resources directly. Underpinning its effectiveness is a foundation of peacetime preparedness, including diversified funding, regional capacity development, and institutionalized training systems that embed resilience into routine practice.

Keywords Network governance · Embedded autonomy · Institutionalized coordination · Organizational resilience · Disaster governance

7.1 From Post-Disaster Collaboration to an Institutionalized Platform: The Emergence and Institutional Embedding of JVOAD

Following a series of large-scale disasters in Japan, the gap between civil society mobilization and the absence of formal coordination mechanisms became increasingly apparent. In response, JVOAD emerged as a hybrid platform bridging this divide. The first part of this section outlines the coordination failures observed between 2011 and 2016, revealing the need for institutional reform. It then traces

Q. Du and R. Shaw, *Social Organizations in Disaster Response and Recovery*, Disaster Risk Reduction, https://doi.org/10.1007/978-981-92-2846-1_7

JVOAD's transformation—from an informal network to a legally recognized intermediary—through its organizational, procedural, and financial embedding. The analysis concludes by identifying three key mechanisms that enabled this transition, offering a contrast to more symbolic models of NGO participation found elsewhere.

7.1.1 Problem Identification and Background: Coordination Gaps in Japan's Disaster Response Landscape

Japan's experience with large-scale disasters in the past two decades has revealed a consistent gap between the high mobilization capacity of civil society and the limited institutional frameworks available for coordinated response. While the number of voluntary organizations and support actors has steadily increased, mechanisms for inter-organizational coordination, information-sharing, and government—nonprofit collaboration remained fragmented or entirely absent prior to 2016.

This gap became most visible during the Great East Japan Earthquake in 2011, when "many volunteers and support organizations gathered in the disaster-stricken areas, but due to the lack of pre-established coordination mechanisms, the overall picture of support was unclear and on-site coordination became difficult" (JVOAD, 2026a). This resulted in duplicated aid in some regions and severe delays in others. Similar breakdowns occurred during the 2015 Kanto-Tohoku Floods and the 2016 Kumamoto Earthquake. In both cases, response efforts were hindered by the absence of centralized information flow and a lack of recognized intermediaries capable of structuring cross-sector collaboration, leading to what was later described as "field-level disorder" and inefficient resource allocation (JVOAD, 2026b).

The recurring nature of such fragmentation pointed to a structural issue: civil society actors had both the will and capacity to contribute, but lacked a formalized institutional interface with government systems. As emphasized in JVOAD's organizational overview, this was not merely a matter of operational inefficiency but of systemic under-preparedness: "there was no platform to connect government, NPOs, and the private sector in a coherent manner before the disaster occurred" (JVOAD, 2026a).

This misalignment between actors reflects a broader challenge discussed in recent disaster governance literature. In hybrid governance settings, the absence of pre-established coordination rules often results in symbolic or reactive inclusion of NGOs, without granting them legitimate operational roles (Özkan & Uzun, 2024). Such exclusion limits adaptive governance, undermines trust, and leaves community-based capacities underutilized during crises.

Japan's case thus reveals a critical paradox: despite its high institutional capacity for emergency response, the country lacked a nationally embedded mechanism that could link civil society's decentralized energy with state-led disaster architecture. The absence of such a mechanism—prior to JVOAD's institutionalization—meant that NGOs were often treated as temporary supplements rather than integrated governance actors. This backdrop sets the stage for understanding how JVOAD emerged as a response to both institutional voids and repeated crisis-driven lessons.

7.1.2 The Progressive Institutionalization of the Coordination Platform

The creation of Japan's JVOAD marked a significant transition from fragmented post-disaster collaboration to an institutionalized civil society platform embedded within the national disaster governance structure. This evolution was not immediate, but the result of accumulated organizational learning and inter-sectoral negotiation following multiple large-scale disasters.

In 2013, civil society actors initiated the "National Disaster Volunteer Support Organizations Network Preparatory Meeting," explicitly modeled on the U.S.-based National VOAD framework (JVOAD, 2026a, 2026b). This informal network sought to create a platform that could facilitate cross-sector communication and shared protocols in anticipation of future disasters. The culmination of this initiative came in October 2016, when JVOAD was formally registered as a certified Nonprofit Organization (認定NPO), marking its transition from a voluntary coalition to a legally recognized entity (JVOAD, 2026a). This legal status granted JVOAD both institutional legitimacy and eligibility for public–private funding streams.

In the years following its formalization, JVOAD achieved embedded governance through a series of memoranda of understanding (MOUs) with national and prefectural governments. These agreements positioned JVOAD as an official coordination hub during disasters, with access to government data and operational channels (JVOAD, 2024a). The organization was granted participation in key national policy platforms, including the Central Disaster Management Council, reflecting its status as a trusted intermediary in state-civil society relations (JVOAD, 2025a). Additionally, JVOAD was commissioned by government bodies to implement tasks such as elderly population mapping and community rebuilding efforts in disaster-affected areas (JVOAD, 2024b).

Financial institutionalization further reinforced JVOAD's platform stability. According to its FY2023 financial report, over 58% of its revenue derived from government-contracted projects, particularly in Ishikawa Prefecture, while the remainder came from private sources such as The Nippon Foundation and designated international donors including Give2Asia (JVOAD, 2024b). This dual-source structure enabled JVOAD to maintain operational independence while securing long-term sustainability.

The pathway from informal coordination to institutionalized governance thus reflects a hybrid model of civil society-led disaster response. Through legal recognition, contractual access to state frameworks, and diversified financial support, JVOAD has positioned itself as a structurally embedded actor within Japan's disaster management system—contrasting sharply with cases where NGOs remain marginal or symbolically included (Özkan & Uzun, 2024).

7.1.3 How Institutional Embedding Was Achieved

The transformation of JVOAD from a reactive coalition into an institutionalized disaster coordination platform illustrates a case of successful embedded governance. This process was not merely technical, but structurally political—emerging through strategic negotiation, legitimation, and trust-building across civil society and state actors.

The post-2011 period in Japan represented a critical juncture in disaster governance. Recurrent large-scale disasters between 2011 and 2016 revealed deep structural gaps in cross-sectoral coordination. Civil society organizations, despite their proactive mobilization, operated in fragmented silos and lacked access to decision-making platforms (JVOAD, 2026a, 2026b). This environment created both urgency and opportunity for institutional reform, enabling civil society networks to claim governance space through demonstrated capacity and negotiated access.

Three mechanisms facilitated JVOAD's institutional embedding. First was legal recognition. By registering as a certified NPO in 2016, JVOAD secured a legal status that granted it not only eligibility for government partnerships but also increased symbolic and procedural legitimacy within Japan's disaster policy architecture (JVOAD, 2026a). This status positioned JVOAD as a public intermediary capable of navigating both state and community expectations.

Second was procedural mechanism-building through formal agreements. JVOAD signed memoranda of understanding with national and local governments, thereby gaining privileged access to operational information and institutional coordination mandates during disaster phases (JVOAD, 2024a). Participation in official policy councils and task forces, such as the Central Disaster Management Council, embedded JVOAD into the governance infrastructure as a recurring actor rather than an ad hoc respondent (JVOAD, 2025a).

Third was financial institutionalization. JVOAD established a hybrid funding structure, with more than 58% of its FY2023 income coming from public contracts—such as support for evacuees in Ishikawa Prefecture—and the rest from designated private grants (JVOAD, 2024b). This fiscal balance reinforced both operational sustainability and political neutrality, allowing JVOAD to maintain autonomy while fulfilling institutional mandates.

Compared to other contexts, such as Turkey—where NGOs are often symbolically included but lack policy access (Özkan & Uzun, 2024)—JVOAD represents a more substantive model of embedded civil society coordination. Through legal positioning, procedural access, and financial integration, JVOAD has evolved into a hybrid coordination platform that functions both within and alongside formal state systems.

7.2 Organizational Networks and Multi-actor Membership: Governing through Decentralized Coordination

To understand JVOAD's distinctive role in Japan's disaster governance, it is necessary to examine how its decentralized network structure enables coordination across sectors without relying on hierarchical authority. This section introduces the conceptual problem of orchestrating collective response among diverse actors, and analyzes how JVOAD's platform-based architecture supports horizontal cooperation. It explores the composition of its multi-actor membership, the institutional routines that sustain coordination, the trust-based mechanisms underpinning relational governance, and the conditions under which civil society gains substantive participation. These dimensions reveal a model of embedded governance rooted in negotiated collaboration and shared operational norms.

7.2.1 Research Problem and Conceptual Framing

In the evolving field of disaster governance, the central challenge is no longer limited to mobilizing resources, but lies increasingly in coordinating a growing number of heterogeneous actors across sectors and jurisdictions. The question of "who orchestrates collective response" has become pivotal, particularly in contexts like Japan where disaster response capabilities are distributed across government agencies, nonprofits, religious organizations, and local communities. JVOAD emerged in response to this coordination vacuum, not as a top-down authority, but as a platform-based governance model rooted in civil society. Rather than attempting to command and control, JVOAD facilitates alignment through horizontal coordination, relational trust, and institutional routines.

This shift reflects a broader governance transition—from centralized command structures to network-based governance models—where power is diffused and coordination relies on shared frameworks and relational capital. The absence of a central coordinating body prior to JVOAD's establishment led to duplicated efforts, information silos, and limited integration across actors (JVOAD, 2026a, 2026b). By creating a formalized network of cross-sectoral actors, JVOAD sought to address the critical need for pre-disaster agreements, role definition, and standard operating procedures.

The challenge, then, is conceptual: How can a civil society-led platform build governance capacity without hierarchical authority? JVOAD addresses this through a deliberately decentralized organizational architecture. Instead of functioning as a directive center, it acts as a non-hierarchical coordination hub that enables multi-directional information flows and negotiated decision-making (JVOAD, 2024a). The goal is not to issue orders, but to build consensus across actors with different mandates and capacities. Coordination is achieved through shared manuals, agreed protocols, and repeated joint training sessions, rather than statutory enforcement (JVOAD, 2025c).

Such an arrangement stands in contrast to models of symbolic NGO participation in disaster response, often seen in more centralized systems. In those contexts, civil society actors may be included nominally, but remain excluded from agenda-setting and strategic planning (Özkan & Uzun, 2024). JVOAD, by contrast, serves as a substantive intermediary—recognized by both government agencies and local stakeholders as a legitimate actor capable of brokering cooperation. Its networked governance structure allows it to mobilize diverse capacities while avoiding the rigidity of bureaucratic hierarchies, marking it as a unique experiment in embedded, civil society-led disaster governance.

7.2.2 Organizational Structure: Decentralized Platform Versus Command-and-Control Model

JVOAD's structural design reflects a deliberate departure from hierarchical, command-and-control models traditionally dominant in disaster governance. Instead, it has adopted a decentralized, networked architecture that enables horizontal coordination across diverse sectors. As of 2025, JVOAD comprises 40 full member organizations spanning a wide spectrum of institutional types. These include specialized disaster relief NPOs (e.g., Peace Boat Disaster Relief [PBV], Rescue Stock Yard [RSY]), religious organizations (e.g., Shinnyo-en, The Salvation Army), academic and medical institutions (e.g., Institute of Disaster Area Support [ACT], Kokushikan University), and nationwide welfare institutions and foundations such as the Japanese Red Cross and the Nippon Foundation (JVOAD, 2025b). This diversity of actors is not merely symbolic; rather, it underpins a flexible coordination ecosystem capable of mobilizing human resources, material supplies, logistics, and psychosocial support from multiple entry points.

What distinguishes JVOAD from traditional disaster coordination models is its emphasis on "coordination-based governance" rather than top-down command. The organization functions not as a central authority issuing directives, but as a facilitative platform that enables consensus formation, role differentiation, and mutual support among members. Such a model relies on pre-existing inter-organizational trust and the institutionalization of communication protocols well before a disaster strikes (JVOAD, 2025c). In contrast to centralized models, where decision-making flows vertically from government to operational actors, JVOAD's structure encourages multi-directional engagement, allowing actors with varied capacities and mandates to participate in co-produced response strategies.

The functional logic of this decentralized platform is supported by two key mechanisms: organizational diversity and pre-disaster relational infrastructure. The wide range of member organizations allows for specialization and distributed responsibility, while pre-disaster networking fosters relational capital and lowers coordination costs during crises (JVOAD, 2025b). This stands in contrast to systems that rely on state-led authority and vertical mandates, which often struggle to integrate community-based or sector-specific expertise.

In this sense, JVOAD offers an institutional response to the challenge of governing complexity under uncertainty. It reframes disaster response not as a matter of control, but of coordination—a platform logic that aligns with contemporary theories of network governance and collaborative resilience. The shift toward such an architecture reflects a broader transformation in disaster management thinking, where legitimacy and efficiency emerge not from centralized command, but from distributed trust, shared norms, and organizational complementarities (Özkan & Uzun, 2024).

7.2.3 Operational Mechanisms: Institutionalizing Coordination as Routine Practice

JVOAD's effectiveness as a coordination platform lies not only in its structural design but also in its institutionalized routines that sustain cross-sector collaboration beyond the moment of disaster. These mechanisms transform coordination from an ad hoc activity into a predictable and routinized practice embedded in the network's organizational life. At the core of this routinization is the JVOAD National Forum (JVOAD 全国フォーラム), a recurring convening mechanism through which member organizations align understandings, share policy proposals, and recalibrate their disaster response capacities (JVOAD, 2024a). These forums serve as institutionalized spaces for deliberation, ensuring that multi-sector actors are synchronized before, during, and after crisis events.

Complementing this deliberative infrastructure is standardized coordination tools that function as shared reference points across the network. The "Guidance for Intermediate Support Organizations in Disasters" (災害中間支援組織の手引き) delineates expected roles, coordination processes, and lessons from past disaster operations (JVOAD, 2024a). This handbook reduces role ambiguity and transaction costs during emergencies by clarifying operational boundaries and fostering interoperability across different organizational cultures. In addition, JVOAD has developed shared information platforms, joint training programs, and pre-disaster simulation exercises in collaboration with regional actors, further embedding coordination as a regular organizational practice (JVOAD, 2025c).

This combination of "relational capital" and "institutional capital" forms the backbone of JVOAD's coordination logic. Relational capital—manifested in repeated face-to-face interaction, familiarity, and mutual trust—is considered essential to quick, adaptive collaboration in chaotic disaster contexts (JVOAD, 2025c). However, JVOAD does not rely solely on informal relationships. It systematically institutionalizes coordination through protocols, manuals, and formalized joint activities.

This dual logic allows the organization to move beyond symbolic declarations of cooperation toward operationally meaningful partnerships.

What emerges is a model of platform governance in which cooperation is not imposed through hierarchy, but sustained through shared knowledge, pre-agreed procedures, and ongoing communicative engagement. Rather than directing resource delivery, JVOAD focuses on coordinating the interface among intermediate support organizations, optimizing resource allocation through information convergence and mutual adjustment (JVOAD, 2024a). The normalization of such routines not only enhances efficiency but also builds a collective identity across the network, embedding resilience within the daily functioning of civil society-led disaster response. In contrast to contexts where NGOs are temporarily activated without sustained mechanisms (Özkan & Uzun, 2024), JVOAD institutionalizes coordination as an organizational norm.

7.2.4 Governance Characteristics: Decentralization and Trust-Based Coordination

JVOAD's governance approach is characterized by a deliberate shift away from hierarchical command structures toward a decentralized, trust-based model of coordination. Rather than acting as a centralized authority, JVOAD functions as a facilitative platform that convenes diverse actors and mediates cooperation among them. This non-hierarchical orientation is foundational to the platform's legitimacy and operational resilience.

One key feature of this model is decentralization. JVOAD does not assume a directive role in disaster response. Instead, it coordinates relationships among intermediary support organizations, ensuring that "支援の提供ではなく、支援の調整" (coordination, not direct delivery) is its core function (JVOAD, 2024c). This orientation reflects a governance logic in which authority is distributed, and interdependence among members becomes the basis for collective action. The organization's operational manuals and coordination handbooks do not prescribe action hierarchically, but instead offer adaptable frameworks based on consensus and mutual adjustment (JVOAD, 2024a).

Equally critical is the emphasis on relational governance. The term "見える関係" (visible relationships) recurs throughout JVOAD's internal reports and training materials, underscoring the importance of interpersonal trust, repeated interaction, and face-to-face communication (JVOAD, 2025c). These relational assets form what the organization refers to as "関係資本" (relational capital), which enables rapid collective mobilization in disaster settings where formal authority is limited or delayed. JVOAD actively cultivates these ties through regular training sessions, pre-disaster workshops, and scenario-based simulations involving both member organizations and government partners (JVOAD, 2025c).

A third defining element is the preference for consensus-based action over command-driven mandates. Coordination meetings during and after disasters are designed not to issue top-down directives, but to form shared understandings ("共通認識の形成") across organizations (JVOAD, 2024c). This deliberative orientation fosters inclusive participation and empowers members to contribute within their respective domains, avoiding jurisdictional conflicts and reinforcing operational trust.

The institutionalization of these coordination norms distinguishes JVOAD's model from more hierarchical systems, where civil society roles are often limited to symbolic presence without substantive input. JVOAD's platform enables a form of embedded governance in which cooperation is achieved through shared procedures, mutual recognition, and structured trust, rather than coercive authority or unilateral control. Such a model reflects a maturing of network-based disaster governance in Japan, anchored in decentralized legitimacy and sustained by continuous relationship-building across sectors (Özkan & Uzun, 2024).

7.2.5 *Embedded Governance and Substantive Participation: JVOAD in Comparative Perspective*

The case of JVOAD illustrates a rare instance of substantive embeddedness of civil society within a national disaster governance system. In contrast to settings where NGOs are symbolically included but structurally marginalized, JVOAD's model reflects sustained institutional integration backed by legal, procedural, and financial mechanisms. This embeddedness ensures that civil society actors are not simply adjunct responders but recognized participants in strategic coordination and policy formation.

One crucial indicator of this embedded governance is the existence of formal participatory structures. JVOAD is not merely an aggregator of NPO activity; it operates under a member-led governance model in which decision-making is distributed among constituent organizations (JVOAD, 2025b). This structure allows for the articulation of civil society priorities through participatory deliberation. As early as 2017, JVOAD began submitting policy proposals to national authorities, including recommendations on evacuee support systems and volunteer mobilization strategies (JVOAD, 2024a). This level of proactive policy engagement reflects more than consultative access—it denotes institutional voice.

Institutional voice is further substantiated through formal agreements with governmental actors. JVOAD has signed memoranda of understanding with both national ministries and prefectural governments, particularly in disaster-prone areas like Ishikawa and Kumamoto (JVOAD, 2024b, 2025a). These agreements grant JVOAD operational legitimacy and access to real-time data sharing platforms, enabling it to coordinate support logistics with official disaster response teams. Furthermore,

JVOAD's recognized status allows it to participate in the Central Disaster Management Council, Japan's highest-level advisory body on disaster policy (JVOAD, 2025a).

This mode of cooperation stands in stark contrast to governance models observed elsewhere. In Turkey, for instance, NGOs are often invited to disaster coordination spaces but lack decision-making authority and procedural integration. Their involvement is frequently limited to "symbolic inclusion," resulting in weak influence over resource allocation or policy design (Özkan & Uzun, 2024). In such contexts, civil society participation is more performative than operational.

7.3 Coordination in Action: Field-Level Mechanisms during the Noto Peninsula Earthquake

The 2024 Noto Peninsula Earthquake served as a critical test for JVOAD's ability to coordinate disaster response in complex, rapidly changing environments. Rather than relying on hierarchical control, the organization facilitated cross-sector collaboration through trust-based mechanisms and adaptive governance routines. This chapter traces JVOAD's operational trajectory from the initial convening of stakeholders and real-time information-sharing to resource mediation and post-crisis institutionalization. The case highlights how platform-based coordination can be effectively implemented through phased, decentralized processes that integrate relational infrastructure and procedural tools in disaster settings.

7.3.1 Framing the Coordination Challenge in Real-Time Disaster Settings

Coordinating disaster response in real-time contexts entails more than the mobilization of aid—it requires managing fluid, multi-actor interactions in volatile, information-scarce environments. The 2024 Noto Peninsula Earthquake revealed the operational complexity of disaster governance, where assistance needs emerged across municipal and sectoral boundaries, necessitating dynamic coordination mechanisms. In such settings, the challenge is not simply who responds, but who facilitates cross-sectoral collaboration in the absence of command authority.

JVOAD does not directly deliver material assistance; rather, its core function lies in enabling coordination across governmental bodies, nonprofits, volunteer groups, and private actors. This platform-oriented approach seeks to mitigate fragmentation and duplication by establishing "支援の調整" (coordination of support) as a central governance principle (JVOAD, 2024c). The organization's role as a convener is especially critical during the early stages of disaster response, where uncertainties dominate and the need for clarity on roles, priorities, and flows of information is most acute.

Unlike top-down coordination models, JVOAD operates through relational and procedural infrastructures that emphasize trust-based governance and adaptive flexibility. Its mechanisms for convening multi-stakeholder coordination meetings, facilitating information flows, and aligning operational priorities demonstrate a shift toward what can be characterized as adaptive governance—an approach that accommodates evolving conditions through decentralized, iterative decision-making processes (JVOAD, 2024c, 2025c).

The empirical question, therefore, is whether such a platform can navigate high-stakes environments effectively without centralized authority. By analyzing JVOAD's field-level coordination during the Noto disaster, this section explores how relational trust, institutional preparedness, and situational adaptability converge to form an operational model of trust-based network governance. It also positions JVOAD's practice in comparative relief contexts, where civil society actors are often structurally marginalized (Özkan & Uzun, 2024), thereby providing a lens to evaluate the institutional distinctiveness and efficacy of Japan's coordination system.

7.3.2 Overview of the Noto Peninsula Earthquake: A Test Case for JVOAD's Coordination Capacity

The Noto Peninsula Earthquake of January 1, 2024, marked a critical stress event for Japan's disaster response systems, testing not only the infrastructure of local governments but also the operational viability of cross-sector coordination mechanisms. The earthquake caused severe destruction across multiple municipalities in Ishikawa Prefecture, generating immediate and large-scale needs for shelter, medical aid, food logistics, and volunteer deployment. With traditional administrative lines quickly overwhelmed, effective coordination across jurisdictional and sectoral boundaries became an urgent priority (JVOAD, 2024c).

In the immediate aftermath, JVOAD activated its pre-existing cooperation framework with the Ishikawa Prefectural Government. Drawing on its national network, it issued an open call to member NPOs, private enterprises, religious institutions, and government bodies to participate in an integrated coordination mechanism. This action was not ad hoc; it was built on formal memoranda of understanding, past training partnerships, and institutional trust accumulated before the crisis (JVOAD, 2025a, 2025c). The approach reflects JVOAD's functional identity as a non-operational but highly engaged intermediary actor.

On January 9, a National Coordination Meeting was held with the Cabinet Office and civil society partners. The meeting established three foundational principles for the response: transparent information-sharing, role clarity among actors, and proactive risk mitigation (JVOAD, 2024c). This coordination logic enabled resource integration without command-and-control mechanisms. Instead of directing operations, JVOAD's role was to facilitate horizontal linkages among actors with diverse capabilities.

Subsequent weeks saw the development of hybrid (online-offline) "Noto Information-Sharing Conferences" co-organized by JVOAD and local intermediary organizations in Ishikawa and Nanao. These covered key operational areas such as volunteer rotation schedules, evacuation shelter management, and debris disposal (JVOAD, 2024c). This ensured alignment between grassroots realities and national-level coordination, reinforcing JVOAD's position as a boundary-spanning facilitator.

Rather than providing relief services itself, JVOAD brokered connectivity between frontline organizations, functioning as a "coordination platform" rather than a delivery agency. This model, rooted in mutual trust and pre-disaster institutional arrangements, allowed for flexibility and responsiveness in a dynamic crisis environment. The Noto case illustrates how civil society-led coordination can scale adaptively in real time when embedded in a structure of shared governance, standing in contrast to systems that marginalize NGOs through symbolic inclusion alone (Özkan & Uzun, 2024).

7.3.3 Coordination in Action: From Information Flow to Resource Integration

JVOAD's coordination during the Noto Peninsula Earthquake unfolded through a staged, adaptive process that moved from initial information alignment to complex resource mediation and eventually to institutional learning. The process exemplified a non-hierarchical governance model where the organization's strength lay not in command capacity but in its ability to build connective infrastructure among disparate actors.

The first phase centered on information-sharing as the foundation for coordinated response. On January 9, 2024—within ten days of the disaster—JVOAD co-convened the first national-level coordination meeting alongside the Cabinet Office and major NPOs. This meeting solidified three organizing principles: resource sharing, role differentiation, and risk mitigation (JVOAD, 2024c). Subsequently, from March through June, JVOAD facilitated a series of hybrid-format "Noto Information-Sharing Conferences" in partnership with local intermediaries such as Ishikawa VOAD and Nanao City Support Center. These conferences focused on real-time operational topics, including volunteer rotation, evacuation shelter management, and disaster waste processing, thereby enabling granular coordination across organizational and geographic boundaries (JVOAD, 2024c).

The second phase involved strategic mediation and resource brokerage. Rather than intervening directly in aid delivery, JVOAD coordinated intermediaries through technical tools and soft infrastructure. It helped implement the "Shien no Omamori"

(しえんのおまもり) project, which provided disaster vehicle ID tags and insurance coverage for civil society responders (JVOAD, 2024c). It also facilitated arrangements for long-distance evacuees from Noto to urban centers such as Kanazawa and Toyama, ensuring consistent care and service linkage across prefectural lines. A centralized database of aid-providing organizations was developed to visualize "who is doing what where," enhancing transparency and minimizing duplication. Additionally, JVOAD worked with TVAC and other national coordination bodies to synchronize volunteer arrival timing and access protocols (JVOAD, 2024c, 2025a).

The third phase focused on codifying these practices into institutional routines. Insights from the Noto case were incorporated into revisions of the "Guidance for Intermediate Support Organizations in Disasters" (JVOAD, 2024a). More importantly, JVOAD launched discussions on formalizing regional disaster support networks through its "Prefecture-Level Network Development Project," aiming to institutionalize the lessons learned during the Noto response (JVOAD, 2024c).

This phased process—from information-sharing to cross-sector mediation and post-crisis institutionalization—demonstrates JVOAD's ability to perform as a trust-based coordination platform under complex, dynamic conditions. It exemplifies the organization's function not as a service provider, but as an embedded broker within Japan's evolving disaster governance architecture.

7.4 Institutional Preparedness: From Fiscal Architecture to Human Capital

JVOAD's disaster coordination capacity is grounded in institutional investments made during non-crisis periods. This includes a mixed funding model that ensures fiscal stability, a decentralized network of regional VOADs that distributes operational capacity, and codified procedures that standardize practices across actors. By embedding resilience into its everyday architecture—through finance, training, and shared protocols—JVOAD moves beyond reactive mobilization, illustrating how civil society-led coordination can be systematized and sustained across time.

7.4.1 Framing the Institutional Basis of Readiness

Disaster response capacity is not solely forged in the immediacy of crisis; it is deeply shaped by the institutional environments built during periods of normalcy. For civil society actors such as JVOAD, the effectiveness of disaster-time coordination hinges on the existence of a robust pre-disaster support infrastructure. Rather than relying on ad hoc mobilization, JVOAD has strategically developed a system of financial, organizational, and human capital support that enables it to function as a reliable intermediary during emergencies.

A critical element of this institutional environment is JVOAD's mixed-source funding model. In FY2024, approximately 60% of its revenue came from national and local government-commissioned projects, such as the Prefectural Disaster Support Network Development Program in Ishikawa and Gifu (JVOAD, 2024b; JVOAD, 2025a). This was complemented by targeted donations from private foundations, including the Nippon Foundation and Toyota Foundation, and consistent support from member organizations through annual dues and cooperation funds. These diverse revenue streams are managed under an independent auditing regime and published annually, ensuring financial transparency and legitimacy (JVOAD, 2024b, 2025a). This structure reflects what can be termed "embedded autonomy": a model that integrates public funding and institutional oversight without compromising civil society independence.

In tandem with financial embedding, JVOAD invests in the development of regional coordination ecosystems. Prefectural-level VOADs are cultivated as the frontline nodes for localized disaster governance. Through the "Prefecture-Level Disaster Support Network Development Program," JVOAD offers technical assistance, funds staff development, and co-designs local training systems with partner organizations (JVOAD, 2025c). In FY2024, particular emphasis was placed on strengthening Ishikawa and Gifu's disaster readiness structures. At the core of this approach is the concept of visible, trust-based relations—encouraging cross-organizational familiarity before crises occur.

Training is further institutionalized through regular joint exercises, coordination simulations, and inter-sectoral workshops, all of which draw upon a standardized action manual, Guidance for Intermediate Support Organizations in Disasters (JVOAD, 2024a). These programs emphasize mutual role recognition, decentralized decision-making, and the importance of relational capital in emergency settings.

7.4.2 Financial Architecture: Diversified Revenue and Embedded Autonomy

The sustainability of JVOAD's coordination capacity during disasters is underpinned by a robust financial architecture that blends institutionalized public funding with diversified private contributions. Rather than relying solely on reactive fundraising or ad hoc donations, JVOAD has established a multi-source revenue structure that reinforces both organizational stability and functional independence. This approach can be characterized as "embedded autonomy"—an operational model that combines state-linked financial legitimacy with non-governmental initiative.

According to financial disclosures, approximately 60% of JVOAD's FY2024 revenue derived from commissioned public-sector projects, including major undertakings such as the "Prefectural Disaster Support Network Development Project" in Ishikawa and Gifu Prefectures (JVOAD, 2025a). These government contracts, while significant in volume, are balanced by targeted grants from private foundations such as the Nippon Foundation, Toyota Foundation, and Give2Asia. These funds are often earmarked for specific programmatic initiatives, such as training disaster coordinators or supporting network infrastructure (JVOAD, 2024b). Membership dues from core and supporting member organizations further diversify the funding pool, reinforcing stakeholder buy-in and organizational accountability.

Crucially, JVOAD has institutionalized rigorous financial governance mechanisms to safeguard independence and transparency. These include third-party audits, public release of annual financial reports, and strict earmarking of designated funds. Government-contracted projects are subject to external review and reporting obligations, ensuring that public funding aligns with mission-driven outcomes rather than steering the organization's agenda (JVOAD, 2025a). This dual structure enables JVOAD to remain responsive to state priorities without being subordinated to them.

Such a financial configuration stands in contrast to the vulnerabilities identified in comparative cases, where NGOs often face exclusion from disaster governance due to unstable funding and perceived lack of professional capacity (Özkan & Uzun, 2024). JVOAD's embedded autonomy allows it to act as a stable intermediary platform capable of long-term planning, capacity development, and cross-sector coordination—even in the absence of immediate crisis triggers.

In essence, financial diversification does more than ensure organizational survival; it operationalizes a preparedness infrastructure that enables sustained engagement across disaster and non-disaster periods. By anchoring its fiscal model in both public legitimacy and private-sector partnership, JVOAD builds the institutional preconditions necessary for resilient, civil society-led disaster governance.

7.4.3 Capacity Development and Human Resource Training: Building an Enabling Environment

A resilient disaster coordination system requires not only institutional funding but also a robust infrastructure for human resource development and organizational capacity building. JVOAD's non-operational but facilitative role depends heavily on its ability to support local intermediaries, cultivate field-level coordination skills, and embed shared protocols across its national network. These efforts are centered on the development of prefectural VOADs and a structured training ecosystem designed to ensure sustained, scalable disaster response capacity.

At the core of JVOAD's human capital strategy lies the "Prefecture-Level Disaster Support Network Development Project," which seeks to strengthen coordination infrastructures in multiple regions by supporting the establishment and expansion of local VOADs. In 2024, particular emphasis was placed on Ishikawa and Gifu prefectures, where JVOAD collaborated with regional actors to prepare local coordination mechanisms, pre-designate partner organizations, and assign trained personnel for potential deployment (JVOAD, 2025c). These efforts included the cultivation of "Disaster Cooperation Supporters" (災害協力支援員), trained individuals embedded in local contexts who are capable of activating cross-sector coordination during emergencies.

Training and simulation exercises serve as the operational backbone of JVOAD's capacity-building strategy. Each year, JVOAD conducts a wide range of joint training programs that target both national and local stakeholders. These include scenario-based coordination drills, NGO-government simulation workshops, and structured debriefings on past disaster experiences (JVOAD, 2025c). Such programs are guided by the official "Manual for Intermediate Support Organizations in Disaster" (災害中間支援組織の手引き), which provides standardized procedures, shared terminology, and role definitions to facilitate interoperability across organizations (JVOAD, 2024a).

More than a technical process, these trainings aim to foster "visible trust-based relations" among actors—an approach grounded in the philosophy of "meeting before disaster, cooperating during crisis." Through repeated engagement in non-disaster periods, JVOAD builds familiarity between individuals and institutions, enhancing not only operational efficiency but also cross-sectoral trust. The trainings serve as a medium for socializing organizational norms, thereby cultivating a shared disaster governance culture that extends beyond any single emergency (JVOAD, 2025c).

7.4.4 Institutional Foundations of Organizational Resilience

The sustainability of JVOAD's coordination model lies not merely in its operational responsiveness, but in the institutional scaffolding that supports its long-term resilience. Three interconnected dimensions—financial embedding, capacity networks, and preparedness infrastructure—jointly underpin this durability, allowing JVOAD to shift from reactive mobilization to anticipatory governance.

Financial embedding is a defining feature of JVOAD's stability. Rather than relying on ad hoc donations or fragmented project-based grants, the organization strategically secures planned, multi-year public commissions from national and local governments. In 2024, nearly 60% of JVOAD's annual income came from such institutional contracts, including government-backed initiatives like the "Prefecture-Level Disaster Support Network Development Project" in Ishikawa (JVOAD, 2025a). Complemented by designated corporate and philanthropic donations—from entities such as the Nippon Foundation, Toyota Foundation, and Give2Asia—and membership fees from affiliated organizations, this diversified revenue base reflects an embedded fiscal architecture that balances independence with accountability (JVOAD, 2024b). Transparency mechanisms, such as annual public audits and itemized financial disclosures, further reinforce trust and legitimacy (JVOAD, 2025a).

Resilience is also structurally distributed through capacity networks. JVOAD does not centralize disaster expertise in its Tokyo office; instead, it empowers regional intermediaries—prefectural VOADs—to become coordination hubs in their own right. These local entities are integrated into national projects, co-develop protocols, and receive tailored training programs that enhance their ability to coordinate under pressure. As part of this, JVOAD has cultivated cross-regional learning platforms that allow expertise to circulate horizontally across the network (JVOAD, 2025c).

This is reinforced by a robust preparedness infrastructure that standardizes practice without rigidifying it. JVOAD's shared handbook for intermediate organizations (JVOAD, 2024a) codifies lessons learned into flexible guidelines, while annual drills, government–NPO simulation exercises, and peer learning workshops translate those protocols into embodied institutional memory. These efforts position "peacetime" not as downtime, but as an institutional incubation period for disaster capacity, gradually deepening coordination literacy across the network (JVOAD, 2025c).

7.4.5 Linking Theory and Practice: Institutional Pathways to Organizational Resilience

The empirical insights drawn from JVOAD's institutional development underscore a crucial dimension of disaster governance: organizational resilience is not merely a function of emergency responsiveness, but a product of embedded institutional arrangements built during non-crisis periods. This understanding challenges dominant critiques that civil society organizations often suffer from marginalization, stemming from financial precarity and limited coordination capacity (Özkan & Uzun, 2024). JVOAD's model presents a counterexample, where pre-disaster investments in funding structures, training ecosystems, and relational infrastructure cultivate not symbolic participation, but functional authority in disaster response.

At the financial level, JVOAD avoids dependence on fragmented or one-off donations. Instead, it has institutionalized a multi-source funding base that includes government-commissioned projects (comprising 60% of its 2024 income), targeted private foundation grants, and structured member contributions (JVOAD, 2024b, 2025a). These income streams are subject to independent audits, public disclosure, and designated-use regulations, forming a fiscal governance model that combines transparency with operational autonomy. This "embedded independence" enables the organization to act with agility in disaster contexts without risking capture by any single funder (JVOAD, 2026b).

Organizational resilience also rests on JVOAD's decentralized capacity architecture. By investing in the formation and professionalization of prefectural VOADs, and by fostering coordination training through joint simulations and knowledge-sharing platforms, JVOAD ensures that response capability is not monopolized at the center, but cultivated across regions (JVOAD, 2024a, 2025c). These systems do not simply transmit skills—they build relationships of trust that are later activated in disaster settings.

Importantly, JVOAD's approach transforms "peacetime" into a period of strategic accumulation—what may be termed a preparedness infrastructure. Handbooks such as the Guidance for Intermediate Support Organizations in Disasters serve not only as operational templates but also as vessels of institutional memory (JVOAD, 2024a). Through annual review cycles, these tools are continuously refined and translated into action through drills and workshops.

The convergence of fiscal embedding, capacity decentralization, and procedural codification constitutes a model of institutionalized resilience, wherein civil society actors are not peripheral responders but central coordinators in Japan's disaster governance system. Rather than treating disaster capacity as a reactive function, JVOAD integrates it into its routine architecture—illustrating a pathway for other contexts where NGOs remain structurally sidelined. This model demonstrates how civil society-led coordination, when anchored in long-term preparedness, can serve as a sustainable and legitimate pillar of disaster governance.

7.5 Conclusion

JVOAD exemplifies how civil society can move beyond reactive mobilization to become a structurally embedded actor in national disaster governance. Emerging from Japan's post-2011 coordination failures, JVOAD institutionalized its role through legal recognition, procedural access, and financial diversification. Its decentralized, trust-based platform enables horizontal coordination across NGOs, faith-based groups, and government agencies, as demonstrated during the 2024 Noto Peninsula Earthquake. Crucially, JVOAD's effectiveness is underpinned by peacetime investments in fiscal infrastructure, human capital, and training systems—turning readiness into routine. By operationalizing "embedded autonomy," JVOAD challenges the notion of NGOs as peripheral responders and offers a transferable model for institutional resilience. The case affirms that civil society leadership, when strategically structured, can play a central, legitimate role in disaster governance.

References

JVOAD. (2024a). *FY2024 annual activity report*. https://jvoad.jp/wp-content/themes/may2024/pdf/jigyo2024.pdf

JVOAD. (2024b). *FY2024 financial report (9th fiscal period)*. https://jvoad.jp/wp-content/themes/may2024/pdf/kessan2024.pdf

JVOAD. (2024c). *Support activities for the 2024 Noto Peninsula Earthquake*. https://jvoad.jp/disaster-name/%e4%bb%a4%e5%92%8c6%e5%b9%b4%e8%83%bd%e7%99%bb%e5%8d%8a%e5%b3%b6%e5%9c%b0%e9%9c%87/

JVOAD. (2025a). *Independent auditor's report for, FY2024*. https://jvoad.jp/wp-content/themes/may2024/pdf/kansa2024.pdf

JVOAD. (2025b). *Member organizations*. https://jvoad.jp/member/

JVOAD. (2025c). *Training and capacity-building activities*. https://jvoad.jp/news-cat/event/

JVOAD. (2026a). *About JVOAD and its activities*. https://jvoad.jp/about/activity/

JVOAD. (2026b). *Organization overview*. https://jvoad.jp/about/outline/

Özkan, O., & Uzun, A. (2024). The democratic touch of disasters: The perception of non-governmental organizations in public institutions in the context of disaster governance. *SSRN*. https://doi.org/10.2139/ssrn.4989771

Part III
Practices of Social Organizations in Emergency Management: Chinese Institutional Context

Chapter 8
Chinese Social Organizations in Disaster Relief: Historical Evolution and Capacity-Building

Abstract Understanding the evolving role of Chinese social organizations in disaster relief requires attention to both historical legacies and institutional reconfigurations. From the embedded communal practices of imperial times to the organizational marginalization under twentieth-century state-building, the foundations of today's disaster governance are deeply shaped by the interplay between moral tradition and bureaucratic control. In the reform era, shifts toward collaborative governance—accelerated by watershed events like the 2008 Wenchuan earthquake—have gradually expanded the space for social participation. Against this backdrop, social organizations have developed distinct forms of capacity across organizational, technical, and collaborative domains, though with marked regional disparities. At the same time, persistent structural constraints highlight the urgency of advancing more integrated and inclusive models of disaster governance.

Keywords Institutional embeddedness · Collaborative governance · Capacity-building · Asymmetry · Policy evolution

8.1 Historical Origins and Institutional Foundations of Social Organizations in Disaster Relief

The evolution of social organizations in China's disaster relief system is deeply embedded in the country's historical, political, and institutional transformations. This section traces a three-phase trajectory: beginning with the moral governance and communal relief practices of imperial China, it then examines how twentieth-century state-building projects marginalized autonomous social actors through corporatist control, and finally analyzes how these legacies continue to shape the hybrid, state-led yet socially embedded governance structure in the present. Understanding this historical continuum reveals both the constraints and embedded potentials that define Chinese civil society's role in disaster governance today.

Q. Du and R. Shaw, *Social Organizations in Disaster Response and Recovery*, Disaster Risk Reduction, https://doi.org/10.1007/978-981-92-2846-1_8

8.1.1 Philanthropy and Communal Relief in Imperial and Late Imperial China

In pre-modern China, disaster relief was characterized by a dispersed, multilayered structure in which local social actors—although lacking formal legal status—played decisive roles in response and recovery efforts. These actors, including local elites, religious institutions, and communal associations, constituted a dense web of quasi-organizational forces that functioned as informal but effective governance entities, particularly in the frequent occurrence of floods, famines, and droughts. Rather than being peripheral, their presence was institutionalized within the normative framework of Confucian governance and shaped enduring traditions of collective moral responsibility.

Local gentry, or shen-shi, were the cornerstone of this grassroots relief structure. Empowered by moral legitimacy and embedded in both the landholding economy and the imperial examination system, these elites mobilized relief through private granaries, ad hoc fundraising, and coordination with officials. In many cases, they stepped in to fill gaps when state capacity was overstretched, especially in rural and frontier regions. Historical records document gentry-led famine relief across regions such as Anhui and Shanxi, where elite households organized food distribution, public works, and orphan care (DuBois, 2015; Fuller, 2015). These efforts operated within the logic of moral governance, where legitimacy was derived not merely from authority but from ethical conduct and social responsibility.

Religious institutions—especially Buddhist monasteries and Daoist temples—added an additional dimension to communal relief. These spaces not only provided physical refuge during disasters but also served as focal points for donation drives, ritual protection, and long-distance charity networks. In particular, temple-based networks enabled the pooling and redistribution of material resources, often crossing regional boundaries (Xiao et al., 2014). The convergence of spiritual duty and charitable work embedded disaster relief in a moral-spiritual cosmology that extended beyond state intervention.

Beyond elite and religious actors, communal institutions such as yicang (community grain granaries), lineage charities, guild halls, and local academies reflected an organized yet non-state infrastructure for social assistance. The yicang system, for example, enabled villages to store surplus grain in good years for redistribution during crisis periods. Evidence from Jiangnan and Lingnan indicates that these mechanisms significantly reduced disaster mortality and promoted local stability (Tao et al., 2025). Similarly, lineage-based organizations administered education subsidies, burial support, and emergency loans—practices that later influenced early modern civic groups in the Republican era (Zhai et al., 2020).

These pre-modern arrangements demonstrate that China's disaster response was neither anarchic nor purely state-centric. Instead, a pluralistic and morally embedded system of co-governance existed, where social actors were recognized as legitimate, if informal, agents of disaster relief. This system laid an early institutional and cultural foundation for contemporary civil society action, even as later state-building projects would reconfigure these dynamics. Understanding these historical continuities is crucial for interpreting both the capacity and constraints of modern Chinese social organizations in disaster governance.

8.1.2 State-Building and the Marginalization of Social Actors in the Twentieth Century

The twentieth century witnessed a fundamental transformation in the governance of disaster relief in China, marked by a shift from a pluralistic system of semi-autonomous social actors to a state-centric model of disaster governance. This transition unfolded through successive stages of state-building that prioritized political consolidation, institutional centralization, and ideological alignment, leaving limited space for independent civic participation. While societal actors continued to exist, their roles in disaster response became increasingly constrained, formalized, and subordinated to state priorities.

In the Republican era (1912–1949), efforts were made to institutionalize civil society's role in disaster response, especially following the devastating Yangtze River floods of 1931 and 1935. These events catalyzed the emergence of philanthropic associations, Christian charities, and urban guilds, many of which collaborated with provincial authorities to provide food, shelter, and medical relief (Zhai et al., 2020). However, warlord fragmentation, fiscal insolvency, and Japanese invasion interrupted these efforts, limiting their sustainability. Attempts to formalize charitable activity through legal registration and coordination mechanisms remained weak and uneven (Jiang, 2013). Relief campaigns frequently relied on elite patronage or international networks rather than systematic domestic governance.

Following the establishment of the People's Republic of China in 1949, a monocentric governance model was gradually institutionalized. Disaster relief was redefined as a core function of the state, administered through vertically integrated bureaucracies and mass mobilization campaigns. During major events such as the 1954 Yangtze flood and the 1976 Tangshan earthquake, the Chinese Communist Party orchestrated centralized responses that emphasized speed, political loyalty, and ideological education alongside material support (DuBois, 2015). In this process, preexisting social organizations were either dissolved or incorporated into the emerging socialist structure, often as "mass organizations" under party supervision.

From the 1950s to the 1980s, the legitimacy of social participation in disaster governance became conditional on organizational dependency—most notably through institutional attachment. Under this system, civil associations were required to affiliate with official sponsors (typically party or government units) to gain legal status, limiting their operational autonomy (Chen, 2015). As a result, grassroots responses were often channeled through government-led structures, while independent initiatives were discouraged or viewed with suspicion. Even humanitarian activities were interpreted through a political lens, with relief framed as an extension of class struggle and socialist solidarity (Wang et al., 2020).

This historical marginalization was not merely a product of authoritarian control, but also a deliberate strategy of state corporatism, designed to integrate social forces into national development while minimizing the risk of political pluralism. Over time, this produced a paradox: while the state built impressive logistical and material capacities for disaster relief, it simultaneously weakened the adaptive potential of society by stifling horizontal networks and non-state innovation.

The result was an institutional configuration where the state's capacity to deliver large-scale aid coexisted with a structurally weakened civil society, laying the foundation for many of the legitimacy and coordination challenges that would later surface in the post-2008 era. These dynamics are essential to understanding the deeply embedded legacy of organizational dependency, which continues to shape how contemporary social organizations seek recognition, resources, and room for maneuver in disaster governance.

8.1.3 Institutional Legacies and the Foundations of Contemporary Relief Systems

Contemporary disaster relief in China reflects a layered institutional structure that blends enduring features of centralized state control with selectively incorporated societal participation. This hybrid system—best described as a form of embedded governance—rests upon historical legacies of hierarchical authority, party-led mobilization, and a constrained model of civil society, yet increasingly incorporates formally recognized social organizations under specific regulatory and operational frameworks.

The institutional embeddedness of social organizations remains structurally shaped by the legacy of state corporatism and regulatory gatekeeping. Although the twenty-first century has witnessed a proliferation of legally registered NGOs and GONGOs (government-organized NGOs), these entities often function under tight administrative oversight and are reliant on state-endorsed project funding, operational legitimacy, or emergency authorization (Chen, 2015; He et al., 2021). This pattern is reinforced by dual management systems, which require registration with the Ministry of Civil Affairs and sponsorship by a professional supervisory agency, limiting independent initiative.

Meanwhile, frequent natural hazards have incentivized the institutionalization of collaborative mechanisms that embed social organizations into government-led disaster response frameworks. Policies such as the Regulations on Natural Disaster Relief (2010) and the revised Emergency Response Law (2024) emphasize the role of "qualified social organizations" in logistics, shelter management, and psychological support. However, these roles are often limited to operational implementation, with minimal involvement in decision-making or strategic planning (Emergency Response Law of the People's Republic of China (Revised 2024), 2024; Ministry of Civil Affairs of the People's Republic of China, 2015). This asymmetry reflects a persistent boundary between state command and social execution, where autonomy is granted but rarely institutionalized.

The functional ambiguity of social organizations further complicates their positioning. As observed in post-Wenchuan dynamics and successive disaster events, many NGOs straddle the line between formal recognition and informal flexibility. Their legitimacy often depends not only on legal status but also on political trust, interpersonal networks with local cadres, or public visibility during crisis moments (Liang et al., 2019; Paltemaa, 2017). This leads to a fragmented ecology where organizations are "recognized but not fully trusted," creating uneven access to coordination platforms and funding streams.

Despite these constraints, the continuity of moral and communal relief traditions remains visible in contemporary forms. Volunteer associations, faith-based charities, and student networks—while now operating under modern regulations—often trace their legitimacy to cultural narratives of collective responsibility and local solidarity (Tao et al., 2025; Xiao et al., 2014). Their embeddedness within communities provides them with rapid response capabilities and social capital, though this potential is not always integrated into formal disaster governance structures.

Overall, the institutional foundations of China's current disaster relief system are marked by both continuity and transformation. While the state retains dominant control over coordination, resource allocation, and public authority, social organizations have carved out a nested role within this architecture—adaptive, responsive, yet structurally limited. This co-opted civil society model defines much of the governance terrain in which Chinese NGOs operate: not as independent counterbalances to state authority, but as auxiliary actors embedded within state-centric frameworks.

8.2 Institutional Shifts and the Resurgence of Social Participation in the Reform Era

The reform era in China has witnessed a gradual reconfiguration of disaster governance, moving from a centralized, state-dominated model toward one characterized by selective inclusion and institutionalization of social participation. This transformation reflects not only the shifting governance philosophy of the state but also responses to practical limitations exposed by major disasters. This section traces the evolution of this shift across three dimensions: the broader governance paradigm shift since 1978, the Wenchuan earthquake as a pivotal turning point, and the formal policy developments that followed. These changes have reshaped the institutional space for social organizations in China's disaster response system.

8.2.1 From Centralized Control to Governance Collaboration: A Paradigm Shift

The evolution of disaster governance in China since the reform era reflects a gradual shift from centralized state control to a more collaborative and multi-actor governance paradigm. This shift has not entailed a weakening of state authority but rather a functional reconfiguration, wherein the state increasingly assumes the role of coordinator and regulator, while selectively incorporating societal actors into disaster response frameworks. This transition was driven by both institutional necessity and pragmatic adaptation to growing disaster risks and governance complexity.

In the early reform years, institutional loosening created new spaces for non-state actors. The dismantling of the commune system and the rise of local administrative discretion offered opportunities for community-based initiatives in areas previously monopolized by the state. However, initial engagements with social actors were informal and fragmented. State capacity remained dominant, and grassroots participation was often framed as supplementary or temporary (Paltemaa, 2017; Wu & Peng, 2017). The governance model was still characterized by top-down mobilization, with limited room for bottom-up initiatives to influence disaster preparedness or policy formation.

As disaster frequency and impact intensified in the late 1990s and early 2000s—exemplified by the 1998 Yangtze River floods—there was increasing awareness that centralized mechanisms alone were insufficient to manage large-scale emergencies effectively. This prompted an institutional rethinking of disaster management, leading to the gradual diffusion of governance responsibilities to include "qualified social organizations," volunteer groups, and community-based networks (Peng, 2017; Yang et al., 2021a, 2021b). The rhetoric of "collaborative governance" (协同治理) began to emerge in policy discussions, accompanied by selective decentralization of operational roles to non-state actors.

This discursive shift was supported by incremental regulatory reforms. The State Council's 2004 "Regulations on the Registration and Management of Social Organizations" allowed for a modest expansion of civil society space, even as the dual registration system continued to constrain autonomy. Meanwhile, local governments, facing mounting fiscal and administrative pressures, began to recognize the functional utility of NGOs and volunteer groups in disaster relief, particularly in service delivery and logistical support (Cui et al., 2022; Wang et al., 2022a, 2022b).

By the mid-2000s, institutional adaptation had become more visible. Pilot programs in provinces such as Guangdong, Sichuan, and Jiangsu introduced joint disaster drills, multi-stakeholder emergency coordination platforms, and volunteer management systems. These mechanisms signaled a softening of the rigid state–society boundary and laid the groundwork for more structured forms of collaboration (Li & Guo, 2022; Zhao et al., 2023a, 2023b). Though often still under tight administrative oversight, these experiments reflected a governance logic that sought resilience through pluralism rather than control alone.

Nevertheless, the paradigm shift remains incomplete. The state retains dominant power in agenda setting, resource distribution, and institutional legitimacy. Most social organizations involved in disaster relief still operate under hierarchical coordination, and policy design continues to privilege state actors. The recognition of social organizations as functional partners—however constrained—represents a meaningful departure from previous exclusionary models. The foundation was thus laid for more significant transformations in the wake of the 2008 Wenchuan earthquake, which would force a reconfiguration of disaster governance on a national scale.

8.2.2 The Wenchuan Earthquake as a Critical Juncture for Social Organizations

The 2008 Wenchuan earthquake, which resulted in over 69,000 fatalities and left millions displaced, marked a pivotal turning point in the trajectory of civil society participation in China's disaster governance. Beyond the scale of the disaster itself, the response it catalyzed exposed both institutional bottlenecks within the state apparatus and the latent potential of societal actors. As such, the event functioned as a "critical juncture" in the historical institutionalist sense—disrupting established governance patterns and opening an unexpected policy window for the incorporation of non-state actors (Menefee & Nordtveit, 2012).

In the immediate aftermath of the earthquake, the sheer scale of humanitarian need overwhelmed traditional state mechanisms, particularly in the early hours and days of the crisis. This capacity strain created both necessity and opportunity for an unprecedented bottom-up mobilization. Reports estimate that over 3,000 domestic NGOs and tens of thousands of grassroots volunteer groups participated in relief operations across affected provinces such as Sichuan, Gansu, and Shaanxi (Yi et al., 2019; Zhai et al., 2020). These organizations offered a range of services—from emergency rescue and supply distribution to psychological counseling and child welfare—functions that had previously been monopolized or tightly regulated by state entities.

The decentralized nature of this response challenged conventional command-and-control frameworks but was not met with outright repression. Instead, local governments displayed what scholars have termed a "tacit acceptance strategy," selectively allowing NGOs to operate while maintaining strategic oversight (Gong, 2021). In many cases, municipalities adopted a "抓大放小" (grasp the big, let go of the small) approach—focusing supervision on larger actors while allowing smaller groups more operational latitude. This shift reflected an emergent pragmatic governance logic that prioritized results over rigid control.

Importantly, the earthquake also triggered a process of institutional feedback and policy learning. Recognizing the efficacy of civil society actors under crisis conditions, central and local authorities initiated retrospective efforts to formalize their inclusion. Several post-disaster reports issued by the Ministry of Civil Affairs emphasized the need to "standardize and support" social organization participation in disaster response (Ministry of Civil Affairs of the People's Republic of China, 2015). Meanwhile, pilot programs were launched to enhance NGO registration processes and improve inter-agency coordination in emergencies (Liang et al., 2019; Yi et al., 2019).

Despite these shifts, tensions remained. Many participating organizations reported difficulties in securing access to disaster zones, lack of legal clarity, and inconsistent treatment by local officials (Yang et al., 2021a, 2021b). Their effectiveness was often contingent on informal ties to local governments or media visibility, rather than institutional guarantees. Nonetheless, the scale and visibility of their engagement permanently altered the state's perception of social organizations—not as marginal or oppositional, but as potentially valuable collaborators under conditions of extreme stress.

As a result, the Wenchuan earthquake reshaped the normative boundaries of disaster governance in China. It created a new legitimacy for bottom-up initiatives, catalyzed institutional reflection within state bodies, and signaled the beginning of a longer-term policy transformation. Subsequent sections will explore how this rupture evolved into more structured and formalized governance arrangements, embedding social organizations more firmly within China's disaster response system.

8.2.3 From Ad Hoc Inclusion to Institutionalization: Policy Evolution Since 2008

The aftermath of the 2008 Wenchuan earthquake marked the beginning of a gradual institutional shift, moving from spontaneous and loosely coordinated civil society participation toward a more formalized integration of social organizations into China's disaster governance system. While the state's dominant role remained intact, the disaster response environment began to witness a cautious but discernible opening, wherein social organizations were no longer viewed merely as ad hoc responders but increasingly as structured collaborators embedded within regulatory frameworks.

This evolution was first reflected in policy formalization. The Regulations on Natural Disaster Relief (2010) provided the initial legal basis for the state to "encourage and regulate" the participation of "social forces" in disaster relief, including logistics, donation management, and temporary shelter services (Regulations on Natural Disaster Relief (State Council Order No. 577), 2010). Although the language remained cautious, this was the first national regulation to explicitly define roles for non-state actors in emergency response, institutionalizing what had previously been tolerated only under exceptional circumstances.

Building on this momentum, the Guiding Opinions on Supporting and Guiding Social Forces in Disaster Relief issued by the Ministry of Civil Affairs in 2015 emphasized the principles of legal registration, autonomous operation, and qualified participation (Ministry of Civil Affairs of the People's Republic of China, 2015). These guidelines marked a partial departure from the earlier attachment-based model (挂靠体制), advocating instead for capacity-building among community-level organizations and professional NGOs capable of engaging in time-sensitive relief operations. The policy also laid out differentiated pathways for "community-based," "professional," and "volunteer-driven" organizations, signaling an attempt to categorize and manage civil participation through formal mechanisms (Chen, 2015; Wu & Peng, 2017).

Further consolidation occurred with the Emergency Response Law, which legally mandated the integration of qualified social organizations into emergency plans at the provincial and municipal levels. The law clarified their roles in disaster monitoring, psychological assistance, evacuation support, and post-disaster reconstruction. Crucially, it provided a normative foundation for pre-disaster coordination and training, enabling these organizations to be included in preparedness efforts rather than merely reactive interventions.

Institutionally, these policies fostered a more nested governance system. Mechanisms such as rapid response platforms, volunteer certification systems, and disaster information-sharing portals were progressively established to facilitate coordination between governmental agencies and registered NGOs (Chen et al., 2025a, 2025b; Yang et al., 2021a, 2021b). These mechanisms not only improved efficiency but also served as channels through which legitimacy and accountability were filtered and maintained. However, access to such platforms often depended on political vetting and alignment with government priorities, limiting their inclusiveness (Chen et al., 2020).

What emerged from this process is a dual-track system of governance: on one hand, the state retains control over strategic planning, funding, and agenda setting; on the other, social organizations are embedded in implementation and service delivery. This model reflects what scholars have termed "state-led institutionalization" (Paltemaa, 2017), wherein civil society is co-opted into governance structures rather than standing apart as an autonomous sector.

In sum, the post-2008 period witnessed a substantive transformation in the policy environment governing disaster relief in China. While autonomy remains constrained, the formalization of roles, development of collaborative platforms, and legal codification of participation have significantly redefined the institutional space for social organizations, transitioning them from peripheral actors to conditional partners in state-led disaster governance.

8.3 Capacity-Building of Chinese Social Organizations: Mechanisms and Evaluation

Capacity-building among Chinese social organizations in disaster relief has emerged as a central issue in understanding their evolving role within state-led governance frameworks. Existing practices and empirical evidence suggest that such capacity is neither singular nor static, but composed of interrelated organizational, technical, and collaborative dimensions that develop through repeated engagement with disasters. Major emergency events have functioned as critical arenas in which these capacities are exercised, reinforced, or constrained, revealing both learning trajectories and persistent regional inequalities. At the same time, growing efforts to evaluate organizational performance through policy, spatial, and multi-criteria tools reflect an expanding demand for accountability, even as fragmented standards and limited data continue to challenge systematic assessment.

8.3.1 Defining and Deconstructing Capacity in Disaster Relief

The capacity of Chinese social organizations to participate meaningfully in disaster relief is a multi-dimensional construct, shaped by their organizational attributes, technical competences, and inter-organizational linkages. Rather than a static quality, capacity must be understood as a dynamic and evolving set of capabilities that develop over time through experience, institutional learning, and policy engagement. Drawing on existing frameworks (Cao et al., 2012; Chen et al., 2025a, 2025b; Wu & Peng, 2017), this section deconstructs capacity into three interdependent dimensions: organizational capacity, technical capacity, and collaborative capacity.

Organizational capacity refers to the internal structural readiness of social organizations to initiate, sustain, and scale disaster response operations. This includes the ability to mobilize financial and human resources, establish functional internal governance, and maintain operational continuity during crises. Research indicates that organizational robustness is often uneven across the sector. Many grassroots NGOs lack stable funding streams, trained personnel, or standardized operating procedures (Gao et al., 2018; Zou & Yuan, 2025). While national-level foundations and well-established organizations such as the One Foundation have demonstrated the ability to deploy teams and resources across multiple regions, smaller actors often depend on personal networks or ad hoc donations. Regulatory environments—such as the dual registration system—also affect organizational sustainability, as legal ambiguity limits access to institutional support and emergency procurement channels (Xia et al., 2021; Zhang, 2012).

Technical capacity involves the tools, knowledge systems, and procedural know-how necessary to engage in effective disaster risk management. It encompasses early warning systems, needs assessment methodologies, information technology infrastructure, and risk mapping abilities. In recent years, capacity in this domain has grown through state-supported training programs and digitalization initiatives, particularly among professionalized NGOs in urban areas (Chen et al., 2025a, 2025b; Wang et al., 2022a, 2022b). However, there remains a technological divide between organizations operating in eastern versus western provinces. Organizations in less developed regions continue to rely on manual data collection and lack access to predictive modeling or real-time communication platforms (Ji et al., 2024). Furthermore, standardized emergency protocols—such as rapid assessment templates or relief distribution systems—are not uniformly adopted across the sector (Zhao et al., 2025a, 2025b).

Collaborative capacity refers to the ability of social organizations to coordinate effectively with state agencies, peer NGOs, and community groups. This includes mechanisms for communication, task-sharing, and joint decision-making. In the context of China's strong state-centered governance, collaboration is both a strategic necessity and a regulatory imperative (Wu & Peng, 2017). Organizations with strong ties to local governments or party-affiliated bodies are more likely to be integrated into official emergency plans and resource allocation processes (Du & Okazaki, 2016). Platforms such as volunteer databases, joint disaster drills, and cross-sectoral working groups have emerged to improve collaborative functioning (Zhao et al., 2025a, 2025b; Zhou et al., 2024). However, coordination often remains personality-driven and lacks formal institutionalization. Inter-organizational trust is fragile, especially when mandates overlap or when access to government funding is competitive.

Taken together, these three dimensions constitute a framework for assessing the operational readiness of Chinese social organizations in disaster relief. Their interactions are recursive: strong organizational infrastructure enables technical learning; technological proficiency enhances credibility in collaborative networks; and effective collaboration, in turn, reinforces legitimacy and resource access. Understanding capacity in this integrated manner provides a necessary foundation for the next section, which examines how these capacities have been deployed, tested, and transformed through participation in major disasters.

8.3.2 Capacity in Action: Case-Based Evidence from Major Disasters

The real-world capacity of Chinese social organizations in disaster relief has been most visibly tested and demonstrated during major disasters. These crisis events not only served as stress tests for organizational preparedness but also revealed structural weaknesses and uneven development across regions. Case-based analysis illustrates how different dimensions of capacity—organizational, technical, and collaborative—have manifested in practice, evolved through experience, and exposed persistent disparities.

The 2008 Wenchuan earthquake was the first large-scale event to mobilize social organizations at an unprecedented scale. Over 3,000 domestic NGOs and volunteer groups responded to the disaster, providing medical care, shelter, and psychological support (Menefee & Nordtveit, 2012; Wu & Peng, 2017). This spontaneous mobilization reflected strong latent organizational and mobilization capacity, especially in urban-based NGOs with prior philanthropic or service delivery experience. However, many organizations struggled with coordination, internal governance, and sustainability. Numerous groups lacked formal registration, clear operational protocols, or mechanisms for coordination with governmental agencies, leading to duplication of efforts and logistical bottlenecks (Chen et al., 2020). Although the Wenchuan case showcased the potential of civil society in disaster relief, it simultaneously revealed the limitations of fragmented organizational structures and the absence of technical and legal frameworks for integration.

In contrast, the response to the 2010 Yushu earthquake in Qinghai Province demonstrated some progress in both technical and collaborative capacity. By this time, a number of professionalized NGOs had begun to adopt needs assessment tools, GPS mapping, and digital reporting systems, allowing more targeted aid delivery (Yang et al., 2021a, 2021b). Coordination with local authorities, particularly in ethnic minority regions, remained a challenge due to linguistic, cultural, and administrative barriers. However, the establishment of temporary coordination platforms by provincial governments—often involving NGOs, faith-based groups, and Red Cross branches—allowed for improved information sharing and resource pooling (Su et al., 2024). The Yushu experience marked a tentative step toward embedding NGOs in a hybrid governance model, although much of the collaborative logic remained experimental and dependent on local discretion.

The 2021 Henan floods offer a more recent and technologically advanced case of social organization engagement. NGOs in cities such as Zhengzhou employed drone surveillance, real-time donation tracking platforms, and public hotline systems to assist in evacuation and distribution efforts (Wang et al., 2024; Zhao et al., 2025a, 2025b). Several organizations were integrated into local emergency command systems, reflecting growing institutional trust and legal clarity. However, the digital divide remained evident: while larger, resource-rich NGOs managed to provide sophisticated support, many grassroots groups in peri-urban and rural areas lacked both access to digital tools and formal coordination channels. As a result, their contributions were limited in scope and visibility, reinforcing the hierarchical differentiation in disaster governance participation (Li et al., 2025; Zhao et al., 2025a, 2025b).

Moreover, a consistent pattern of regional disparity emerged across all cases. Organizations based in eastern and coastal provinces tended to exhibit stronger organizational and technical capacity due to better infrastructure, funding access, and training opportunities (Ji et al., 2024). In contrast, western and central regions—often more disaster-prone—were characterized by a thinner civil society ecosystem, limited human resources, and lower NGO density. These areas were also more reliant on government-led responses, with social organizations playing only a peripheral role (Gao et al., 2018; Liu et al., 2021). This spatial imbalance has resulted in uneven coverage during disasters and the emergence of what some scholars call "rescue blind zones."

8.3.3 *Evaluation and Measurement: Tools for Capacity Assessment*

As Chinese social organizations assume more prominent roles in disaster relief, the need for systematic evaluation of their capacities has grown increasingly urgent. Capacity assessment serves not only as a diagnostic tool for identifying organizational strengths and weaknesses but also as a policy instrument for resource allocation, regulatory oversight, and long-term capability development. In recent years, several methodological frameworks have been proposed and piloted in China to evaluate the disaster-related performance of social organizations. However, challenges of standardization, data integration, and institutional trust continue to hinder the effectiveness of these tools.

Among the most widely discussed models is the Policy Modeling Consistency (PMC) approach, which has been applied to assess the alignment between organizational strategies and overarching policy goals. The PMC model allows evaluators to quantify the consistency between a social organization's disaster relief agenda and national priorities, particularly in areas such as social mobilization, emergency responsiveness, and compliance with government regulations (Li & Guo, 2022). This model has been adopted in several municipal-level assessments, particularly in eastern provinces where local governments seek to professionalize and rank participating NGOs. While useful in revealing policy gaps, PMC evaluations tend to focus heavily on documentation and formal alignment, often overlooking operational flexibility or informal community linkages.

Another growing trend is the use of Geographic Information Systems (GIS) and spatial analysis tools to evaluate capacity from the perspective of coverage and responsiveness. These tools enable real-time mapping of relief deployment, volunteer distribution, and supply logistics, offering granular insights into spatial disparities in service provision. For example, evaluations after the 2021 Henan floods utilized GIS data to identify areas underserved by both state and non-state actors, revealing that nearly 30% of rural townships in the affected regions had no NGO presence within the first 48 h (Zhao et al., 2025a, 2025b). While GIS offers a valuable spatial lens, its application is often limited by data accessibility, particularly for grassroots organizations lacking technical infrastructure or government data-sharing permissions.

In addition, Multi-Criteria Decision-Making (MCDM) models have gained traction for integrating both quantitative and qualitative indicators in evaluating organizational performance. These models typically combine inputs such as response time, service diversity, beneficiary feedback, and coordination efficiency to generate a composite performance score (Chen et al., 2025a, 2025b). MCDM methods have been applied in pilot evaluations in cities like Chengdu and Shanghai, enabling comparative assessments across different NGO types and intervention stages. However, concerns remain about subjectivity in indicator weighting and the limited transferability of these models to rural or resource-constrained contexts.

Despite these innovations, several structural limitations persist. First, the lack of unified national standards for capacity assessment has resulted in fragmented practices, with local authorities and academic institutions applying divergent methodologies. Second, data fragmentation severely restricts longitudinal tracking; many NGOs fail to maintain consistent records or lack the capacity for systematic data collection (Wang et al., 2022a, 2022b). Third-party evaluations—critical for ensuring objectivity—remain rare due to limited legal frameworks, funding constraints, and bureaucratic resistance to independent oversight (Hou & Wu, 2020). As a result, performance assessments tend to be either internally driven or commissioned by supervising agencies, limiting critical scrutiny.

Moreover, evaluation tools often fail to capture relational and adaptive capacities, such as the ability to build trust with local communities, respond to rapidly evolving needs, or collaborate flexibly across sectors (Wu & Peng, 2017; Zhao et al., 2025a, 2025b). These dimensions, while less quantifiable, are crucial for effective disaster response, particularly in complex and uncertain environments.

8.4 Constraints and Future Prospects for Social Organizations in Disaster Governance

Persistent constraints continue to shape the role of Chinese social organizations in disaster governance, revealing a complex interplay between institutional design and practical participation. Despite growing policy recognition, hierarchical state–society relations, ambiguous legal status, and uneven policy implementation confine social organizations largely to auxiliary roles. These structural limitations are reinforced by legitimacy deficits, unstable resource channels, and stark regional and urban–rural disparities in organizational capacity, collectively constraining responsiveness and long-term engagement. At the same time, recent legal reforms and localized collaborative experiments point toward alternative governance trajectories. Advancing toward more effective disaster governance will depend on institutional embedding, sustained cross-sector coordination, and mechanisms that support continuous participation across all phases of disaster management.

8.4.1 *Institutional Marginalization and Asymmetric State–Society Relations*

Despite the increasing visibility of Chinese social organizations in disaster relief, their role remains structurally constrained by an overarching state-centric governance framework. Although official policy rhetoric emphasizes social participation in emergency response, institutional arrangements often relegate social organizations to auxiliary or temporary roles, limiting their ability to institutionalize participation in disaster governance. This disjuncture between policy endorsement and operational practice reflects deeper asymmetries in state–society relations.

The dominant governance logic in disaster management in China is hierarchical, with central and local governments exercising primary control over emergency planning, resource allocation, and response coordination (DuBois, 2015; Wu & Peng, 2017). Within this structure, social organizations are frequently positioned as supportive actors—mobilized post hoc to supplement state efforts rather than embedded as co-governance partners. This mode of engagement has been described as "临时征用" (temporary conscription), wherein social organizations are activated in times of crisis but excluded from sustained planning and institutional design (He et al., 2021).

Evidence from disaster events such as the Wenchuan (2008) and Yushu (2010) earthquakes illustrates this pattern. While thousands of NGOs were involved in on-the-ground relief, few were granted formal roles within emergency command systems or policy-making bodies. Instead, coordination was often ad hoc, based on personal networks or local discretion, reinforcing their peripheral status (Jiang, 2013; Paltemaa, 2017). Even in more recent disasters such as the Henan floods (2021), many grassroots organizations reported being excluded from formal emergency briefings or logistical planning processes, despite providing frontline assistance (Liu et al., 2021; W. Wang et al., 2024).

At the policy level, legal documents such as the Emergency Response Law and Regulations on Natural Disaster Relief affirm the role of "social forces" but lack detailed implementation guidelines on how these actors are to be systematically integrated into governance structures. This vagueness allows for discretionary enforcement by local governments, leading to significant variation in cooperation mechanisms across provinces and municipalities. In practice, some localities develop temporary NGO coordination platforms during crises, while others exclude non-state actors entirely from formal decision-making channels (Zhou et al., 2024).

The asymmetry is further reinforced by the dual registration system, which requires NGOs to secure a professional supervisory unit before legal recognition—an often prohibitive barrier, especially for grassroots or advocacy-oriented organizations (Gong, 2021; Yi et al., 2019). Without formal recognition, many organizations are excluded from public procurement contracts, joint drills, or emergency coordination lists. This exclusion not only limits operational legitimacy but also restricts access to funding and information flows, undermining both preparedness and responsiveness.

8.4.2 Legitimacy Challenges and Responsiveness Limitations

A core barrier undermining the effectiveness of Chinese social organizations in disaster governance lies in their precarious legitimacy status and constrained emergency responsiveness. While policy discourses often call for "diversified social participation" in disaster response, institutional realities frequently deny social organizations the formal recognition, resource access, and operational readiness required for timely and sustained intervention.

At the institutional level, legal identity remains an unresolved challenge. The current registration system imposes significant procedural hurdles, requiring social organizations to obtain sponsorship from a supervising agency before formal recognition. This "dual management" model has excluded many grassroots NGOs—particularly those engaged in advocacy or rights-based relief—from legal registration, thereby marginalizing them in official response systems (Gong, 2021; Yi et al., 2019). Even among registered organizations, the scope of permissible activities is often narrowly defined, and bureaucratic delays can hinder rapid mobilization during emergencies (DuBois, 2015). Without a clear legal identity, many organizations face limitations in fundraising, resource procurement, and coordination with government agencies.

These legitimacy constraints are compounded by unstable and fragmented resource channels. Most social organizations rely heavily on private donations, which fluctuate dramatically across disaster cycles. For example, during the 2008 Wenchuan earthquake, donations surged to over 76 billion RMB, yet more than 90% were directed to government-led channels, leaving many civil society actors underfunded despite their frontline roles (Fuller, 2015; Wang et al., 2020). Subsequent events have shown similar patterns: large, media-covered disasters may temporarily boost NGO funding, but smaller-scale or rural events receive limited philanthropic attention, further eroding organizational continuity (Paltemaa, 2017; Su et al., 2024).

In terms of responsiveness, the operational flexibility of many organizations is hindered by inadequate institutional infrastructure. Most grassroots NGOs lack permanent staff, professional training, and emergency stockpiles, making them ill-equipped for rapid deployment (Chen et al., 2025a, 2025b; Li et al., 2025). Heavy dependence on volunteers, while valuable for community engagement, results in fluctuating response capacity and difficulties in ensuring procedural consistency or accountability (Zhao et al., 2023a, 2023b). Moreover, information asymmetry—particularly in disaster-prone rural regions—prevents many organizations from receiving timely alerts or coordinating with official early warning systems (Ji et al., 2024).

Attempts to professionalize the sector have gained momentum, with capacity-building initiatives supported by local civil affairs bureaus and international donors (Hou & Wu, 2020). However, these efforts often prioritize larger, urban-based NGOs, exacerbating the marginalization of rural or issue-specific organizations. Additionally, national laws such as the Emergency Response Law mention social forces only in broad terms, without establishing mechanisms for role definition, performance standards, or integration into incident command structures.

8.4.3 Regional Disparities and Capacity Inequality

The spatial distribution of disaster relief capacity among Chinese social organizations is marked by profound regional inequalities. These disparities manifest not only in the concentration of organizational resources in economically advanced regions, but also in uneven access to information, infrastructure, and institutional support across urban–rural and east–west divides. As a result, organizations based in under-resourced areas often find themselves marginalized in both preparedness and emergency response phases.

Empirical data reveal that social organizations are disproportionately concentrated in eastern coastal provinces such as Guangdong, Jiangsu, and Zhejiang, which benefit from more favorable policy environments, philanthropic cultures, and economic infrastructure (Gao et al., 2018; Li et al., 2025). In contrast, western and central provinces—including many disaster-prone areas such as Sichuan, Yunnan, and Gansu—have fewer registered NGOs, lower levels of professionalization, and limited financial or technological resources (Ji et al., 2024; Wu & Peng, 2017). For instance, after the 2021 Henan floods, several counties in western parts of the province reported no local NGO presence within the first 72 h, despite extensive damage and displacement (Zhao et al., 2025a, 2025b).

The urban–rural divide further complicates organizational capacity. Urban-based organizations tend to benefit from higher media visibility, digital connectivity, and established government linkages, whereas rural NGOs operate in fragmented ecosystems with weaker support networks (Wang et al., 2022a, 2022b; Zhao et al., 2025a, 2025b). Disparities in access to early warning systems and real-time disaster data frequently delay rural responses, while bureaucratic bottlenecks and limited funding pipelines impede their ability to scale up operations. Moreover, recruitment and retention of skilled personnel in rural areas remain a persistent challenge, with many organizations relying on part-time or ad hoc volunteer networks (Liu et al., 2021).

Regional inequality is not merely a logistical challenge but also a structural governance issue. In areas where local governments lack incentives or mechanisms to incorporate civil society actors into disaster plans, organizational participation is often limited to reactive or symbolic roles (Du & Okazaki, 2016; Liang et al., 2019). In some disaster-prone zones, especially in western ethnic minority regions, this has resulted in what scholars describe as "participation deserts" or "rescue blind zones," where neither governmental nor non-governmental forces are sufficiently embedded to ensure effective disaster governance (He et al., 2021; Jiang, 2013).

8.4.4 *Toward Integrated and Collaborative Governance: Future Directions*

As Chinese social organizations continue to seek a meaningful role in disaster governance, structural transformation toward integrated, multi-actor collaboration becomes imperative. Despite fragmented progress over the past two decades, emerging institutional frameworks and policy discourses signal a growing consensus on the necessity of embedding social organizations into the full disaster governance cycle—from preparedness and response to recovery and reconstruction.

The legal and regulatory basis for this transition has gradually expanded. The Emergency Response Law emphasizes the need for coordinated efforts among governments, social organizations, and communities, introducing clauses that encourage institutionalized participation of non-state actors in local emergency planning and drills. Similarly, the Ministry of Civil Affairs proposed the creation of permanent coordination platforms at the provincial and municipal levels, designed to institutionalize cross-sector engagement in disaster scenarios. However, the implementation of these mandates remains uneven, hindered by local bureaucratic inertia and the absence of enforceable accountability mechanisms (Du & Okazaki, 2016; Zhou et al., 2024).

One key strategy moving forward is the development of collaborative governance models that actively link government agencies with social organizations and community groups. Research on recent disaster responses highlights the value of "tri-sector coordination" models, particularly in urban areas where policy support and resource availability are stronger (Cui et al., 2022; Paltemaa, 2017). Pilot programs in cities such as Chengdu and Shenzhen have demonstrated how integrated platforms—such as disaster response command centers or digital volunteer networks—can facilitate real-time coordination, task-sharing, and joint decision-making (Hou & Wu, 2020; Zhao et al., 2025a, 2025b). However, these remain localized and often lack replication pathways in rural or under-resourced regions.

Another promising direction is the promotion of inter-organizational coalitions and NGO consortia. These alliances allow smaller organizations to pool resources, share technical expertise, and present unified proposals to governmental actors (Su et al., 2025; Zhao et al., 2025a, 2025b). Cross-regional mutual aid mechanisms have also been explored, particularly after the 2021 Henan floods, where organizations from different provinces coordinated supply chains and information channels through pre-existing networks (Wang et al., 2024). However, the lack of stable funding and regulatory clarity continues to constrain the scaling of such initiatives.

Ultimately, transitioning toward an inclusive and collaborative disaster governance system requires more than top-down policy articulation. It necessitates legal empowerment, fiscal commitment, and the establishment of mechanisms for ongoing participation—beyond ad hoc mobilization—throughout the disaster governance continuum (Cao et al., 2012). This includes standardized entry points for NGOs in emergency command systems, regularized capacity assessments, and co-developed protocols that recognize the complementary value of social organizations in managing uncertainty and complexity. Without such structural shifts, the potential of social organizations in disaster relief will remain underutilized.

8.5 Conclusion

The evolution of Chinese social organizations in disaster relief reflects a dynamic interplay between historical legacies, institutional transformations, and capacity development. From their moral roots in communal relief to their marginalization under state-building, these organizations have gradually reemerged through reform-era governance shifts and post-disaster opportunities. Capacity-building efforts across organizational, technical, and collaborative domains have expanded their functional scope, yet persistent challenges—ranging from legal ambiguity and resource constraints to regional disparities—continue to limit their structural integration. Despite these barriers, emerging policy frameworks and localized innovations suggest a path toward more inclusive and institutionalized collaboration. Fully realizing the potential of social organizations in disaster governance will depend on sustained legal, fiscal, and procedural reforms that embed them as essential actors throughout the disaster management cycle.

References

Cao, L., Luo, H. F., & Wang, X. B. (2012). Building and application of GIS platform in the provincial disaster relief and emergency command system. *Advanced Materials Research, 518–523*, 5684–5687. https://doi.org/10.4028/www.scientific.net/AMR.518-523.5684

Chen, H., Lin, X., Guo, Y., & Qi, X. (2025a). Heterogeneity and spatiotemporal evolution characteristics of regional emergency logistics response capacity: A case of China. *Frontiers in Public Health, 13*, 1461354. https://doi.org/10.3389/fpubh.2025.1461354

Chen, Q. (2015). Climate shocks, state capacity and peasant uprisings in North China during 25–1911 CE. *Economica, 82*(326), 295–318. https://doi.org/10.1111/ecca.12114

Chen, Y., Tadikamalla, P. R., Shang, J., & Song, Y. (2020). Supply allocation: Bi-level programming and differential evolution algorithm for natural disaster relief. *Cluster Computing, 23*(1), 203–217. https://doi.org/10.1007/s10586-017-1366-6

Chen, Y., Yi, R., Li, J., & Chen, A. (2025b). Coping or adapting? Exploring regional priority of capacity building for disaster resilience. *Journal of Cleaner Production, 525*, Article 146502. https://doi.org/10.1016/j.jclepro.2025.146502

Cui, P., Ge, Y., Li, S., Li, Z., Xu, X., Zhou, G. G. D., Chen, H., Wang, H., Lei, Y., Zhou, L., Yi, S., Wu, C., Guo, J., Wang, Q., Lan, H., Ding, M., Ren, J., Zeng, L., Jiang, Y., & Wang, Y. (2022). Scientific challenges in disaster risk reduction for the Sichuan-Tibet Railway. *Engineering Geology, 309*, Article 106837. https://doi.org/10.1016/j.enggeo.2022.106837

Du, F., & Okazaki, K. (2016). Building improvement responses to multi-hazard risk in the historic Dali Dong Village, Guizhou, China. *International Journal of Disaster Risk Reduction, 19*, 64–74. https://doi.org/10.1016/j.ijdrr.2016.08.014

DuBois, T. D. (2015). Before the NGO: Chinese charities in historical perspective. *Asian Studies Review, 39*(4), 541–553. https://doi.org/10.1080/10357823.2015.1081870

Emergency Response Law of the People's Republic of China (Revised 2024) (2024). https://www.gov.cn/yaowen/liebiao/202406/content_6960130.htm

Fuller, P. (2015). Decentring international and institutional famine relief in late nineteenth-century China: In search of the local. *European Review of History: Revue Européenne D'histoire, 22*(6), 873–889. https://doi.org/10.1080/13507486.2015.1048189

Gao, L., Wu, Q., Li, Y., Ding, D., Hao, Y., Cui, Y., Kang, Z., Jiao, M., Liang, L., Ferrier, A., Ning, N., & Sun, H. (2018). How prepared are hospitals' emergency management capacity? Factors influencing efficiency of disaster rescue. *Disaster Medicine and Public Health Preparedness, 12*(2), 176–183. https://doi.org/10.1017/dmp.2016.25

Gong, L. (2021). China's disaster diplomacy in Southeast Asia. *China Review, 21*(4), 77–107. https://www.jstor.org/stable/48635893

He, H., Delang, C. O., Zhou, J., Li, Y., & He, W. (2021). Simulation of social resilience affected by extreme events in ancient China. *Climatic Change, 166*(3–4), 42. https://doi.org/10.1007/s10584-021-03134-9

Hou, C., & Wu, H. (2020). Rescuer, decision maker, and breadwinner: Women's predominant leadership across the post-Wenchuan earthquake efforts in rural areas, Sichuan China. *Safety Science, 125*, Article 104623. https://doi.org/10.1016/j.ssci.2020.104623

Ji, Z., Zheng, H., & Zou, J. (2024). Integrated approach for flood shelter site selection and emergency response accessibility analysis at the provincial scale: A case study in Henan, China. *Journal of Water and Climate Change, 15*(10), 5291–5308. https://doi.org/10.2166/wcc.2024.475

Jiang, L. (2013). Implementation of disaster reduction measures and enhancement of integrated risk governance in China: Summary of the remarks at the high-level dialogue, official statement, and plenary session of the fourth session of the global platform on disaster risk reduction, 21–23 May 2013, Geneva, Switzerland. *International Journal of Disaster Risk Science, 4*(2), 101–104. https://doi.org/10.1007/s13753-013-0011-0

Li, Y., Cheng, W., Zhang, Y., Stanny, M., & Du, G. (2025). Mapping urban-rural divide of disaster vulnerability and implications for community resilience building in China. *Habitat International, 165*, Article 103580. https://doi.org/10.1016/j.habitatint.2025.103580

Li, Z., & Guo, X. (2022). Quantitative evaluation of China's disaster relief policies: A PMC index model approach. *International Journal of Disaster Risk Reduction, 74*, Article 102911. https://doi.org/10.1016/j.ijdrr.2022.102911

Liang, Y., Cheng, J., Ruzek, J. I., & Liu, Z. (2019). Posttraumatic stress disorder following the 2008 Wenchuan earthquake: A 10-year systematic review among highly exposed populations in China. *Journal of Affective Disorders, 243*, 327–339. https://doi.org/10.1016/j.jad.2018.09.047

Liu, Y., Zhu, J., Shao, X., Adusumilli, N. C., & Wang, F. (2021). Diffusion patterns in disaster-induced internet public opinion: Based on a Sina Weibo online discussion about the 'Liangshan fire' in China. *Environmental Hazards, 20*(2), 163–187. https://doi.org/10.1080/17477891.2020.1758608

Menefee, T., & Nordtveit, B. H. (2012). Disaster, civil society and education in China: A case study of an independent non-government organization working in the aftermath of the Wenchuan earthquake. *International Journal of Educational Development, 32*(4), 600–607. https://doi.org/10.1016/j.ijedudev.2011.10.002

Ministry of Civil Affairs of the People's Republic of China. (2015). *Guiding opinions on supporting and guiding social forces to participate in disaster relief*. https://www.gov.cn/gongbao/content/2016/content_5046085.htm

Paltemaa, L. (2017). China's changing disaster management regime. *China Information, 31*(3), 284–303. https://doi.org/10.1177/0920203X17726611

Peng, L. (2017). Crisis crowdsourcing and China's civic participation in disaster response: Evidence from earthquake relief. *China Information, 31*(3), 327–348. https://doi.org/10.1177/0920203X17735668

Regulations on Natural Disaster Relief (State Council Order No. 577) (2010). https://www.gov.cn/flfg/2010-07/14/content_1661409.htm

Su, H., Zhao, X., Wang, L., & Li, Y. (2024). How rural community resilience evolves after a disaster? A case study of the eastern Qinghai-Tibetan Plateau China. *Applied Geography, 165*, Article 103238. https://doi.org/10.1016/j.apgeog.2024.103238

Su, Y., Zeng, Q., & Shu, Q. (2025). The impact of disasters on adaptive collective action among farmers: Evidence from China's border regions. *Systems, 13*(12), 1065. https://doi.org/10.3390/systems13121065

Tao, L., Chen, X., Su, Y., Tian, F., & Chen, S. (2025). Social response behaviors reveal the enhancement and loss of drought resilience: Example of Sichuan, China in the Qing Dynasty. *Climate Risk Management, 50*, Article 100743. https://doi.org/10.1016/j.crm.2025.100743

Wang, F., Xie, Z., Pei, Z., & Liu, D. (2022a). Emergency relief chain for natural disaster response based on government-enterprise coordination. *International Journal of Environmental Research and Public Health, 19*(18), 11255. https://doi.org/10.3390/ijerph191811255

Wang, H., Ye, H., Liu, L., & Li, J. (2022b). Evaluation and obstacle analysis of emergency response capability in China. *International Journal of Environmental Research and Public Health, 19*(16), 10200. https://doi.org/10.3390/ijerph191610200

Wang, J., Bu, K., Yang, F., Yuan, Y., Wang, Y., Han, X., & Wei, H. (2020). Disaster risk reduction knowledge service: a paradigm shift from disaster data towards knowledge services. *Pure and Applied Geophysics, 177*(1), 135–148. https://doi.org/10.1007/s00024-019-02229-w

Wang, W., Zhu, X., Lu, P., Zhao, Y., Chen, Y., & Zhang, S. (2024). Spatio-temporal evolution of public opinion on urban flooding: Case study of the 7.20 Henan extreme flood event. *International Journal of Disaster Risk Reduction, 100*, 104175. https://doi.org/10.1016/j.ijdrr.2023.104175

Wu, F., & Peng, L. (2017). The rise of non-governmental disaster relief in China. *SSRN Electronic Journal*. https://doi.org/10.2139/ssrn.2916221

Xia, C., Nie, G., Li, H., Fan, X., Yang, R., & Zeng, X. (2021). Comparative analysis of the earthquake disaster risk of cities in Eastern China based on lethal levels—A case study of Yancheng City, Suqian City and Guangzhou City. *Geomatics, Natural Hazards and Risk*. https://www.tandfonline.com/doi/abs/https://doi.org/10.1080/19475705.2021.2009580

Xiao, L., Fang, X., Zhang, Y., Ye, Y., & Huang, H. (2014). Multi-stage evolution of social response to flood/drought in the North China Plain during 1644–1911. *Regional Environmental Change, 14*(2), 583–595. https://doi.org/10.1007/s10113-013-0516-5

Yang, W., Fang, J., & Liu-Zeng, J. (2021a). Landslide-lake outburst floods accelerate downstream hillslope slippage. *Earth Surface Dynamics, 9*(5), 1251–1262. https://doi.org/10.5194/esurf-9-1251-2021

Yang, Z., Liu, G., Wang, L., Liu, S., & Fu, X. (2021b). Evolution of hydro-mechanical behaviours and its influence on slope stability for a post-earthquake landslide: Implications for prolonged landslide activity. *Bulletin of Engineering Geology and the Environment, 80*(11), 8803–8822. https://doi.org/10.1007/s10064-021-02463-4

Yi, J.-H., Wang, J., & Wang, G.-G. (2019). Using monarch butterfly optimization to solve the emergency vehicle routing problem with relief materials in sudden disasters. *Open Geosciences, 11*(1), 391–413. https://doi.org/10.1515/geo-2019-0031

Zhai, X., Fang, X., & Su, Y. (2020). Regional interactions in social responses to extreme climate events: A case study of the north china famine of 1876–1879. *Atmosphere, 11*(4), 393. https://doi.org/10.3390/atmos11040393

Zhang, H. (2012). What has China learnt from disasters? evolution of the emergency management system after SARS, Southern Snowstorm, and Wenchuan Earthquake. *Journal of Comparative Policy Analysis: Research and Practice, 14*(3), 234–244. https://doi.org/10.1080/13876988.2012.687621

Zhao, F., Zhou, Q., Liu, L., Liu, F., Ma, W., Li, H., Chen, Q., & Liu, Y. (2025a). Evaluation of disaster resilience and optimization strategies for villages in the Hengduan Mountains Region China. *Sustainability, 17*(22), 10176. https://doi.org/10.3390/su172210176

Zhao, S., Dai, F., Deng, J., Wen, H., Li, H., & Chen, F. (2023a). Insights into landslide development and susceptibility in extremely complex alpine geoenvironments along the western Sichuan-Tibet Engineering Corridor China. *CATENA, 227*, Article 107105. https://doi.org/10.1016/j.catena.2023.107105

Zhao, W., Wang, J., Xu, Y., Chen, S., Zhang, J., Tang, S., & Wang, G. (2023b). Community resilience assessment and identification of barriers in the context of population aging: A case study of Changchun city China. *Sustainability, 15*(9), 7185. https://doi.org/10.3390/su15097185

Zhao, X., Wang, H., Morikawa, S., Wang, S., Tang, J., Sugiyama, S., & Sriwarnasinghe, S. M. (2025b). Mapping non-profits participation in disaster mitigation: A data-driven study of functional diversity, spatial patterns and driving factors of China's emergency management nonprofit organizations. *International Journal of Disaster Risk Reduction, 118*, Article 105252. https://doi.org/10.1016/j.ijdrr.2025.105252

Zhou, J., Nie, G., & Liu, Y. (2024). Multi-criteria comparison analysis of spatial equity for emergency shelters in old and new urban districts: A case study in Wuhua District and Chenggong District of Kunming City, China. *International Journal of Disaster Risk Reduction, 108*, Article 104547. https://doi.org/10.1016/j.ijdrr.2024.104547

Zou, M., & Yuan, Y. (2025). China's comprehensive disaster reduction. *International Journal of Disaster Risk Science, 1*(1), 24–32. https://doi.org/10.3974/j.issn.2095-0055.2010.01.005

Chapter 9
Wenchuan Earthquake: The First Year of Social Organizations in Disaster Governance

Abstract The 2008 Wenchuan Earthquake marked a historical turning point in the evolution of China's disaster governance and civil society development. This chapter traces how the disaster catalyzed the emergence, negotiation, consolidation, and institutionalization of social organizations in the governance process. It begins by examining the spontaneous mobilization of grassroots NGOs and citizen volunteers in the immediate aftermath of the quake. It then analyzes their evolving interactions with the state—ranging from pragmatic cooperation to embedded control—before turning to how these organizations developed internal capacities, sectoral networks, and collaborative governance mechanisms. Finally, the chapter explores how civic engagement matured into routinized and partially institutionalized participation, highlighting a gradual but uneven transformation from ad hoc relief efforts to an increasingly pluralized disaster governance regime.

Keywords Spontaneous mobilization · State–society negotiation · Professionalization and institutionalization · Governance pluralization

9.1 The Emergence of Civil Society in Crisis: Spontaneous Mobilization and the Rise of Social Organizations

The 2008 Wenchuan Earthquake marked a pivotal moment in the evolution of civil society in China, revealing a dynamic shift in the state–society relationship within disaster governance. This chapter examines how the disaster catalyzed spontaneous civic mobilization, leading to the emergence of grassroots NGOs, volunteer networks, and citizen-led initiatives. It analyzes the enabling role of digital media, the evolution of informal coalitions, and the complex interactions between non-state actors and local authorities. By tracing the transition from initial chaos to tentative organizational maturation, the chapter highlights both the possibilities and constraints of civic participation in a state-centric governance environment.

Q. Du and R. Shaw, *Social Organizations in Disaster Response and Recovery*, Disaster Risk Reduction, https://doi.org/10.1007/978-981-92-2846-1_9

9.1.1 A Critical Juncture: Disaster as Opportunity for Civil Society

The 2008 Wenchuan Earthquake marked a watershed moment in China's state–society dynamics, particularly in the domain of disaster governance. The catastrophe, which caused nearly 70,000 deaths and displaced millions, not only exposed the limitations of state-led emergency response but also catalyzed the spontaneous emergence of civil society actors—including grassroots NGOs, volunteer networks, and citizen-led initiatives—on an unprecedented scale. Rather than being a mere reaction to a disaster, the mobilization that followed constituted a critical juncture, fundamentally disrupting existing governance logics and creating new openings for non-state actors to participate in public affairs.

Multiple studies identify the Wenchuan Earthquake as a moment when the boundary between the state and society became permeable. The disaster created a rupture in China's traditionally state-dominated governance architecture, enabling what had previously been latent or suppressed forms of associational life to surface (Sun, 2017). Civil society actors, including both registered and informal organizations, began to assert themselves in disaster relief, often operating in areas where government capacity was overstretched or absent.

The term "NGO explosion" has been used to describe the remarkable proliferation of organizations that entered the disaster zone within days of the earthquake. Within the first two weeks, thousands of volunteers and over 200 social organizations were reported on the ground in Sichuan province. Many were grassroots entities operating without formal registration, mobilized through personal networks, online calls-to-action, and the collective emotional response to the disaster. This mobilization was not only geographically widespread but also temporally sustained, laying the groundwork for longer-term recovery initiatives (Lu et al., 2021).

Importantly, digital communication technologies played a catalytic role in this transformation. Blogs, forums, and social media platforms such as Weibo functioned as real-time coordination and amplification tools (Gao, 2020; Wu & Chang, 2018). These tools enabled rapid information exchange, the mapping of needs and resources, and cross-regional collaboration among previously unconnected actors. In many cases, public discourse online also served to hold local governments accountable and to legitimate the actions of NGOs in the eyes of the broader public.

However, while the Wenchuan Earthquake unlocked the potential of civil society in disaster response, it also revealed the fragility of this emerging sector. Many of the organizations lacked operational experience, legal status, or stable funding mechanisms, which led to coordination failures and resource duplication (Lu et al., 2021). Additionally, the state's initial permissiveness gradually gave way to heightened regulation, as authorities moved to reassert control over the narrative and logistics of disaster recovery.

9.1.2 Spontaneous Mobilization: Volunteers, Networks, and Public Sentiment

The Wenchuan Earthquake of 2008 marked an unprecedented moment of bottom-up civic mobilization in post-reform China. In the hours and days following the disaster, tens of thousands of volunteers—including individuals, student associations, corporate employees, religious groups, and grassroots NGOs—spontaneously traveled to the disaster zone, forming an emergent and largely uncoordinated field of humanitarian engagement (Kang, 2017; Lu et al., 2021).

This wave of voluntary action was not only massive in scale, but also striking in its heterogeneity and speed. Within a week of the earthquake, over 200,000 volunteers had converged in Sichuan, often without affiliation or formal training (Kang, 2017). Many were driven by nationalistic empathy, shared trauma, or exposure to real-time images and narratives disseminated through online platforms. While major national organizations such as the Red Cross were initially present, it was the informal and grassroots actors who quickly filled logistical and emotional gaps on the ground (Huang et al., 2011).

Crucially, this surge of engagement was made possible by emerging digital technologies and media platforms. Public forums, SMS networks, QQ groups, and blogs became essential tools for organizing supplies, locating missing persons, and disseminating calls for help (Gao, 2020). This digital infrastructure enabled a unique triadic mobilization mechanism—citizens, civil society organizations, and media—which significantly bypassed conventional state channels (Wu & Chang, 2018). The internet created a "self-organizing space" in which affective solidarity and practical coordination could thrive despite institutional constraints (Yin & Wang, 2010).

At the same time, this spontaneous mobilization exposed structural tensions. Many volunteers lacked disaster response expertise, and informal groups often duplicated services or strained local logistics systems (Lu et al., 2021). Conflicts occasionally arose between NGOs and local officials over access, legitimacy, or control of resources. Some accounts describe a state of "benevolent chaos," where enthusiasm outpaced coordination mechanisms (Menefee & Nordtveit, 2012).

Despite these shortcomings, the Wenchuan case demonstrated the latent capacity of society to respond rapidly and collectively in times of crisis. This civic energy laid the groundwork for more organized forms of disaster response in subsequent years and shifted public perception of civil society from a marginal actor to a necessary component of national resilience infrastructure (Ji & Pang, 2021; Sun, 2017).

9.1.3 The Rise of Grassroots NGOs and Citizen-Led Initiatives

The Wenchuan Earthquake catalyzed the emergence of grassroots NGOs and citizen-led initiatives that had previously occupied a marginal or latent position within

China's disaster governance framework. Prior to the earthquake, China's civil society sector was relatively constrained, both in terms of legal recognition and operational space. Most registered NGOs were either government-organized NGOs (GONGOs) or operated under heavy administrative oversight, with truly grassroots actors often operating informally or under the radar (Kang, 2017; Liu & Wang, 2019).

The immediate post-quake period, however, saw an explosive increase in the number and visibility of such organizations. In Sichuan alone, more than 200 local NGOs and hundreds of informal civic groups became active in relief and recovery operations within weeks of the disaster (Lu et al., 2021). These organizations filled critical service gaps, ranging from logistics coordination to psychosocial support, particularly in remote and underserved areas where state reach was limited or delayed (Menefee & Nordtveit, 2012; D. Xu et al., 2018a, 2018b).

Importantly, many of these grassroots organizations operated without formal registration, functioning instead as semi-institutionalized networks or ad hoc coalitions. This "gray zone" of civic engagement enabled rapid, flexible responses but also revealed regulatory ambiguities and legal vulnerabilities (Sun, 2017). For example, several volunteer collectives initially emerged from university-based networks or social media forums and gradually evolved into more structured NGOs during the response phase (Gao, 2020).

In Chengdu, a coalition of local philanthropic organizations formed a joint disaster response platform within days of the earthquake, pooling personnel, donations, and technical resources to conduct coordinated field operations (Lu et al., 2021). Meanwhile, youth-based networks in Beijing and Shanghai remotely mobilized supplies and coordinated volunteer deployment through digital platforms and peer networks, reflecting a geographically distributed yet socially cohesive form of engagement (Gao, 2020; Wu & Chang, 2018).

Despite their institutional fragility, these grassroots actors demonstrated strong adaptive capacity and local knowledge, allowing them to navigate the complex terrain of post-disaster governance. Their emergence marked a shift from occasional civic involvement to sustained engagement, laying the groundwork for long-term organizational development and policy advocacy in the disaster field (Ji & Pang, 2021; Kang, 2017).

9.1.4 Technology and Media as Mobilization Catalysts

The 2008 Wenchuan Earthquake marked a critical turning point not only in the scale of civil society engagement but also in the role of technology and media as enablers of mass mobilization. In the immediate aftermath of the disaster, digital communication technologies—especially blogs, forums, and early social media platforms—played a central role in bypassing traditional state-controlled information channels and enabling horizontal coordination among citizens, volunteers, and emerging civil organizations (Gao, 2020; Yin & Wang, 2010).

Information flows, which had traditionally been monopolized by state media, were rapidly decentralized. Individuals on the ground began posting real-time updates, photos, and resource needs via digital platforms. These user-generated reports enabled both the rapid dissemination of situational awareness and the spontaneous organization of volunteer teams across the country (Wu & Chang, 2018). Rather than passively receiving top-down instructions, citizens became active agents in shaping disaster response narratives and logistics, signaling a shift in the discursive and operational space of governance.

Digital technologies also facilitated the emergence of what could be described as a "civic information infrastructure." Volunteer mapping projects, crowd-sourced donation platforms, and online coordination hubs—many initiated by NGOs or tech-savvy individuals—allowed for the tracking of resource allocation and identification of underserved areas (Lu et al., 2021). In this sense, digital media were not simply tools of expression, but platforms of governance that contributed directly to crisis management functions such as needs assessment and supply chain coordination (Sun, 2017).

Crucially, these developments challenged the previous hierarchy of authority in disaster governance. While the state maintained central control over formal response protocols, the visibility and reach of digital communication allowed civil society actors to assert their presence and legitimacy in new ways. Independent media accounts and online platforms became mechanisms for public scrutiny, accountability, and legitimacy production, creating pressure on local governments to respond more transparently (Yin & Wang, 2010).

At a deeper level, the widespread adoption of technology revealed an evolving governance ecology in which non-institutional actors leveraged communication tools to co-produce public value during emergencies. This digital-enabled civic engagement not only amplified grassroots voices but also expanded the boundaries of permissible participation in an otherwise tightly managed public sphere (Gao, 2020; Sun, 2017).

9.1.5 From Chaos to Organization: Initial Limitations and Lessons

The initial burst of civic participation following the Wenchuan Earthquake, while unprecedented in scale and visibility, also exposed the fragility of China's emergent civil society infrastructure. Many grassroots NGOs and volunteer collectives entered the disaster zone with immense enthusiasm but limited professional experience, leading to challenges in coordination, duplication of effort, and inefficient resource distribution (Lu et al., 2021). Lacking disaster-specific training or standardized operational protocols, numerous actors struggled with project planning, beneficiary targeting, and logistics management, especially in remote or high-need areas.

These shortcomings were not merely technical but systemic. Most organizations operated outside formal regulatory frameworks and had little or no prior interaction with local authorities. As such, their integration into the official emergency response system was largely improvised, often relying on ad hoc personal relationships or informal arrangements. This absence of institutionalized mechanisms for collaboration hindered both the legitimacy and sustainability of non-state interventions (Liu & Palmer, 2021; Ni & Zhan, 2017).

The government's response to this spontaneous mobilization evolved from tolerance to increasing scrutiny. While local officials initially welcomed support from civic actors, especially given the scale of destruction, their attitudes gradually shifted toward control and containment as governance pressures mounted. Scholars have noted a visible transition from "open access" to "selective engagement" policies, whereby only a subset of "government-friendly" NGOs were granted operational space (Kang, 2017; Tang & Wang, 2020). Others faced bureaucratic obstruction, licensing difficulties, or were gradually marginalized in the reconstruction phase.

Moreover, competing priorities between state actors and civil society groups—such as speed versus inclusivity, visibility versus sustainability—often resulted in tension. For instance, official discourse emphasized rapid rebuilding and social stability, while some grassroots organizations prioritized community participation and psychosocial support, which were viewed as secondary by local governments (Huang et al., 2011; Menefee & Nordtveit, 2012).

Nevertheless, these early limitations catalyzed a longer-term process of learning and adaptation. Several networks that emerged during the earthquake—such as inter-NGO alliances and professional volunteer platforms—subsequently institutionalized their models, developed training standards, and advocated for more transparent partnership frameworks in future crises (Lu & Xu, 2015; J. Xu et al., 2018a, 2018b). While the Wenchuan experience revealed systemic gaps, it also laid the groundwork for a more structured and resilient civil society presence in disaster governance in China.

9.2 Between the State and the People: Interaction and Negotiation in Disaster Response

The Wenchuan Earthquake marked a rare moment in China's disaster governance where the boundaries between state and civil society became unusually porous. In the face of overwhelming devastation, the government temporarily relaxed regulatory controls, allowing NGOs and grassroots actors to enter the disaster zone and contribute to relief efforts. This initial openness, however, soon gave way to a strategy of embedded governance, wherein civil society actors were incorporated into state-led structures through hierarchical coordination. As NGOs navigated these shifting conditions, they engaged in strategic compliance while also testing the limits of state-defined boundaries, particularly in areas like community reconstruction and psychosocial support. Over time, this evolving relationship settled into a model of managed collaboration, where NGOs gained formal recognition but remained confined to technocratic, state-sanctioned roles. This section examines these layered dynamics of interaction, negotiation, and control, revealing a governance trajectory shaped by both institutional pragmatism and authoritarian constraint.

9.2.1 The Initial Openness: Pragmatic Governmental Tolerance in Crisis

In the immediate aftermath of the Wenchuan Earthquake, the Chinese state demonstrated an unusual degree of openness toward civil society actors, creating what many scholars regard as a moment of pragmatic tolerance rather than institutional transformation. Faced with the overwhelming scale of destruction and limited state capacity to respond across all affected areas, government authorities at both central and local levels temporarily relaxed conventional controls over NGO registration, operational boundaries, and funding channels (Huang et al., 2011). This permissiveness enabled a range of non-state actors—including grassroots NGOs, informal volunteer collectives, and religious organizations—to enter the disaster zone and provide critical services in coordination, shelter, medical care, and psychosocial support (Kang, 2020; Xu et al., 2018a, 2018b).

The operational space granted to NGOs during this phase was characterized not by a shift in governance ideology but by crisis-induced necessity. This was especially true for service-oriented and assistance-based organizations whose activities were seen as complementary to state priorities. These organizations often occupied the role of auxiliary actors, functioning within domains where the state lacked reach or technical specialization. Several reports describe how previously dormant or informal groups were allowed to operate openly, sometimes with the tacit support of local officials who faced immense logistical challenges in managing displaced populations and infrastructure damage (Liu & Wang, 2019; Lu & Xu, 2015).

Nonetheless, this window of openness was not institutionally grounded but rather reflective of a contingent and instrumental logic. As several researchers argue, the temporary regulatory vacuum constituted a form of "provisional tolerance" rather than genuine structural reform (Menefee & Nordtveit, 2012). While NGOs were permitted to act, their legal ambiguity and dependence on local discretion made their access to operational space highly variable. In some areas, coordination was facilitated through ad hoc government–NGO liaison mechanisms, while in others, NGOs had to navigate unclear boundaries and shifting expectations (Kang, 2020; Ni & Zhan, 2017).

In essence, the early stage of the Wenchuan response was marked by an "adaptive pragmatism" on the part of the state—an openness born from necessity rather than from an intentional strategy of civic inclusion. While this allowed social organizations to step into critical roles, it also set the stage for later tensions as the government began to reassert control and institutional boundaries were reestablished. The flexibility of this moment thus represented both an opportunity for civic engagement and a reminder of its fragility in an authoritarian governance context.

9.2.2 Power Dynamics and Embedded Governance: Control Through Co-optation

Following the initial period of tolerance, the Chinese state moved swiftly to reassert organizational control over the post-earthquake relief field by constructing formalized coordination mechanisms and embedding civil society actors within state-led governance structures. In many affected areas, the government established official platforms—such as joint rescue coordination centers and temporary liaison offices—to channel and manage the participation of NGOs and volunteer groups (Liu & Palmer, 2021). These institutional arrangements enabled authorities to integrate external actors into a broader reconstruction agenda while maintaining oversight and steering capacity.

This emerging mode of "embedded governance" did not reflect a horizontal partnership model but instead relied on hierarchical incorporation and administrative supervision. NGOs were treated less as autonomous policy stakeholders and more as functional extensions of the state's crisis response apparatus. Their legitimacy was increasingly tied to their ability to align with local bureaucratic logics, rather than to community accountability or sectoral expertise. As noted in several case studies, government actors exercised selectivity in their partnerships—prioritizing technically competent, politically neutral, or previously vetted NGOs, while marginalizing groups seen as unfamiliar, critical, or community-driven (Kang, 2020; Liu & Wang, 2019).

These power asymmetries were often manifested through patron–client or principal–agent relations, wherein state authorities played the role of gatekeepers, distributing access, resources, and recognition to compliant actors. NGOs that demonstrated administrative fluency and deference to local governance norms were more likely to receive logistical support, media visibility, and future contracting opportunities. Conversely, those that operated independently or engaged in rights-based or participatory approaches encountered obstacles such as delays in registration, site access restrictions, or project suspensions (Kang, 2020; Liu & Wang, 2019).

In this context, the state retained ultimate control over the reconstruction narrative, timelines, and priorities, while NGOs were selectively incorporated as "technical executors" rather than co-governors. This logic of incorporation reveals a pragmatic yet instrumental state strategy: allowing civil society presence insofar as it reinforces state-defined objectives, but limiting its political autonomy. Rather than signaling a shift toward pluralistic governance, the embedded governance model that emerged in Wenchuan represented a calibrated form of co-optation—where participation was permitted, but only within administratively sanctioned boundaries (Liu & Palmer, 2021; Xu et al., 2018a, 2018b).

9.2.3 Contestation and Negotiation: Strategic Compliance and Boundary-Pushing

While the post-Wenchuan context enabled unprecedented engagement of NGOs in disaster response, the space was neither normatively stable nor politically neutral. Interactions between state and civil society actors were marked by continuous negotiation, in which NGOs adopted strategic postures to navigate the ambiguities of administrative acceptance, while subtly contesting or expanding the boundaries of permissible action. This dynamic interplay reveals a mode of negotiated governance, in which institutional power was not merely imposed, but also reshaped through tactical compliance and selective resistance.

In order to secure operational legitimacy, many NGOs aligned themselves with official frameworks and avoided politically sensitive agendas. This strategic compliance entailed the adoption of state-sanctioned rhetoric, the prioritization of "technical" service delivery over rights-based advocacy, and deliberate attempts to build rapport with local officials (Liu & Wang, 2019; Xu et al., 2018a, 2018b). For example, several education- and health-focused NGOs rebranded their post-disaster programs as "capacity building" rather than "psychosocial recovery," in order to remain within the acceptable vocabulary of local governance (Kang, 2020; Xu et al., 2018a, 2018b).

At the same time, NGO actors did not passively accept their subordinate role. Some organizations engaged in boundary-pushing behaviors that subtly challenged the contours of state control. These efforts were most visible in sectors such as community reconstruction, trauma counseling, and participatory education, where NGOs leveraged their field access and professional expertise to advocate for bottom-up approaches that deviated from the dominant "stability-first" reconstruction narrative (Lu & Xu, 2015; Menefee & Nordtveit, 2012). Initiatives that encouraged community self-organization, public deliberation, or critical pedagogy—though rarely oppositional in form—nonetheless expanded the discursive and practical space for civil engagement.

Such negotiation often took place in institutional interstices—zones of policy ambiguity or administrative neglect—where NGOs could act with a degree of autonomy. The fragmented nature of local governance, combined with limited disaster management capacity, created openings for experimental practices, particularly in rural or underserved areas (Ni & Zhan, 2017; Sun, 2017). These practices did not confront state authority directly, but rather tested its limits through pragmatic adaptation and local embeddedness.

The outcome was neither full incorporation nor clear resistance, but a hybrid form of governance, where civil society actors selectively adapted to state logics while preserving limited forms of autonomy. This negotiated space was inherently unstable, contingent on political discretion, local bureaucratic cultures, and NGOs' own capacity to perform institutional fluency. Nonetheless, it underscored the agency of non-state actors in shaping the contours of disaster governance—even within authoritarian settings.

9.2.4 *Evolving Relationship Models: From Conflict to Managed Collaboration*

As the emergency phase of the Wenchuan Earthquake response transitioned into long-term reconstruction, the relationship between the state and civil society gradually shifted from initial contestation to a more structured, yet asymmetrical, mode of collaboration. This evolution was marked by the state's increasing efforts to institutionalize NGO participation through formal mechanisms such as service contracting, accreditation systems, and performance evaluations, thereby consolidating a model of managed collaboration or "controllable cooperation" (Liu & Palmer, 2021; Xu et al., 2018a, 2018b).

In the reconstruction phase, multiple local governments actively engaged selected NGOs in project implementation, particularly in areas such as housing reconstruction, education, and community rehabilitation. These organizations were often those deemed "low-risk," "service-oriented," or already embedded in pre-existing state networks (Kang, 2020; Lu & Xu, 2015). For instance, the Ya'an NGO Service Center, initially a grassroots coordination platform, was formalized into a semi-official body that mediated between government needs and NGO capacities, operating under government supervision while retaining limited operational flexibility (Xu et al., 2018a, 2018b).

To standardize cooperation, local governments introduced procurement-based governance models that framed NGOs as contractors providing technical services. These arrangements entailed fixed deliverables, state-defined goals, and performance auditing systems, thereby transforming previously autonomous civic actors into technical executors rather than independent policy stakeholders (Liu & Wang, 2019; Zhang & Tang, 2021). While this institutionalization granted NGOs a degree of legal recognition and fiscal support, it simultaneously restricted their agenda-setting power and advocacy potential (Sun, 2017; Zhang & Tang, 2021).

This shift also reflected the consolidation of hierarchical coordination mechanisms. Coordination platforms, often led by government bureaus or GONGOs, served to filter NGO participation, ensure alignment with state priorities, and monitor field activities. In many instances, cross-sector partnerships were limited to transactional collaborations rather than genuine co-governance, with NGOs expected to remain politically neutral and technically efficient (Menefee & Nordtveit, 2012; Wu & Chang, 2018).

Despite these constraints, the managed collaboration model did offer certain gains for civil society. It enabled the professionalization of some NGOs, the establishment of inter-organizational networks, and increased public visibility (Lian & Shaw, 2025; Lu et al., 2021). However, the broader governance logic remained state-centric. Civil society actors were incorporated as auxiliary agents within a state-dominated architecture, rather than as equal partners in governance innovation. This dynamic reinforced a top-down mode of disaster management in which institutional inclusion did not equate to political empowerment.

9.3 Building Collaborative Mechanisms: Organizational Practices and Capacity Development

In the wake of the Wenchuan Earthquake, social organizations were compelled not only to respond to immediate humanitarian needs but also to experiment with new ways of working across organizational and institutional boundaries. As emergency interventions unfolded, fragmented initiatives gradually gave way to coordinated NGO networks, tripartite interfaces linking the state, civil society, and communities, and more routinized approaches to resource management and project implementation. Through these practices, social organizations accumulated operational experience, professional capacity, and relational capital. Over time, such experiential learning was consolidated and diffused across subsequent disaster contexts, enabling a shift from episodic mobilization toward more stable, semi-institutionalized forms of participation in China's evolving disaster governance landscape.

9.3.1 From Fragmented Actions to Coordinated Networks

In the immediate aftermath of the Wenchuan Earthquake, civil society's involvement was characterized by fragmented, decentralized, and spontaneous initiatives. A wide variety of NGOs, volunteer collectives, and professional associations entered the disaster zones with limited coordination, often operating in silos due to the absence of pre-existing collaborative frameworks (Tang & Wang, 2020). This initial fragmentation led to duplication of efforts, uneven service coverage, and limited information exchange, particularly in remote areas where logistical and communication barriers further complicated cooperation.

Over time, however, a shift toward more organized and specialized forms of collaboration began to emerge. NGOs gradually coalesced into thematic and regional networks, such as coalitions focused on psychological counseling, educational continuity, and reconstruction planning. These alliances promoted knowledge sharing, delineated service boundaries, and pooled technical expertise, facilitating more efficient resource deployment (Tang & Chen, 2025). For instance, in the mental health sector, cross-organizational working groups developed shared intervention protocols and referral systems to streamline care and reduce overlap.

This transition from fragmented action to networked coordination was also facilitated by intermediary platforms and shared infrastructures. Notably, the establishment of regional NGO hubs—such as the Sichuan NGO Development Center—served as physical and institutional spaces for convening actors, archiving best practices, and developing standard operating procedures. These platforms played a critical role in institutionalizing inter-organizational learning and promoting collective legitimacy in the eyes of local governments (Lu et al., 2021; Xu et al., 2018a, 2018b). Through regular coordination meetings, joint capacity-building workshops, and common funding applications, civil society actors began to co-construct a governance space that was previously fragmented and fluid.

Importantly, these emerging networks were not merely operational tools; they also carried normative implications. By creating peer accountability mechanisms and promoting shared values of transparency and participation, NGO coalitions contributed to the early institutionalization of disaster response protocols outside of state structures. Over time, this helped transform ad hoc engagement into semi-professionalized governance practices that could be reactivated during subsequent disasters, such as in Ya'an or Lushan (Tang & Wang, 2020).

9.3.2 Tripartite Mechanisms: State–NGO–Community Interfaces

In the aftermath of the Wenchuan Earthquake, collaborative governance evolved beyond the state–NGO dyad to encompass a more complex tripartite configuration involving local governments, civil society organizations, and community members. This structure—gradually emerging from practice rather than pre-existing design—enabled more adaptive and inclusive forms of recovery governance, especially in service delivery, psychosocial support, and community reconstruction. Multiple studies have documented the gradual formation of this "triangular" cooperation model, whereby the state offered political legitimacy and partial resources, NGOs brought technical expertise and mobilization capacities, and community residents actively participated in program design and implementation (Jiang, 2013; Lu & Xu, 2015).

In concrete terms, NGOs often acted as brokers, translating government mandates into locally meaningful interventions while simultaneously channeling grassroots voices upward. This intermediary role allowed NGOs to promote what some scholars describe as "deliberative governance" in disaster contexts, whereby affected communities were not merely aid recipients but co-producers of recovery solutions (Ji & Pang, 2021). For example, in post-disaster education initiatives, NGOs designed participatory school rebuilding programs that engaged teachers, parents, and local officials in decision-making processes, leading to more legitimate and culturally resonant outcomes. Similarly, in psychosocial interventions, organizations developed community-based trauma recovery activities that reflected local norms and emphasized social cohesion (Tang & Chen, 2025).

The success of these tripartite mechanisms hinged on the NGO's ability to navigate dual accountabilities: upward to government funders and regulators, and downward to communities. While this balancing act often constrained their autonomy, it also incentivized procedural innovations. In some cases, community feedback mechanisms and participatory planning sessions became institutionalized as part of project routines (Ji & Pang, 2021; Jiang, 2013). These practices enhanced project sustainability and contributed to a broader shift from top-down delivery to consultative engagement.

Nonetheless, these collaborative interfaces remained uneven. Community participation varied widely depending on local leadership dynamics, state attitudes toward civic engagement, and the NGO's organizational capacity (Ni & Zhan, 2017). Moreover, while tripartite coordination fostered short-term gains in trust and effectiveness, it did not always translate into long-term institutional change. Still, the emergence of these cooperative arrangements marked a key inflection point in China's disaster governance—signaling a tentative move toward more socially embedded, interactive forms of recovery that combined technical delivery with participatory legitimacy.

9.3.3 Navigating Resources and Implementation Challenges

In the aftermath of the Wenchuan Earthquake, NGOs and CSOs faced significant resource and implementation challenges that shaped their operational capacities and long-term development trajectories. Despite the urgency of disaster recovery, many organizations operated in environments marked by stark asymmetries in funding, institutional access, and administrative support. The predominant reliance on external funding—particularly international donations—exacerbated the volatility of NGO operations. As highlighted in several studies, domestic philanthropic infrastructure in China remained underdeveloped at the time, rendering grassroots organizations highly susceptible to fluctuations in overseas donor priorities and short-term project cycles.

Compounding these financial uncertainties was the opaque and often selective distribution of governmental resources. While some NGOs were invited into local coordination platforms or reconstruction initiatives, others—particularly those without pre-existing ties to state actors—struggled to access subsidies or logistical support. This discrepancy created a landscape in which organizations with strong bureaucratic alignment thrived, while more independent or rights-based groups encountered systemic exclusion (Ji & Pang, 2021; Ni & Zhan, 2017). The result was an uneven operational field where effectiveness hinged not only on professional capacity but also on political embeddedness.

To navigate this challenging terrain, many NGOs undertook internal reforms aimed at enhancing organizational professionalism and credibility. These included the establishment of project management systems, financial accountability protocols, and standardized monitoring and evaluation mechanisms. Some organizations even adopted self-imposed auditing procedures to improve transparency and signal reliability to both donors and government agencies (Ji & Pang, 2021). Such institutionalization efforts represented a shift from informal voluntarism to semi-professionalized service delivery.

Sustainability emerged as a parallel concern. The prolonged nature of reconstruction—spanning years rather than months—demanded continuity in staffing, donor relations, and community engagement. However, staff turnover, burnout, and unstable funding made it difficult for many organizations to maintain programmatic consistency. To address this, some NGOs cultivated long-term partnerships with universities, professional associations, and international agencies, thereby diversifying their resource base and extending technical support (Liu & Palmer, 2021; Lu & Xu, 2015).

9.3.4 Toward Institutional Learning and Practice Diffusion

The post-Wenchuan earthquake period marked a critical phase in which NGOs and CSOs not only engaged in direct service delivery but also gradually internalized new forms of organizational learning. This institutional learning—defined as the capacity to adapt practices through reflective evaluation and contextualization—was central to the transformation of Chinese civil society from reactive actors to strategic participants in disaster governance (Tang & Wang, 2020). Importantly, such learning was not purely endogenous but emerged through an ongoing translation of international models into localized operational logics, influenced by political constraints and administrative realities (Lu & Xu, 2015; Sun, 2017).

A key driver of this institutional learning was the cross-contextual adaptation of global disaster recovery frameworks. Several NGOs incorporated lessons from the Sphere standards and community-based disaster risk reduction (CBDRR) protocols into project design, but these were localized through iterative engagements with township officials, local norms, and resource realities (Liu & Palmer, 2021; Ni & Zhan, 2017). In psychological support and participatory education programs, for example, practices shifted from imported curriculum models to hybridized formats co-designed with village cadres and community teachers, reflecting both contextual relevance and political sensitivity (Ji & Pang, 2021; J. Xu et al., 2016).

Another manifestation of practice diffusion was the development of replicable operational templates. Collaborative frameworks piloted in Wenchuan—such as tripartite coordination between governments, NGOs, and residents—were later adapted in other disaster zones including the 2013 Lushan and 2014 Ludian earthquakes. In these cases, models such as community rehabilitation hubs, NGO service centers, and inter-organizational coalitions were transplanted with modifications, demonstrating the routinization of emergency experiences into scalable models (Lu et al., 2021; Tang & Chen, 2025).

This gradual normalization of NGO involvement also contributed to the professionalization of the sector. Organizations invested in training, internal monitoring mechanisms, and inter-organizational learning platforms—transforming short-term relief practices into long-term institutional memory (Jiang, 2013; Tang & Wang, 2020). Moreover, policy dialogues initiated by select NGOs began to reflect this shift, positioning them not just as implementers but as knowledge holders within the broader reconstruction ecosystem (Gao, 2020; Wan et al., 2023).

9.4 From Participation to Institutionalization: The Transformation of Disaster Governance

The Wenchuan Earthquake marked a turning point in China's disaster governance by catalyzing the transition of social organizations from peripheral responders to institutionalized governance actors. This chapter traces this transformation across four interlinked dimensions: the emergence of policy learning and state recognition of NGO roles; the deepening of social organization involvement in long-term reconstruction and disaster risk reduction (DRR); the development of professional norms and sectoral coordination; and the gradual pluralization of governance through structured civic engagement. These shifts reflect the embedding of non-state actors into China's evolving disaster governance regime under a logic of controlled inclusion and collaborative adaptation.

9.4.1 Policy Learning and Institutional Recognition

The 2008 Wenchuan Earthquake became a watershed for state–society relations in disaster governance, not only in terms of operational response but also in prompting institutional introspection within the Chinese state. As NGOs and CSOs rapidly filled urgent service gaps in relief and early recovery, their visibility and effectiveness led to a re-evaluation of their legitimacy by both central and local authorities. What began as a reactive accommodation gradually evolved into an active process of policy learning, wherein the state recognized the need to incorporate non-state actors into future governance architectures (Tang & Wang, 2022; Zhang & Tang, 2021).

This shift was neither uniform nor immediate. However, empirical studies indicate that the earthquake catalyzed a discursive and bureaucratic acknowledgment of social organizations' complementary functions in areas such as community mobilization, psychosocial support, and livelihood restoration (Jiang, 2013; Wu & Chang, 2018). This emerging recognition was translated into policy terms through several administrative channels. Post-disaster planning documents, such as Sichuan's reconstruction guidelines and national-level policy frameworks, began explicitly including social organizational participation as part of multi-stakeholder governance (Lian &

Shaw, 2025). Such inclusion marked a move from informal tolerance to formal policy alignment, albeit within state-defined boundaries.

Importantly, recognition was not unconditional. Institutional incorporation was accompanied by increasing regulatory clarity, particularly regarding NGO registration, funding eligibility, and scope of activities. This process of "elastic embeddedness" combined political receptiveness with administrative containment—enabling NGO access to governance spaces while maintaining vertical accountability structures (Tang & Wang, 2022). For instance, disaster response was one of the few domains where unregistered or temporarily registered NGOs were granted exceptional operational flexibility during the emergency, but this flexibility was gradually narrowed in the reconstruction phase (Huang et al., 2011; Ji & Pang, 2021).

The state's policy learning was thus characterized by a dual logic: instrumental openness paired with institutional selectivity. While some NGOs were included in consultative processes and implementation partnerships, others—especially those with rights-based or advocacy orientations—faced continued marginalization. Nonetheless, compared to pre-Wenchuan dynamics, the earthquake triggered a substantial recalibration of state–society interactions, embedding disaster-related NGO participation into a more routinized and governable framework (Ni & Zhan, 2017; Sun, 2017).

9.4.2 Social Organizations in Long-Term Reconstruction and DRR

Following the emergency phase of the Wenchuan Earthquake, social organizations gradually expanded their engagement from short-term relief to long-term reconstruction and disaster risk reduction, marking a functional shift beyond ad hoc humanitarianism. Initial interventions, often centered on material aid and psychosocial support, evolved into more structured roles in rebuilding infrastructure, restoring livelihoods, and fostering community resilience (Pu & Chang-Richards, 2024; Tang & Wang, 2020). This trajectory reflected not only the increasing capacity of NGOs and CSOs, but also the evolving governance landscape that allowed for deeper institutional cooperation.

In the context of urban and rural reconstruction, social organizations played a bridging role between state plans and community needs. For example, several NGOs facilitated participatory housing design in resettlement areas, ensuring cultural relevance and usability, particularly for elderly and disabled residents (Wang et al., 2017). In rural villages such as Xuankou and Yingxiu, community-based reconstruction projects implemented by NGOs involved local stakeholders in rebuilding schools, health centers, and livelihoods, often in cooperation with township governments and international donors (Lin et al., 2017; Lu & Li, 2020). These experiences highlighted a shift toward collaborative governance, where civil actors were no longer peripheral service providers, but essential co-constructors of post-disaster recovery.

Beyond infrastructure, social organizations contributed to institutionalizing DRR practices at the grassroots level. Projects that focused on enhancing local risk awareness, hazard mapping, and emergency preparedness drills were developed in cooperation with village committees and local education departments (Liu & Palmer, 2021; Xu et al., 2016). Some NGOs also introduced training programs on first aid, trauma counseling, and earthquake simulation exercises, especially in schools and community centers (Huang et al., 2011). These efforts not only addressed immediate vulnerabilities but also embedded DRR into routine community development.

Furthermore, cross-sectoral collaborations gained momentum in the reconstruction phase. Partnerships between NGOs, universities, and research institutions enabled the application of scientific tools—such as risk assessment models and participatory evaluation frameworks—to strengthen evidence-based decision-making (Lyu et al., 2022; Wan et al., 2023). In turn, these alliances promoted capacity building within the NGO sector, enhancing its technical credibility and policy relevance.

9.4.3 Toward Norms, Standards, and Professionalization

The decade following the Wenchuan Earthquake witnessed a gradual but significant evolution in the internal practices and sectoral coordination of social organizations. As their involvement in disaster governance deepened, NGOs and CSOs began to move beyond ad hoc responses toward more standardized, professionalized forms of operation. This transformation was not only a response to external demands for accountability and efficiency, but also a reflection of internal learning processes and growing sectoral maturity (Zhang & Tang, 2021).

One notable development was the emergence of shared norms and technical standards. In the immediate aftermath of the earthquake, many grassroots NGOs lacked clear protocols for project implementation, volunteer management, or coordination with government agencies. However, through repeated engagement in disaster relief and reconstruction activities, organizations developed internal guidelines related to financial transparency, needs assessment, and impact evaluation. These practices were gradually diffused across the sector through peer learning, workshops, and donor requirements (Liu & Wang, 2019; Lu & Xu, 2015).

Sectoral networks also played a key role in institutionalizing professional norms. Platforms such as the "Sichuan NGO Disaster Response Coordination Group" and regional alliances for emergency response facilitated the sharing of resources, standardized training programs, and the co-development of operational frameworks (Tang & Wang, 2022). These meso-level structures functioned as informal intermediaries between state and society, enhancing horizontal coordination and building trust with local authorities. Moreover, they contributed to the formation of what some scholars term a "governance interface," where policy feedback, operational capacity, and legitimacy were co-produced across state–society boundaries (Ji & Pang, 2021; Ni & Zhan, 2017).

Professionalization also entailed a shift in organizational identity. Many NGOs that had emerged spontaneously after the earthquake began formal registration processes, recruited full-time staff, and adopted bureaucratic structures akin to those of public institutions (Kang, 2020). This bureaucratization was not merely a response to state regulation—it was also a strategic adaptation to participate in formal policy arenas, qualify for government contracts, and gain donor confidence. In doing so, NGOs redefined themselves not only as service providers but as credible governance actors.

9.4.4 Civic Engagement and Governance Pluralization

The Wenchuan Earthquake not only catalyzed the operational expansion of social organizations, but also fostered the conditions for a deeper transformation in China's governance architecture. Despite ongoing political constraints, the experience of civic mobilization during and after the disaster opened new possibilities for participatory governance at the local level. What began as an exceptional moment of crisis response gradually evolved into more sustained forms of civic engagement and multilevel cooperation (Sorace, 2014).

This pluralization of governance was first visible in the way community-based organizations, NGOs, and informal volunteer networks collaborated with township and village administrations in the delivery of services. Local authorities, facing capacity deficits, became increasingly receptive to external assistance, particularly in sectors such as education, public health, psychological support, and rural reconstruction (Liu & Wang, 2019; Lu & Xu, 2015). In many cases, NGOs embedded themselves in local institutional landscapes, not only providing services but also engaging in agenda-setting processes and community planning efforts (Lin et al., 2017).

A paradigmatic shift from "government control" to "governance coordination" was thus set in motion. Some localities institutionalized consultative platforms that brought together state agencies, social organizations, and community representatives for joint decision-making and resource allocation (Ji & Pang, 2021; Ni & Zhan, 2017). These hybrid governance mechanisms, though still often informal and contingent, reflected a nascent form of collaborative governance in disaster recovery and risk reduction.

Importantly, civic engagement after Wenchuan was no longer confined to episodic volunteerism or emergency relief. Organizations that had initially emerged for disaster response extended their work into long-term community resilience initiatives, public policy advocacy, and social capacity building (Pu & Chang-Richards, 2024; Wang et al., 2017). In the process, they not only professionalized internally but also contributed to the emergence of a more participatory governance ethos at the grassroots level.

Nevertheless, pluralization did not imply democratization in the Western sense. The central state maintained its supervisory and gatekeeping role, shaping the scope, and form of acceptable civic participation (Kang, 2020; Tang & Wang, 2020). Within this controlled space, civic actors found strategic ways to embed themselves, promote bottom-up accountability, and institutionalize their roles in disaster governance systems.

9.5 Conclusion

The aftermath of the 2008 Wenchuan Earthquake revealed a profound reconfiguration of state–society relations in China's disaster governance landscape. What began as spontaneous civic mobilization gradually evolved into more structured forms of participation, negotiation, and institutional adaptation. Social organizations, once peripheral to public governance, became increasingly embedded through mechanisms of policy learning, collaborative reconstruction, and professionalization. Their presence not only addressed immediate gaps in relief and recovery but also contributed to the formation of new norms, networks, and governance interfaces. While the process was shaped by selective incorporation and regulatory containment, it nevertheless marked a shift toward more pluralized and routinized civic engagement. The Wenchuan experience thus offers a critical lens for understanding how crises can open contingent spaces for institutional change within authoritarian contexts.

References

Gao, H. (2020). The power of the square: Public spaces and popular mobilization after the Wenchuan earthquake. *Journal of Chinese Political Science, 25*(1), 89–111. https://doi.org/10.1007/s11366-019-09643-w

Huang, Y., Zhou, L., & Wei, K. (2011). 5.12 Wenchuan earthquake recovery government policies and non-governmental organizations' Participation. *Asia Pacific Journal of Social Work and Development*, *21*(2), 77–91. https://doi.org/10.1080/21650993.2011.9756108

Ji, M., & Pang, H. (2021). Survival or crisis? Funding distribution from grant-making foundations to grassroots social organizations in China. *Communist and Post-Communist Studies, 54*(4), 137–156. https://doi.org/10.1525/j.postcomstud.2021.54.4.137

Jiang, L. (2013). Implementation of disaster reduction measures and enhancement of integrated risk governance in China: Summary of the Remarks at the High-Level Dialogue, Official Statement, and Plenary Session of the Fourth Session of the Global Platform on Disaster Risk Reduction, 21–23 May 2013, Geneva, Switzerland. *International Journal of Disaster Risk Science, 4*(2), 101–104. https://doi.org/10.1007/s13753-013-0011-0

Kang, Y. (2020). Dispersed domination through patron-Clientelism: The evolution of the local state–NGO relationship in post-disaster Sichuan. *Journal of Contemporary China, 29*(124), 598–613. https://doi.org/10.1080/10670564.2019.1677367

Kang, Y. (2017). The development of grassroots Chinese NGOs following the Wenchuan earthquake of 2008: Three Case Studies, Four Modi Vivendi. *VOLUNTAS: International Journal of Voluntary and Nonprofit Organizations*, *28*(4), 1648–1672. https://doi.org/10.1007/s11266-017-9837-1

Lian, F., & Shaw, R. (2025). Civil societies and disaster risk reduction in china: policy and literature analysis. *Sustainability, 17*(23), 10644. https://doi.org/10.3390/su172310644

Lin, L., Wang, Y., & Liu, T. (2017). Perception of recovery of households affected by 2008 Wenchuan earthquake: A structural equation model. *PLoS ONE, 12*(8), Article e0183631. https://doi.org/10.1371/journal.pone.0183631

Liu, Q., & Palmer, D. A. (2021). Chinese NGOs at the interface between governmentality and local society: An actor-oriented perspective. *China Information, 35*(2), 158–178. https://doi.org/10.1177/0920203X20942094

Liu, Q., & Wang, R. Y. (2019). Peasant resistance beyond the state: Peasant–NGO interactions in post Wenchuan earthquake reconstruction China. *Journal of Contemporary China, 28*(115), 151–166. https://doi.org/10.1080/10670564.2018.1497917

Lu, Y., & Li, R. (2020). Rebuilding resilient homeland: An NGO-led post-Lushan earthquake experimental reconstruction program. *Natural Hazards, 104*(1), 853–882. https://doi.org/10.1007/s11069-020-04194-3

Lu, Y., & Xu, J. (2015). NGO collaboration in community post-disaster reconstruction: Field research following the 2008 Wenchuan earthquake in China. *Disasters, 39*(2), 258–278. https://doi.org/10.1111/disa.12098

Lu, Y., Zhan, C., Li, R., & Su, M. (2021). An NGO disaster relief network for small and medium-scale natural hazards in China. *Natural Hazards, 106*(3), 2689–2709. https://doi.org/10.1007/s11069-021-04560-9

Lyu, J., Bao, Y., Xie, H., Li, H., Hu, Y., & He, X. (2022). Spatiotemporal variation of the loose deposits in Baisha River Basin after the 2008 Wenchuan earthquake. *Journal of Mountain Science, 19*(9), 2544–2555. https://doi.org/10.1007/s11629-022-7519-y

Menefee, T., & Nordtveit, B. H. (2012). Disaster, civil society and education in China: A case study of an independent non-government organization working in the aftermath of the Wenchuan earthquake. *International Journal of Educational Development, 32*(4), 600–607. https://doi.org/10.1016/j.ijedudev.2011.10.002

Ni, N., & Zhan, X. (2017). Embedded government control and Nonprofit revenue growth. *Public Administration Review, 77*(5), 730–742. https://doi.org/10.1111/puar.12716

Pu, G., & Chang-Richards, A. Y. (2024). Identifying critical factors affecting the livelihood recovery following disasters: A comparative analysis of China and New Zealand. *International Journal of Disaster Risk Reduction, 114*, Article 104958. https://doi.org/10.1016/j.ijdrr.2024.104958

Sorace, C. (2014). China's vision for developing Sichuan's post-earthquake countryside: Turning unruly peasants into grateful urban citizens. *The China Quarterly, 218*, 404–427. https://doi.org/10.1017/S0305741014000642

Sun, T. (2017). Earthquakes and the typologies of state–society relations in China. *China Information, 31*(3), 304–326. https://doi.org/10.1177/0920203X17720337

Tang, G., & Chen, B. (2025). Unpacking the dynamics of nonprofit purpose-oriented networks in disaster relief: A tale of two cases. *Public Administration and Development*, pad.70044. https://doi.org/10.1002/pad.70044

Tang, G., & Wang, F. (2020). What factors contribute to nonprofit collaboration? An analysis of response and recovery efforts after the 2008 Wenchuan Earthquake China. *Safety Science, 125*, Article 104624. https://doi.org/10.1016/j.ssci.2020.104624

Tang, G., & Wang, F. (2022). What contributes to the sustainability of self-organized non-profit collaboration in disaster relief? A fuzzy-set qualitative comparative analysis. *Public Management Review, 24*(3), 466–488. https://doi.org/10.1080/14719037.2020.1834608

Wan, G., Wang, H., Geng, X., & Huang, K. G. (2023). Natural disasters and corporate philanthropy: A double movement perspective. *Management and Organization Review, 19*(1), 98–137. https://doi.org/10.1017/mor.2022.40

Wang, S., Tang, W., Qi, D., Li, J., Wang, E., Lin, Z., & Duffield, C. F. (2017). Understanding the role of built environment resilience to natural disasters: lessons learned from the Wenchuan earthquake. *Journal of Performance of Constructed Facilities, 31*(5), 04017058. https://doi.org/10.1061/(ASCE)CF.1943-5509.0001062

Wu, W.-N., & Chang, S.-M. (2018). Collaboration mechanisms of Taiwan Nonprofit organizations in disaster relief efforts: drawing lessons from the Wenchuan Earthquake and typhoon Morakot. *Sustainability, 10*(11), 4328. https://doi.org/10.3390/su10114328

Xu, J., Xu, D., & Lu, Y. (2016). Resident participation in post-Lushan earthquake housing reconstruction: A multi-stage field research method-based inquiry. *Environmental Hazards, 15*(2), 128–147. https://doi.org/10.1080/17477891.2016.1142417

Xu, D., Hazeltine, B., Xu, J., & Prasad, A. (2018a). Public participation in NGO-oriented communities for disaster prevention and mitigation (N-CDPM) in the Longmen Shan fault area during the Wenchuan and Lushan earthquake periods. *Environmental Hazards, 17*(4), 371–395. https://doi.org/10.1080/17477891.2018.1491382

Xu, J., Xu, D., Lu, Y., & Wang, Q. (2018b). A bridged government–NGOs relationship in post-earthquake reconstruction: The Ya'an service center in Lushan earthquake. *Natural Hazards, 90*(2), 537–562. https://doi.org/10.1007/s11069-017-3056-3

Yin, L., & Wang, H. (2010). People-centred myth: Representation of the Wenchuan earthquake in China Daily. *Discourse and Communication, 4*(4), 383–398. https://doi.org/10.1177/1750481310381581

Zhang, H., & Tang, G. (2021). Paired assistance policy and recovery from the 2008 Wenchuan earthquake: A network perspective. *Disasters, 45*(1), 126–157. https://doi.org/10.1111/disa.12413

Chapter 10
Yushu Earthquake: Social Organizations in Ethnic and High-Altitude Contexts

Abstract The 2010 Yushu Earthquake revealed not only the devastation of a natural disaster, but also the fragility and complexity of governance in peripheral, high-altitude, and ethnically distinct regions. This chapter explores how disaster shock interacted with structural constraints to produce a temporary governance vacuum, into which civil society actors—NGOs, monasteries, and volunteer networks—rapidly inserted themselves. It unfolds in four analytical sections: the conditions that enabled civil society emergence, the diverse roles social organizations assumed in emergency and recovery phases, their strategies of cultural adaptation and local embedding, and finally, the gaps, dilemmas, and lessons for inclusive disaster governance. Through this layered analysis, the chapter highlights the evolving, contested, and contextually grounded nature of social response in disaster governance landscapes like Yushu.

Keywords Governance vacuum · Civil society mobilization · Cultural embeddedness · Inclusive governance

10.1 Governance Challenges and Civil Society Emergence in Peripheral Regions

The 2010 Yushu Earthquake revealed not only the physical devastation of a high-altitude disaster, but also the underlying governance fragilities in ethnically distinct and infrastructurally marginalized regions. This section examines how pre-existing structural constraints shaped institutional responses, how disaster shock produced a temporary governance vacuum, and why civil society actors emerged as early responders. By analyzing their emergence through the lenses of flexibility, embeddedness, and moral mobilization, this section further transitions from the conditions of appearance to the practices of action, laying the groundwork for understanding how non-state actors operated within—and sometimes beyond—the fragmented post-disaster governance field.

Q. Du and R. Shaw, *Social Organizations in Disaster Response and Recovery*, Disaster Risk Reduction, https://doi.org/10.1007/978-981-92-2846-1_10

10.1.1 Governance Vulnerabilities in Peripheral Ethnic Regions

The Yushu Earthquake exposed not only the immediate risks of a natural disaster but also the structural constraints inherent to governance in peripheral, high-altitude, and ethnically distinct regions. Geographical remoteness, combined with infrastructural deficits, imposed severe limitations on emergency preparedness and institutional responsiveness.

But before analyzing the response dynamics in Yushu, it is necessary to briefly introduce Organization Z, whose situation reports are referenced throughout this chapter and several subsequent ones. Organization Z is a volunteer-based disaster information coordination group that emerged from grassroots collaboration during the 2008 Wenchuan Earthquake and was formally established amid the 2010 Yushu Earthquake. Over the past fifteen years, the organization has participated in a wide range of disaster responses both within China and internationally, including—but not limited to—operations in South Asia, East Asia, the Middle East, and other disaster-affected contexts.

Organization Z focuses on the collection, verification, and dissemination of disaster-related information under conditions of uncertainty, time pressure, and limited institutional capacity. Its coordination model has evolved alongside shifts in digital communication technologies, moving from early reliance on QQ groups and online forums to more recent use of WeChat, Lingxi forms, and collaborative platforms such as Tencent Docs and Shimo. These tools enable geographically dispersed volunteers to engage in distributed data processing, real-time reporting, and iterative peer verification across different phases of disaster response.

In this chapter, as in subsequent ones, references to Z's reports help illuminate how social organizations navigated informational, logistical, and institutional challenges on the ground. Its documentation provides a longitudinal empirical lens through which bottom-up coordination practices, adaptive information work, and the interaction between civil society actors and fragmented governance environments can be systematically examined across multiple disaster settings.

To contextualize how such bottom-up coordination unfolded, it is important to examine the physical and institutional environment in which the Yushu response took place. Yushu, located over 4000 m above sea level, is characterized by extreme climatic conditions, seasonal road closures, and dispersed settlements. These physical conditions did not merely delay response; they shaped the very nature of what response was institutionally feasible (Feng et al., 2013; Liu et al., 2011). Even under normal circumstances, administrative response times were protracted and material mobilization relied heavily on cross-provincial dispatching (Organisation Z, 2010a). In disaster scenarios, these limitations were amplified, creating significant temporal and logistical gaps.

In addition to geographic constraints, the social structure of Yushu was deeply embedded in religious and cultural networks, particularly those associated with Tibetan Buddhism. Monasteries, monks, and local ritual leaders held considerable authority and were key actors in both everyday governance and crisis response (Sun et al., 2019). This meant that formal governance structures had to operate through, rather than independently of, existing religious institutions. In early response phases, monasteries served as centers for shelter and information coordination, often more trusted than state institutions (Organisation Z, 2010a; Sun & Qi, 2023). Such dynamics reflect the salience of "embedded governance," where informal institutions mediate access to state resources and legitimacy in culturally distinct regions (Ma et al., 2022).

Crucially, these conditions should not be mistaken for an absence of state presence. Rather, governance in Yushu reflects uneven distributions of state capacity. While national-level directives were issued rapidly, execution faltered across the central-local-communal chain. The cascading nature of state response—central mandates, provincial logistics, township implementation—meant that delays and information distortion were frequent (Jiang, 2013; Wei et al., 2013). In such settings, disaster governance was less about state absence than about the erosion and fragmentation of institutional coordination at lower levels.

10.1.2 Disaster Shock and the Formation of a Temporary Governance Vacuum

The Yushu Earthquake struck with instantaneous force, yet formal institutional responses unfolded in a staggered, procedural manner. This temporal mismatch between the immediacy of impact and the phased structure of official relief operations was especially pronounced in high-altitude regions like Yushu, where logistical limitations further slowed intervention. Although national-level mobilization was initiated rapidly, the first 24 to 48 h saw significant gaps in external presence, as road damage, altitude constraints, and resource bottlenecks impeded access (Organisation Z, 2010a; Wei et al., 2013). Local needs such as trauma care, corpse retrieval, and temporary shelter emerged immediately, but official mechanisms required time to coordinate, authorize, and deploy, creating a window of institutional latency.

Compounding this temporal gap was the disruption of communication infrastructures. Earthquake-induced damage to information networks made it difficult for village-level actors to relay urgent needs upward, while incoming aid teams often operated with fragmented or outdated knowledge of on-the-ground conditions (Organisation Z, 2010a; Zhou et al., 2013). In some cases, aid arrived in forms ill-suited to local customs or needs, highlighting the consequences of uncoordinated dispatches. Relief distribution lacked spatial targeting, with over-supply in some areas and absence in others. The information environment was not chaotic per se but unintegrated—characterized by dispersed efforts without a stable coordination center (Deng et al., 2019).

These dynamics produced what can be understood as a temporary governance vacuum. Not defined by the absence of the state, this vacuum reflected a moment in which formal authority had not yet operationalized control across the affected terrain, and non-state actors entered the gap. Civil society organizations, local monasteries, and volunteer groups began to operate with relative autonomy, especially in areas yet to be reached by the centralized command structure (Organisation Z, 2010b). This condition was not a void but a contingent space of institutional underreach—stage-specific and fillable—where the logics of flexibility, embeddedness, and moral urgency temporarily superseded those of hierarchy and protocol.

10.1.3 Why Civil Society Actors Emerged First

In the early hours following the Yushu Earthquake, civil society actors emerged as primary responders—not because they possessed superior institutional power, but because their structure and positioning enabled faster, more flexible intervention. Unlike formal governmental agencies that required cross-level authorization and procedural compliance, many grassroots NGOs, volunteer collectives, and religious networks operated with flat hierarchies and minimal entry barriers. Decision-making was localized and adaptive, allowing teams to mobilize resources and personnel based on immediate assessments rather than formal instructions. For instance, several grassroots organizations began search and relief operations within hours of the earthquake, even before the arrival of national-level rescue forces (Organisation Z, 2010a; Peng, 2017).

This rapid activation was also rooted in the social embeddedness of many civil society actors. In Tibetan communities like Yushu, social life is closely structured around religious institutions, kinship networks, and customary authority. Monasteries and monks, long central to daily governance and community rituals, played a key role in both early warning dissemination and logistical coordination (Sun et al., 2019). Many NGOs and volunteer teams entering the area leveraged pre-existing relationships with local religious figures, which facilitated not only trust-building but also access to hard-to-reach villages and distribution points (Organisation Z, 2010a). Embeddedness thus functioned as an operational asset, enabling smoother entry and faster alignment with local expectations and norms.

Beyond structural advantages, moral mobilization was a crucial driver of civil society emergence. The emotional shock triggered by the disaster stimulated an immediate public response, especially among urban-based volunteers and donors. The narratives of suffering circulated online and through media channels prompted spontaneous donation drives, peer-to-peer volunteer recruitment, and the establishment of ad hoc relief teams (Deng et al., 2019; Peng, 2017). The affective dimension of disaster—grief, urgency, solidarity—translated into real-time resource flows, often faster than formalized aid.

These dynamics do not suggest that civil society actors replaced the state. Rather, they filled a temporal and spatial void by being faster, closer, and more socially proximate. Their early involvement was enabled not by institutional superiority but by a unique alignment of organizational agility, local legitimacy, and moral urgency—qualities especially critical in the context of a high-altitude, ethnically distinct disaster zone.

10.1.4 Transition: From Emergence to Action

The immediate presence of civil society actors in the early aftermath of the Yushu Earthquake was a response shaped by urgency and proximity. The capacity to appear rapidly did not automatically translate into sustained or effective intervention. Many organizations, especially volunteer collectives and smaller NGOs, faced significant limitations in resources, logistical coordination, and operational professionalism once the relief effort extended beyond the first 72 h. Despite early access to affected zones, they often lacked standardized procedures for needs assessment, data management, or equitable distribution, which became increasingly critical as the situation evolved (Organisation Z, 2010b; Peng, 2017).

Logistical challenges were further exacerbated by geographic and environmental constraints. High-altitude terrain and infrastructural fragility slowed the delivery of essential supplies and hindered communication between actors. Some grassroots groups withdrew within days due to altitude sickness, inadequate preparation, or operational exhaustion (Liu et al., 2015; Organisation Z, 2010c). Others continued to operate, but with growing inefficiencies due to lack of coordination with governmental actors or other civil society teams. While flexibility allowed for initial responsiveness, the absence of clear division of labor and information-sharing platforms often led to redundant efforts or overlapping service areas (Deng et al., 2019).

Moreover, differences in professional capacity among civil society actors led to a divergence in roles. Larger NGOs with medical or educational specializations were able to transition into structured service delivery, while less experienced teams remained focused on basic distribution or emotional support. This stratification highlighted an important distinction between appearance and sustained impact, raising questions about what constitutes effective disaster governance in peripheral and culturally distinct regions (Ren et al., 2016; Zhou et al., 2013).

To move beyond the explanatory logic of "why they appeared," it becomes essential to analyze "what they did and how." This includes interrogating the actual practices adopted on the ground, the types of services provided, the degree of alignment with local needs, and the nature of interactions—both cooperative and conflictual—with state institutions and local communities. In this regard, operational effectiveness was not determined solely by the speed of entry but by the ability to adapt, coordinate, and embed within a complex sociopolitical and environmental landscape.

10.2 Social Organizations' Roles in Emergency Relief and Recovery

In the aftermath of the 2010 Yushu Earthquake, social organizations played a pivotal role in filling critical gaps left by delayed or overstretched state mechanisms. This section examines how a diverse array of actors—ranging from professional NGOs to religious networks and digitally coordinated volunteer groups—mobilized across different phases of the disaster response. It explores four key dimensions: the typological diversity of these organizations, the specific emergency functions they assumed, the mechanisms through which they mobilized and coordinated action, and their transition from short-term relief to long-term recovery and reconstruction. These facets reveal a layered, adaptive, and culturally embedded civil society response.

10.2.1 Typology and Diversity of Social Organizations in the Yushu Earthquake Response

The Yushu Earthquake response featured an unusually diverse constellation of social organizations, reflecting a pluralized and multi-scalar civil society ecosystem. Rather than being dominated by a single organizational type, the relief landscape was composed of overlapping actor clusters, each driven by different mandates, resources, and capacities.

First, professional NGOs constituted a core segment of the relief apparatus. Organizations such as the China Foundation for Poverty Alleviation (CFPA), China Heart, and Blue Sky Rescue Team played a pivotal role in organizing large-scale relief operations. These groups brought specialized capabilities—such as medical care, logistics, and technical equipment—and demonstrated a degree of operational continuity from previous disasters (Liu et al., 2015; Organisation Z, 2010a, 2010c). Their participation marked a maturation of China's non-profit sector, transitioning from ad hoc relief to semi-institutionalized disaster response systems (Peng, 2017).

Second, religious and community-based organizations, particularly those rooted in Tibetan Buddhism, served as culturally embedded actors in the Yushu response. Monasteries not only offered physical shelter but also became central logistical hubs, leveraging their legitimacy and spatial familiarity to organize aid distribution. Monks and lay volunteers coordinated burial services, established refugee centers, and mediated communication between local communities and incoming NGOs (Organisation Z, 2010a; Sun & Qi, 2023). These actors blurred the lines between spiritual authority and humanitarian response, acting as moral intermediaries in a culturally distinct environment.

Third, volunteer coalitions—both spontaneous and institutional—provided essential manpower and adaptive labor. University-based volunteer teams, local youth groups, and informal collectives mobilized rapidly through peer networks and social media. Often lacking formal credentials, these actors nonetheless filled key gaps in transportation, communication, and food delivery in the early phases of the crisis (Deng et al., 2019; Organisation Z, 2010b). Their decentralized nature allowed for quick mobilization, though coordination challenges persisted over time.

Finally, digital and networked platforms facilitated novel forms of participation. Microblog-based campaigns, online donation drives, and message board coordination allowed urban supporters to contribute financially and logistically despite being geographically removed from the disaster site. These platforms lowered barriers to entry, democratized access to information, and created feedback loops between donors, NGOs, and recipients (Zhou et al., 2013).

10.2.2 Key Functions and Modalities of Emergency Actions

Social organizations in the Yushu Earthquake response undertook a wide range of emergency functions that complemented, and in some phases preceded, formal state-led operations. These actions were not uniform but varied by organizational capacity, timing, and degree of local embeddedness, forming a layered response system across the acute and early recovery phases.

In the critical first 72 h, search and rescue (SAR) activities were largely initiated by local residents, monks, and village networks rather than professional rescue units. Given the immediate collapse of housing and limited access for external teams, monks and villagers conducted initial rescue efforts using basic tools and local knowledge of settlement layouts (Organisation, 2010a). Professional NGOs such as the Blue Sky Rescue Team entered soon after, bringing specialized equipment, standardized rescue procedures, and trained personnel. Joint operations emerged in which local actors provided guidance and access while professional teams handled high-risk extractions, illustrating an emergent division of labor shaped by urgency and terrain (Liu et al., 2015; Shen et al., 2015).

As the response moved beyond immediate rescue, the distribution of relief supplies became a central task. National and regional NGOs, including the China Foundation for Poverty Alleviation and the Hua Xia Philanthropy Foundation, coordinated large-scale deliveries of tents, food, drinking water, and medical kits. In recognition of high-altitude conditions, some organizations adjusted supply composition to include cold-resistant tents, thermal clothing, and high-calorie food (Organisation,). Nevertheless, logistical challenges were evident. Repeated deliveries to accessible town centers contrasted with shortages in remote villages, reflecting coordination gaps and information asymmetries. Language barriers and unfamiliarity with local customs further complicated distribution efficiency (Organisation Z, 2010e, 2010f; Ren et al., 2016).

Emergency medical and psychological care constituted another critical modality. Medical NGOs and volunteer doctors cooperated with temporary field hospitals and air evacuation systems to treat severe trauma and altitude-related complications (X. Liu et al., 2012). As mortality stabilized, attention gradually shifted toward psychological support, particularly for children who had lost family members. Volunteer counselors and NGO-led teams initiated trauma counseling sessions in schools and temporary shelters, with some programs continuing beyond the acute phase (Organisation Z, 2010e, 2010f, 2010g, 2010h).

Education continuity and child protection efforts emerged as relief priorities once basic survival needs were addressed. NGOs supported the establishment of temporary learning spaces, often in collaboration with monasteries and local teachers. Instruction in the Tibetan language was maintained to ensure cultural continuity, while targeted interventions addressed the needs of orphaned, disabled, and migrant children, groups identified as particularly vulnerable in the post-disaster environment (Jia & Zhang, 2013; Organisation Z, 2010g; Sun et al., 2019).

10.2.3 Mobilization Mechanisms and Organizational Strategies

The mobilization of social organizations during the Yushu Earthquake response was marked by a combination of rapid digital coordination, religious legitimacy, and adaptive logistical innovation. These mechanisms were essential in a context where formal institutional infrastructures were either delayed or disrupted.

One critical mechanism was the spontaneous formation of digital networks, particularly through platforms such as Weibo. Within hours of the earthquake, microblogging sites became hubs for information exchange, volunteer recruitment, and donation drives. Volunteer collectives were formed through hashtag-based campaigns, allowing decentralized groups to self-organize, coordinate transport, and target hard-hit areas (Zhou et al., 2013). This form of peer-to-peer mobilization often outpaced centralized systems, highlighting how digital trust networks could substitute for formal authority in the initial phase (Peng, 2017).

Equally important was the embedded role of religious networks. Monasteries in Yushu, already central to local governance and social cohesion, quickly transformed into operational bases. They served not only as sites for material storage and distribution but also as communication nodes connecting villages and external responders (Organisation,). Monks assumed dual roles—offering spiritual care while coordinating aid—thereby legitimizing the presence of both local and external actors. Their embeddedness provided both access and authority, allowing smoother integration with local cultural norms (Sun & Qi, 2023).

Adaptation to environmental and logistical constraints formed the third pillar of effective mobilization. High-altitude terrain and fragmented road networks necessitated the use of alternative transport methods, including 4WD vehicles, horses, and pack animals, often managed by local coordination units (Organisation Z, 2010f). In addition, language and cultural barriers between Han-speaking NGOs and Tibetan communities were mitigated through local translators and guides who ensured communication and conflict resolution in the field (Wang et al., 2012; Xin et al., 2025).

10.2.4 From Emergency Relief to Recovery and Local Reconstruction

The transition from emergency response to post-disaster recovery in Yushu revealed stark differences among social organizations in terms of continuity, adaptability, and institutional integration. While the initial phase saw a surge of spontaneous civil society involvement, only a subset of actors maintained a long-term presence and evolved their strategies toward sustainable reconstruction and capacity-building.

Some professional NGOs demonstrated a clear transition from short-term aid to more structured developmental engagement. For instance, China Heart extended its activities beyond initial medical outreach to support rural health training programs in Yushu's surrounding townships, focusing on building local capacity for primary care delivery (Organisation Z, 2010i; Peng et al., 2021). Similarly, organizations like the China Foundation for Poverty Alleviation shifted their operations from direct relief logistics to long-term livelihood support, such as rebuilding schools and supporting income-generating activities for earthquake-affected households (Organisation Z, 2010i). These strategic shifts reflect an organizational capacity to move from "relief delivery" toward "social reconstruction," often involving reallocation of internal resources and re-negotiation of local roles.

The effectiveness of these long-term interventions was closely tied to the cultivation of local partnerships and trust-based governance mechanisms. In Yushu, social legitimacy was not automatically granted to external actors; it had to be negotiated through ongoing relationships with village committees, local religious authorities, and community-based organizations. NGOs that successfully embedded themselves in these networks—through co-governance practices, participatory planning, and local hiring—were better positioned to implement sustained recovery initiatives (Sun et al., 2019). In some cases, partnerships that began informally during the emergency phase evolved into semi-institutionalized mechanisms for joint decision-making on reconstruction priorities (Ma et al., 2022).

This post-crisis institutionalization was not uniform. While some national NGOs gained formalized presence through project offices or registered branches, smaller volunteer groups either withdrew or collapsed under the weight of operational fatigue and funding limitations (Organisation Z, 2010i). However, the legacy of these groups persisted through knowledge transfer, new local organizations, and heightened civic awareness within affected communities.

10.3 Cultural Adaptation and Local Embeddedness in High-Altitude, Ethnic Contexts

The Yushu Earthquake response unfolded within an intricate ecological and cultural landscape, where high-altitude conditions and Tibetan Buddhist traditions deeply shaped humanitarian dynamics. This section examines how social organizations navigated these contextual specificities through six interrelated dimensions: the physical-geographical and ethno-religious characteristics of Yushu; the central role of monasteries in coordinating aid; the recalibration of relief materials to align with local customs; organizational adaptations to altitude-induced logistical challenges; the use of language and cultural brokers to bridge communication gaps; and the strategic value of cultural sensitivity in enhancing operational legitimacy, trust, and efficiency throughout both emergency relief and recovery phases.

10.3.1 Contextual Specificities of the Yushu Earthquake Area

The particular geographic, cultural, and social composition of Yushu Prefecture critically shaped the dynamics of disaster response following the 2010 earthquake. Located in southwestern Qinghai Province, Yushu sits at an average elevation exceeding 4000 m above sea level. The region's high-altitude terrain, coupled with extreme temperature fluctuations, thin air, and limited transport infrastructure, imposed severe constraints on relief logistics and volunteer operations. Many non-local responders reported symptoms of acute mountain sickness, forcing some NGOs to adjust deployment timelines or restructure supply chains with local support (Organisation Z, 2010e; Xin et al., 2025). Transportation to remote villages often required relay systems involving 4WD vehicles, motorbikes, and pack animals, particularly where road access was blocked or non-existent (Liu et al., 2012).

Cultural and social conditions further complicated external intervention. Yushu is a Tibetan Autonomous Prefecture, where over 90% of residents identify as ethnically Tibetan and practice Tibetan Buddhism (Sun et al., 2019). The centrality of monasteries in local governance, social cohesion, and moral authority meant that conventional top-down models of disaster coordination proved inadequate without religious mediation. In many cases, monks—not administrative officials—acted as primary sources of information, logistics organizers, and emotional stabilizers during the initial response period (Ma et al., 2022). Additionally, traditional beliefs, ritual calendars, and specific mourning practices shaped community preferences around burial, shelter, and material assistance, requiring aid actors to adjust their approaches to avoid friction.

Taken together, Yushu's ecological extremity and deeply embedded religious-cultural structures created both obstacles and opportunities for aid delivery. Relief actors who failed to account for this contextual specificity often encountered distrust, inefficiency, or outright rejection, whereas those who adapted to local realities found pathways to more effective and accepted interventions (Organisation Z, 2010e). The following sections explore how such adaptation unfolded across organizational strategies and operational choices.

10.3.2 Religious Mediation and Temple-Based Coordination

In the Yushu Earthquake response, religious institutions—particularly Tibetan Buddhist temples and monastic networks—functioned not only as spiritual centers but as pivotal logistical and relational nodes within the humanitarian landscape. In many localities, monks were the first to mobilize, providing physical shelter, basic medical aid, and emotional reassurance to displaced populations. Their embeddedness within the community enabled them to rapidly assess needs and coordinate internal responses before external actors arrived (Organisation Z, 2010f). This form of embedded legitimacy allowed temples to operate as culturally authoritative intermediaries, mediating between affected communities and unfamiliar outside organizations.

Temples offered more than symbolic capital—they served as material depots, communication nodes, and volunteer coordination hubs. Some were repurposed as distribution centers for food, tents, and medicine, while others became venues for multilingual translation and information exchange between aid workers and local residents (Organisation Z, 2010g). In areas with limited state infrastructure or damaged government buildings, religious institutions filled the vacuum of formal governance by facilitating spatial order and maintaining community cohesion. Their credibility helped mitigate rumors, defuse panic, and ensure a relatively smooth inflow of humanitarian resources.

Social organizations that proactively partnered with monasteries often experienced greater local cooperation. By aligning aid delivery with temple calendars or involving monks in needs assessments, these groups gained cultural legitimacy and access to social capital otherwise unavailable to unfamiliar outsiders (Sun & Qi, 2023). Some NGOs adapted their operational protocols—such as seeking ritual approval before site entry or using monks as co-signatories for supply lists—to reinforce local acceptance. This strategic co-governance, while informal, contributed to operational efficiency and reduced cultural friction.

10.3.3 Culturally Appropriate Aid and Material Recalibration

Aid delivery in the aftermath of the Yushu Earthquake revealed the limitations of standardized relief models when applied to ethnically distinct and culturally embedded communities. Early shipments of pre-packaged food, instant noodles, canned meat, and unfamiliar hygiene products were met with skepticism, resistance, and in some cases, outright rejection by Tibetan residents. Standardized provisions clashed with local dietary norms rooted in pastoralist lifestyles and religious codes. Many families preferred tsampa (roasted barley flour), butter tea, and yak-derived dairy products—foods not only nutritionally appropriate for high-altitude living but also imbued with symbolic and spiritual significance (Organisation Z, 2010e).

Initial misalignments in material aid prompted recalibration efforts among responsive NGOs. For example, some organizations revised their supply chains to incorporate culturally accepted foods such as barley-based nutritional supplements and vacuum-packed butter blocks. Others partnered with local monasteries and community leaders to design distribution strategies that honored ritual norms and facilitated community ownership over relief goods. In one case, powdered milk was widely refused unless accompanied by Tibetan-language preparation instructions and the endorsement of local lamas—underscoring the need for both linguistic and moral translation in effective humanitarian response (Organisation Z, 2010f, 2010g).

Material recalibration extended beyond food to include daily-use items such as hygiene kits and clothing. Many Tibetan women expressed discomfort with donated sanitary supplies that did not reflect traditional preferences or lacked proper explanation. In response, culturally attuned organizations introduced modified hygiene packages and relied on female monks or local health workers to provide explanations and training (Organisation Z, 2010h). These iterative adjustments demonstrate that cultural sensitivity is not an ancillary concern but a core operational principle—one that affects both the efficiency and legitimacy of post-disaster aid in ethnically diverse, highland regions such as Yushu.

10.3.4 High-Altitude Constraints and Organizational Adaptation

The physical geography of Yushu—situated at over 4000 m above sea level—posed acute challenges for disaster response operations. Thin air, extreme temperature fluctuations, and rugged terrain not only endangered the health of incoming responders but also impeded the speed, scale, and reliability of aid delivery. Many volunteers and NGO personnel reported symptoms of altitude sickness within hours of arrival, including dizziness, nausea, and shortness of breath, severely limiting their operational capacity (Liu et al., 2012). To mitigate this, several organizations prioritized local recruitment and short-cycle training of residents who were physiologically adapted to the high-altitude environment. These community-based teams became essential in sustaining logistics chains and executing fieldwork over extended periods (Organisation Z, 2010g).

Material logistics had to be reconfigured in response to ecological constraints. Relief supplies such as oxygen tanks, stoves, and medical kits required special packaging to withstand freezing temperatures and pressure instability. Certain medications and perishable items lost efficacy if not stored in insulated containers, prompting some medical NGOs to create alpine-specific medical packs for highland deployment

(Xin et al., 2025). Similarly, tents and shelters distributed during the early phase were redesigned with thermal linings and wind-resistant structures to better serve families exposed to sub-zero nights in temporary encampments (Organisation Z, 2010g).

The fragmented transport infrastructure necessitated multi-modal delivery strategies. In zones where roads were impassable due to landslides or ice, trucks had to transfer goods to motorbikes, yaks, or even human porters for the final leg. This reliance on staged relay networks delayed delivery timelines and introduced coordination complexity, especially for time-sensitive materials like food and medicine. However, NGOs that incorporated local logistical knowledge—such as village-level pathfinding or weather forecasting—were better able to navigate these physical constraints and sustain outreach in peripheral settlements (Liu et al., 2012; Xin et al., 2025).

10.3.5 Language, Communication, and Cultural Brokerage

Communication emerged as a critical fault line during the early phases of the Yushu Earthquake response. Most national-level NGOs and rescue teams arrived without Tibetan-speaking staff or prior exposure to local dialects, leading to frequent misunderstandings, delays in service delivery, and even rejection of aid by local residents (Peng, 2017). The language gap, compounded by divergent worldviews and socio-cultural norms, meant that literal translation often failed to enable meaningful exchange. In numerous cases, volunteers were unable to explain the usage of medical or hygiene items, resulting in confusion and distrust among affected communities (Organisation Z, 2010h).

To overcome this barrier, several organizations adopted intermediary mechanisms grounded in cultural brokerage. Effective actors rapidly hired local interpreters and guides—not merely as translators, but as facilitators of social trust and mediators of symbolic meaning. Monks, in particular, assumed a hybrid role, functioning simultaneously as spiritual leaders, cultural gatekeepers, and logistical coordinators. Their involvement in communication activities—such as broadcasting updates, explaining relief protocols, or organizing joint briefings—helped reduce anxiety and legitimize external interventions (Organisation Z, 2010h; Sun & Qi, 2023).

Visual communication also played a vital role in bridging linguistic divides. Organizations like the Blue Sky Rescue Team adapted their training manuals into Tibetan, adding pictograms and step-by-step visual instructions to support volunteer onboarding (Peng, 2017). Bilingual signage was deployed in aid distribution sites and medical tents to guide residents, while some teams used pre-recorded audio messages in local dialects for mobile dissemination in remote settlements.

Ultimately, communication was not merely a technical task but a socio-cultural negotiation. Organizations that invested in translation infrastructure, participatory dialogue, and local narrative framing gained stronger acceptance and smoother coordination with Tibetan communities (Jia & Zhang, 2013; Organisation Z, 2010h).

10.3.6 *Cultural Sensitivity as an Efficiency Driver*

In the culturally rich and religiously devout context of Yushu, culturally attuned interventions proved to be not only ethically appropriate but also operationally effective. Social organizations that respected local beliefs, religious rhythms, and community hierarchies were able to embed themselves more quickly and sustainably into post-disaster recovery processes. This embeddedness translated into higher participation, smoother coordination, and greater trust from affected populations (Sun et al., 2019).

Organizations that aligned their operational tempo with ritual calendars, avoided disruptions during prayer times, and sought permission before entering sacred areas encountered fewer on-the-ground conflicts (Organisation Z, 2010i). By contrast, groups that used loudspeakers near temples or managed deceased bodies without involving monks faced community backlash and were at times denied access to local coordination platforms (Jia & Zhang, 2013). These missteps were not merely symbolic; they materially hindered operational effectiveness by provoking resistance, delaying aid delivery, and obstructing volunteer deployment.

Cultural sensitivity was also a critical factor in longer-term recovery. Educational and psychosocial programs that integrated Tibetan Buddhist practices—such as prayer-based mourning rituals or lama-led group counseling—were more accepted and better attended than purely biomedical models (Organisation Z, 2010i). In school reconstruction efforts, NGOs that engaged local artisans, preserved religious iconography, and consulted with monastic leaders ensured higher rates of student return and parental support (Sun et al., 2019).

Rather than treating cultural adaptation as a logistical burden, effective organizations viewed it as a strategic asset. These actors recognized that legitimacy in a post-disaster Tibetan context is relational, not only bureaucratic. Accordingly, culturally embedded approaches were not just more ethical—they were more efficient, enabling sustained collaboration, reduced conflict, and enhanced local ownership of reconstruction processes (Jia & Zhang, 2013; Organisation Z, 2010i).

10.4 Gaps, Dilemmas, and Lessons for Inclusive Disaster Governance

The Yushu Earthquake response unveiled profound dilemmas in disaster governance, especially within ethnically distinct and high-altitude contexts. This section identifies four interlinked governance challenges: mismatches between aid supply and local needs, the systematic neglect of vulnerable groups, fragmented coordination among aid actors, and the marginalization of local voices in planning and decision-making. These deficits not only undermined operational efficiency but eroded community trust. Drawing on empirical evidence, the final sub-section distills lessons for building inclusive, adaptive, and culturally grounded governance systems—emphasizing the need for coordination platforms, participatory mechanisms, and flexible, locally informed aid strategies in complex humanitarian settings.

10.4.1 Fragmented Aid and Mismatch Between Needs and Supply

The immediate humanitarian response to the 2010 Yushu Earthquake was marked by an overwhelming influx of donations and aid actors. However, this surge of goodwill exposed significant inefficiencies due to fragmentation and misalignment between available supplies and the actual needs of affected communities. While tents, food, clothing, and medicines poured in from across China, much of the aid failed to consider the ecological, cultural, and logistical specificities of the high-altitude Tibetan region.

Two parallel problems persisted across many camps: over-supply of culturally mismatched items and critical shortages of essential goods. Instant noodles, bottled water, and synthetic clothing—though abundant—were often unsuitable for the plateau's climate or rejected due to dietary and cultural preferences. Large quantities of food were left unopened or discarded due to taste aversion and incompatibility with local habits. In contrast, urgently needed supplies such as women's hygiene products, warm blankets, infant formula, and altitude-appropriate shelters remained scarce. Striking imbalances emerged—for example, some locations faced food surpluses while others lacked basic maternal or infant care items (Organisation Z, 2010f, 2010g, 2010h).

This uneven distribution pattern was exacerbated by a lack of unified needs assessments and centralized coordination mechanisms. Donations often arrived unrequested and without prior consultation with frontline actors or local stakeholders. Some NGOs operated with independent logistics and criteria, resulting in duplication of services in certain villages and complete neglect in others (Liu et al., 2012). The absence of a shared data platform or interoperable distribution framework led to redundant supply chains and delivery bottlenecks (Zhou et al., 2013).

Furthermore, standardized aid kits lacked cultural and climatic relevance. For instance, mass-distributed cotton quilts and low-altitude tents offered insufficient insulation for nighttime temperatures, which dropped below − 10 °C in many temporary settlements (Organisation Z, 2010g). Aid actors who failed to recalibrate their provisioning—both in content and delivery—undermined the overall efficacy of relief efforts.

10.4.2 Exclusion of Vulnerable Groups in Emergency Planning

Despite the massive influx of humanitarian resources following the Yushu Earthquake, vulnerable groups—such as the elderly, women, children, disabled individuals, and orphans—remained systematically underrepresented in emergency planning and aid delivery. The absence of targeted assessments for these populations resulted in significant gaps in protection, accessibility, and service provision across resettlement sites and temporary shelters.

Protective mechanisms for unaccompanied children and elderly individuals were notably absent in several resettlement areas, where no clear registration systems or targeted care protocols had been established (Organisation Z, 2010g). In multiple camps, children were seen wandering without supervision, and elderly residents lacked access to essential medical support or mobility assistance. These vulnerabilities were exacerbated by the failure to identify and register at-risk individuals during early needs assessments, resulting in their systematic exclusion from aid prioritization frameworks (Ni et al., 2018).

Shelter design further reflected the lack of attention to gender-specific needs. Many temporary camps were constructed without separate sanitation facilities or secure sleeping arrangements for women and girls, increasing the risk of harassment and undermining residents' safety and dignity (Jia & Zhang, 2013). Standardized aid kits similarly failed to accommodate differentiated needs across gender and age; women frequently lacked access to appropriate sanitary products, while older adults struggled to use unfamiliar hygiene supplies without assistance (Organisation Z, 2010g).

Similarly, disabled individuals were rarely considered in infrastructure planning. Temporary shelters and distribution sites lacked accessible pathways, assistive equipment, or dedicated support personnel. This exclusion significantly reduced their ability to access food, medical services, or participate in community decision-making processes (Ni et al., 2018).

10.4.3 Weak Inter-Organizational Coordination and Role Ambiguity

The Yushu Earthquake response highlighted significant fragmentation in inter-organizational coordination, as various actors—government agencies, domestic NGOs, international organizations, and religious groups—operated with overlapping mandates and unclear lines of responsibility. In the absence of a unified command structure or shared information platform, aid delivery became disjointed, and organizational roles remained ambiguous throughout the critical early stages.

Widespread confusion in shelter allocation and distribution responsibilities led to cases where "multiple organizations arrived at the same site without coordination, while adjacent villages received no assistance for days" (Organisation Z, 2010i). This unevenness resulted not only in duplicated efforts—such as over-supply of generic food in accessible zones—but also in the formation of so-called aid deserts, where peripheral communities were left underserved despite being severely affected (Ren et al., 2016; Zhou et al., 2013).

The lack of centralized coordination also impaired logistics. In several cases, NGOs reported delays due to conflicting instructions from different levels of government or because access permits were inconsistently issued between jurisdictions (Deng et al., 2019). Meanwhile, religious actors—especially monasteries that had organically assumed leadership in local relief—were often excluded from formal decision-making, despite being pivotal in mobilization and community mediation (Organisation Z, 2010h, 2010i).

Operational fragmentation further extended to data management and needs assessment. Different organizations deployed varying criteria for prioritizing aid recipients, with minimal data sharing. This led to redundant surveys, gaps in vulnerable group identification, and inefficient resource use (Liu et al., 2012). As a result, logistical bottlenecks intensified, and coordination meetings—where they occurred—were reactive rather than strategic.

10.4.4 Local Voices and Knowledge Marginalized in Governance Design

Despite the scale of mobilization following the Yushu Earthquake, disaster governance frameworks were largely designed and implemented through externally driven institutional logics, with limited incorporation of local voices and indigenous knowledge systems. Planning processes were predominantly shaped by government agencies and large NGOs operating under standardized emergency protocols, while Tibetan community leaders, monks, and grassroots women's groups were rarely involved in decision-making beyond symbolic consultation (Sun & Qi, 2023).

Local religious leaders played a critical yet often underutilized role in shaping culturally appropriate and socially accepted aid distribution. Drawing on their deep familiarity with settlement layouts and community dynamics, monks advocated for using temples as central hubs for relief allocation to minimize disorder, promote fairness, and align with local norms. Despite the practical and cultural relevance of these suggestions, they were frequently overlooked in favor of more administratively convenient but less context-sensitive channels. This sidelining weakened community trust and diminished the effectiveness of aid delivery, even when material resources were adequate (Organisation Z, 2010h).

Language barriers further compounded the marginalization of local voices. Many aid procedures, registration forms, and feedback mechanisms were issued exclusively in Mandarin, rendering them inaccessible to monolingual Tibetan speakers. As a result, affected residents struggled to articulate needs, lodge complaints, or influence aid priorities. The absence of systematic translation and interpretation not only slowed information flow but also structurally excluded victims from governance processes intended to serve them (Ge et al., 2023; Organisation Z, 2010h).

Women's perspectives were particularly underrepresented. Although women bore disproportionate caregiving burdens in temporary shelters, their participation in needs assessments and recovery planning remained minimal. Local women's groups and informal caregivers were seldom consulted, leading to blind spots in shelter safety, hygiene provision, and child protection measures (Organisation, 2010g). This pattern reflects a broader tendency to treat affected populations as passive recipients rather than knowledgeable agents.

The marginalization of local knowledge in Yushu was not due to its absence, but to institutional designs that privileged speed and standardization over participation and contextual intelligence. The case demonstrates that excluding local voices is not merely a normative failure but a practical one, undermining coordination, legitimacy, and responsiveness. More inclusive governance designs—rooted in linguistic accessibility, religious mediation, and community participation—are essential for effective disaster response in ethnically distinct and peripheral regions (Organisation Z, 2010h; Sun & Qi, 2023).

10.4.5 Lessons for Inclusive, Adaptive, and Culturally Grounded Governance

The Yushu Earthquake response revealed not only gaps in coordination and inclusion, but also highlighted emergent practices that offer direction for more resilient and context-sensitive disaster governance. In peripheral and ethnically diverse regions such as Yushu, effective governance requires more than logistical efficiency—it demands institutional designs that accommodate cultural particularities, recognize local agency, and foster multi-level collaboration.

A first lesson concerns the necessity of coordination platforms that transcend institutional silos. During the Yushu response, overlapping mandates and unstructured partnerships between state agencies, NGOs, religious institutions, and local leaders led to inefficiencies and resource misallocation (Deng et al., 2019; Organisation Z, 2010i). Creating multi-actor coordination hubs—especially at the meso-level between township and prefecture—could allow for clearer task distribution, real-time information exchange, and adaptive strategy adjustment during complex emergencies (Ren et al., 2016).

Second, participatory governance structures that involve vulnerable groups throughout the disaster cycle—from risk assessment to aid feedback—are essential. In the Yushu case, women, the elderly, and disabled populations were largely passive recipients of aid, rather than contributors to decision-making (Jia & Zhang, 2013; Ni et al., 2018). Institutionalizing consultative mechanisms, such as village-level councils including female caregivers or local monks, would enable more nuanced assessments and increase legitimacy of interventions.

Flexibility in aid design emerged as another critical factor. Standardized procedures and supply kits often clashed with local needs and customs. In contrast, organizations that adjusted their strategies based on environmental, dietary, and spiritual considerations gained community trust and achieved higher effectiveness (Organisation Z, 2010f, 2010g, 2010h, 2010i; Sun et al., 2019). For example, the NGO China Heart collaborated with temple networks to co-design aid distribution systems that reflected local norms of fairness and respected religious spaces. As a result, tensions diminished and beneficiary satisfaction increased (Organisation Z, 2010i; Peng, 2017).

10.5 Conclusion

The Yushu Earthquake serves as a powerful lens into the complexities of disaster governance in peripheral, high-altitude, and ethnically distinct regions. This chapter traced how structural vulnerabilities and institutional latency created a temporary governance vacuum, enabling the rapid emergence of social organizations. Through their diverse roles in relief and recovery, these actors demonstrated both the potential and the limitations of civil society in navigating fragmented systems and cultural complexity. The response also revealed critical gaps—particularly in inter-agency coordination, cultural appropriateness, and the inclusion of vulnerable groups and local voices. As disaster governance becomes increasingly decentralized and multi-actor, the Yushu case underscores the urgent need for flexible, participatory, and culturally grounded frameworks that can bridge institutional divides and empower affected communities.

References

Deng, Y., Su, G., Gao, N., & Sun, L. (2019). Perceptions of earthquake emergency response and rescue in China: A comparison between experts and local practitioners. *Natural Hazards, 97*(2), 643–664. https://doi.org/10.1007/s11069-019-03663-8

Feng, T., Hong, Z., Fu, Q., Ma, S., Jie, X., Wu, H., Jiang, C., & Tong, X. (2013). Application and prospect of high-resolution remote sensing and geo-information system in estimating earthquake casualties. *Earthquake Hazards*. https://doi.org/10.5194/nhessd-1-7137-2013

Ge, J., Tang, H., & Ji, C. (2023). Self-incremental learning for rapid identification of collapsed buildings triggered by natural disasters. *Remote Sensing, 15*(15), 3909. https://doi.org/10.3390/rs15153909

Jia, M., & Zhang, Z. (2013). Critical mass of women on bods, multiple identities, and corporate philanthropic disaster response: Evidence from privately owned Chinese firms. *Journal of Business Ethics, 118*(2), 303–317. https://doi.org/10.1007/s10551-012-1589-7

Jiang, L. (2013). Implementation of disaster reduction measures and enhancement of integrated risk governance in China: Summary of the remarks at the high-level dialogue, official statement, and

plenary session of the fourth session of the global platform on disaster risk reduction, 21–23 May 2013, Geneva, Switzerland. *International Journal of Disaster Risk Science, 4*(2), 101–104. https://doi.org/10.1007/s13753-013-0011-0
Liu, J., Fan, Y., & Shi, P. (2011). Response to a high-altitude earthquake: The Yushu Earthquake example. *International Journal of Disaster Risk Science, 2*(1), 43–53. https://doi.org/10.1007/s13753-011-0005-8
Liu, X., Tang, B., Yang, H., Liu, Y., Xue, C., & Zhang, L. (2015). The technical efficiency of earthquake medical rapid response teams following disasters: The case of the 2010 Yushu Earthquake in China. *International Journal of Environmental Research and Public Health, 12*(12), 15390–15399. https://doi.org/10.3390/ijerph121214991
Liu, X., Zhang, L., Liu, Y., Kang, P., & Liu, Z. (2012). Critical care aeromedical evacuation staff in batang airport after the Yushu Earthquake at high altitude. *Aviation, Space, and Environmental Medicine, 83*(4), 436–440. https://doi.org/10.3357/ASEM.3148.2012
Ma, T., Swallow, B., Zhong, L., Xu, K., Sang, W., & Jia, L. (2022). Local perspectives on social-ecological transformation: China's Sanjiangyuan National Park. *Environment, Development and Sustainability, 26*(1), 1809–1829. https://doi.org/10.1007/s10668-022-02786-6
Ni, W., Shu, J., & Song, M. (2018). Location and emergency inventory pre-positioning for disaster response operations: Min-max robust model and a case study of Yushu Earthquake. *Production and Operations Management, 27*(1), 160–183. https://doi.org/10.1111/poms.12789
Organisation, Z. (2010a). *Situation report No. 1: Qinghai Yushu earthquake relief (April 14, 2010).*
Organisation, Z. (2010b). *Situation report No. 2: Qinghai Yushu earthquake relief (April 14, 2010).*
Organisation, Z. (2010c). *Situation report No. 3: Qinghai Yushu earthquake relief (April 14, 2010).*
Organisation, Z. (2010d). *Situation report No. 4: Qinghai Yushu earthquake relief (April 14, 2010).*
Organisation, Z. (2010e). *Situation report No. 5: Qinghai Yushu earthquake relief (April 14, 2010).*
Organisation, Z. (2010f). *Situation report No. 6: Qinghai Yushu earthquake relief (April 14, 2010).*
Organisation, Z. (2010g). *Situation report No. 7: Qinghai Yushu earthquake relief (April 14, 2010).*
Organisation, Z. (2010h). *Situation report No. 8: Qinghai Yushu earthquake relief (April 14, 2010).*
Organisation, Z. (2010i). *Situation report No. 9: Qinghai Yushu earthquake relief (April 14, 2010).*
Peng, C., Yin, Z., Zhang, X., Shao, H., Pang, M. (2021). China University of Geosciences (Beijing), Beijing 100029, China,,China Institute of Geo-Environment Monitoring, China Geological Survey, Ministry of Natural Resources, Beijing 100081, China, &,China University of Geosciences, Wuhan 430074, China. (2021). A comparative study of the main factors controlling geohazards induced by seismic events in Western China since the Wenchuan earthquake. *China Geology, 5*(0), 1–15. https://doi.org/10.31035/cg2022009
Peng, L. (2017). Crisis crowdsourcing and China's civic participation in disaster response: Evidence from earthquake relief. *China Information, 31*(3), 327–348. https://doi.org/10.1177/0920203X17735668
Ren, P., Xu, Z., & Gu, J. (2016). Assessments of the effectiveness of an earthquake emergency plan implementation with hesitant analytic hierarchy process. *International Journal of Information Technology & Decision Making, 15*(06), 1367–1389. https://doi.org/10.1142/S0219622016500383
Shen, W., Jiang, L., Zhang, M., Ma, Y., Jiang, G., & He, X. (2015). Practical experiences and lessons of medical response to major disasters in China. *Hong Kong Journal of Emergency Medicine, 22*(5), 270–280. https://doi.org/10.1177/102490791502200503
Sun, L., & Qi, W. (2023). Tibetan Buddhist belief and disaster resilience: A qualitative exploration of the Yushu area China. *Disasters, 47*(3), 788–805. https://doi.org/10.1111/disa.12563
Sun, L., Su, G., Tian, Q., Qi, W., Liu, F., Qi, M., & Li, R. (2019). Religious belief and Tibetans' response to earthquake disaster: A case study of the 2010 Ms 7.1 Yushu earthquake, Qinghai Province, China. *Natural Hazards, 99*(1), 141–159. https://doi.org/10.1007/s11069-019-03733-x
Wang, Z., Xu, B., Wu, L., Shen, Y., Li, Z., & Li, H. (2012). Geohazards observations and natural disaster assessment in China. *IEEE International Geoscience and Remote Sensing Symposium, 2012*, 950–953. https://doi.org/10.1109/IGARSS.2012.6351394

Wei, B., Su, G., & Liu, F. (2013). Public response to earthquake disaster: A case study in Yushu Tibetan Autonomous Prefecture. *Natural Hazards, 69*(1), 441–458. https://doi.org/10.1007/s11069-013-0710-2

Xin, X., Wang, S., & Zhang, T. (2025). Truck-drone supported humanitarian relief logistics network design: A two-stage distributionally robust optimization approach. *Transportation Research Part C: Emerging Technologies, 178*, Article 105231. https://doi.org/10.1016/j.trc.2025.105231

Zhou, Y., Yang, L., Van De Walle, B., & Han, C. (2013). Classification of Microblogs for Support Emergency Responses: Case Study Yushu Earthquake in China. *2013 46th Hawaii International Conference on System Sciences*, 1553–1562. https://doi.org/10.1109/HICSS.2013.129

Chapter 11
Lushan Earthquake: Early Practices of Government–Society Collaboration

Abstract The Lushan Earthquake marked more than a humanitarian emergency—it served as a pivotal moment in the evolving dynamics of state–society collaboration in China's disaster governance. Compared to earlier disasters, notably Wenchuan, the response revealed a shift: government actors showed greater openness, NGOs operated with increasing professionalism, and new mechanisms for information flow and coordination began to take shape. This chapter traces how these changes unfolded across institutional settings, operational practices, and social outcomes. From the emergence of informal coordination platforms to differentiated roles between state and civil actors, and from improved psychosocial support to persistent inequalities in service delivery, the Lushan case offers insight into both the possibilities and limitations of collaborative governance under crisis conditions.

Keywords State–society collaboration · NGO professionalism · Coordination platforms · Service inequality · Institutional constraints

11.1 Institutional and Operational Context: How Collaboration Became Possible

The Lushan Earthquake response marked a critical juncture in the evolution of government–society collaboration in China's disaster governance. This section examines the institutional and operational conditions that made such collaboration possible. It traces the maturation of the governance environment, including improved emergency protocols and policy shifts toward civil society inclusion. It then analyzes the growing organizational capacity of NGOs, highlighting their professionalization, functional specialization, and networked operations. Finally, it explores the operational dynamics on the ground, where real-time information flow, increased field access, and emergent coordination platforms enabled new forms of interaction between state and social actors under emergency conditions.

Q. Du and R. Shaw, *Social Organizations in Disaster Response and Recovery*, Disaster Risk Reduction, https://doi.org/10.1007/978-981-92-2846-1_11

11.1.1 Governance Environment

By the time the Lushan Earthquake struck on April 20, 2013, China's disaster governance framework had entered a new phase of institutional maturity. Compared to the 2008 Wenchuan Earthquake, the country's emergency response system had become more structured, formalized, and capable of coordinating multi-level interventions. The national emergency response was triggered almost immediately, with the State Council Earthquake Relief Headquarters launching a Level I response within hours of the quake (Organisation Z, 2013a). Over 9000 military personnel, along with specialized medical and engineering teams, were rapidly deployed, reflecting a higher degree of state readiness and command coordination. Local governments also demonstrated improved capacity and responsiveness, operating within a clearer division of responsibilities under the Emergency Response Law, which had gradually shaped the procedural and legal foundation for disaster governance (Sun, 2017; Wu & Peng, 2017). The decentralization of operational tasks—such as shelter management, supply logistics, and medical services—enabled local authorities to act swiftly while aligning with national directives.

This institutional environment was further shaped by policy shifts that redefined the role of social actors in disaster response. In contrast to the Wenchuan experience, where many NGOs operated informally or faced state resistance, the Lushan Earthquake marked a turning point in official attitudes toward civil society participation. Government agencies, including the Ministry of Civil Affairs and Sichuan provincial authorities, increasingly acknowledged the contributions of social organizations and encouraged their involvement through public communications, designated coordination platforms, and simplified registration processes (Rong et al., 2015; Zhang et al., 2016). These policy adaptations signaled a deliberate move toward inclusion, though within a managed framework where government still retained agenda-setting authority. During the first week of the earthquake response, NGOs became increasingly active in areas traditionally dominated by state agencies, including logistics, psychosocial support, the setup of child-friendly spaces, and needs assessments. Their engagement was not limited to parallel efforts but increasingly aligned with official mechanisms through participation in government-led coordination platforms and joint briefings (Organisation Z, 2013a, 2013b, 2013c, 2013d, 2013e, 2013f). This marked a notable shift in state–civil society interaction, reflecting a gradual move toward structured, if still uneven, collaborative arrangements.

At the same time, the growing legitimacy and professionalization of NGOs laid the groundwork for meaningful collaboration. Many of the organizations involved in the Lushan response had emerged or matured in the post-Wenchuan era, with more defined governance structures, specialized teams, and adherence to humanitarian standards (Lu & Li, 2020; Zhao et al., 2015). No longer solely grassroots volunteers, these actors included professionalized entities with sector-specific expertise—in areas such as child protection, mental health, or disability services—whose operational approaches aligned more easily with government expectations. Participation in real-time information sharing, joint needs assessments, and formal activity logging marked a shift from spontaneous engagement to structured integration (Organisation Z, 2013c, 2013d).

11.1.2 Organizational Capacity of Social Actors

By the time the Lushan Earthquake struck in 2013, Chinese social organizations—especially NGOs—had undergone significant transformations in scale, structure, and strategic orientation. Compared to the spontaneous and fragmented participation observed during the 2008 Wenchuan Earthquake, the Lushan response reflected a shift from volunteer-driven enthusiasm to more organized, professionalized, and networked action. This evolution in the organizational capacity of social actors was a critical enabling factor for effective government–society collaboration in the emergency response.

The years between Wenchuan and Lushan saw a steady increase in the number of disaster-oriented NGOs in China. Many grassroots organizations founded during or after the 2008 disaster had become institutionalized, developing stable organizational structures, securing registration or partnerships, and gaining experience through prior interventions. According to field assessments, the NGOs involved in Lushan demonstrated higher levels of preparedness, including pre-arranged deployment plans, internal coordination protocols, and trained personnel (Zhao et al., 2015). Instead of ad hoc volunteer mobilization, many organizations operated with defined roles such as child protection, needs assessment, shelter management, or psychological support. The increasing functional differentiation among NGOs contributed to a more efficient division of labor and clearer coordination interfaces with state agencies.

This transformation was also marked by a discernible shift from "impulse-driven" voluntary action to "organization-based" and "project-based" interventions. Unlike in earlier disasters where volunteers often acted independently, by 2013 many actors were embedded within organizational frameworks that provided legitimacy, logistical support, and strategic guidance (Xu et al., 2016). These NGOs had begun adopting standard operating procedures, needs assessment tools, and activity reporting mechanisms aligned with international humanitarian practices. Their ability to articulate professional identities and deliver measurable outcomes enhanced their credibility in the eyes of both the government and affected communities.

The Lushan Earthquake also witnessed the emergence of inter-organizational networks and cross-regional alliances that expanded both the capacity and reach of social actors. Several NGOs participated in collaborative mechanisms such as regional coalitions and issue-based working groups, enabling resource pooling and rapid information sharing (Lu & Li, 2020). For instance, some organizations from Beijing, Guangzhou, and Chengdu formed coordinated task forces to jointly operate in hard-hit townships like Baoxing and Lushan County. These networks facilitated a horizontal mode of coordination among NGOs while also improving vertical communication with government departments. Such coalition-based approaches proved especially useful in accessing remote areas, as NGOs often shared transportation and logistical resources under difficult terrain and infrastructure conditions (Lu et al., 2023).

Crucially, the maturation of social actors also meant they were increasingly prepared to align their operations with government priorities, thereby reducing friction and enabling collaborative governance. NGOs did not merely act as independent service providers but often embedded themselves in local coordination frameworks or served as implementing partners for official recovery projects. Some organizations developed relationships with township governments or local bureaus, using these ties to gain operational access and to jointly deliver services (Zhao et al., 2015).

11.1.3 Operational Dynamics and the Emergence of Coordination Practices

The operational environment during the Lushan Earthquake created favorable conditions for emergent collaboration between governmental actors and social organizations. One of the most critical factors was the establishment of a real-time information flow system that significantly improved the speed and accuracy of situational awareness. In the early phase of the emergency response, daily updates were disseminated by coordination platforms such as the "NGO Joint Office for Ya'an Earthquake Relief" (NGO联合救灾办公室), which gathered and distributed detailed data on damages, needs, and activities. By April 21—only one day after the earthquake—over 150 NGOs had been reported active in the field, with updated figures on casualties, displaced populations, and shelter conditions made available on a daily basis (Organisation Z, 2013a, 2013b, 2013c).

The second enabling factor was the evolving governmental stance on access and collaboration. Compared with the Wenchuan Earthquake in 2008, there was a noticeable shift in how the state responded to the presence of social actors. By April 23, field access to NGO teams was no longer limited to logistics support at the periphery; grassroots teams such as psychological support providers, needs assessment groups, and emergency education teams were explicitly listed in situational summaries, with their locations and tasks publicly documented (Organisation Z, 2013d, 2013e, 2013f). The inclusion of social actors in official updates marked an operational shift in openness. Although coordination was still informal in the early days, by April 24, local governments began organizing regular meetings with NGO representatives in Lushan and Baoxing counties, signaling an increasing recognition of civil society's value in response capacity (Organisation Z, 2013f).

Third, a preliminary platform-based coordination mechanism started to emerge during this period. Although not institutionalized, several collaborative models developed in practice. NGO coalitions and joint platforms, such as the "Sichuan NGO Disaster Relief Joint Action" (四川民间组织灾后联合行动), played a role in matching needs and services. These informal structures used tools such as shared spreadsheets, public WeChat groups, and mobile mapping to facilitate real-time coordination of shelter services, supply distribution, and psychosocial support (Organisation Z, 2013e, 2013f, 2013g). While limitations in data accuracy and duplication persisted, this was a key moment where information technology and grassroots organizing converged into a rudimentary coordination infrastructure.

This operational environment thus reflected a convergence of rapid information dissemination, institutional tolerance for civil society presence, and grassroots coordination innovation. These factors, while far from perfect, represented a transitional phase in China's disaster governance, where previously peripheral actors were informally but meaningfully integrated into the frontline response. Rather than resulting from top-down design, this collaboration evolved organically, shaped by urgency, capacity, and mutual accommodation.

11.2 Collaborative Practices in Action: Platforms, Information Sharing, and Role Differentiation

The Lushan Earthquake response demonstrated how collaboration between state and societal actors was not only initiated but operationalized through distinct yet interrelated mechanisms. This section examines three critical dimensions of such collaboration. First, emergent platform mechanisms—both physical and digital—enabled coordination among heterogeneous actors. Second, information sharing practices, including needs assessments and community-level feedback, facilitated situational awareness despite persistent asymmetries. Third, role differentiation between government, NGOs, and professional organizations allowed for a more efficient division of labor across hard and soft services. These practices reflected a shift toward a hybrid, pluralistic mode of disaster governance in China.

11.2.1 Platform Mechanisms

During the response to the Lushan Earthquake, a set of emergent platform mechanisms played a foundational role in enabling coordination between governmental agencies and a growing ecosystem of social organizations. Unlike the ad hoc and fragmented structures observed during earlier disasters, this period marked a transition toward the preliminary institutionalization of joint coordination platforms. These platforms, although largely informal, enabled the sharing of information, division of labor, and synchronization of efforts across heterogeneous actors (Lu & Li, 2020; Xu, Wang, et al., 2018).

One of the most prominent coordination structures was the "NGO Joint Office for Ya'an Earthquake Relief" (NGO联合救灾办公室), which served as a temporary coordination node in the early days of the response. Established within the first week following the earthquake, this joint office brought together various NGOs operating in Lushan and Baoxing counties and acted as an interface with local government units. By April 24, this office was already hosting regular coordination meetings, compiling field reports, and facilitating communication among NGOs and local bureaus (Organisation Z, 2013f). While it had no formal authority, its role in centralizing information flows and linking actors in the field gave it practical relevance in a complex operational environment (Wu & Peng, 2017).

In addition to physical coordination hubs, digital and data-sharing platforms increasingly supported field operations. NGOs developed and applied shared spreadsheets, mobile-based mapping tools, and WeChat groups to track service coverage, identify unmet needs, and reduce duplication. Tools such as those used on April 25 and 27 were employed to record and update shelter availability across affected townships, helping ensure more balanced allocation of tents, food, and psychosocial support (Organisation Z, 2013g, 2013j). Although these systems lacked real-time automation or formal government integration, their grassroots-driven application marked a shift in operational practice and anticipated the emergence of more structured digital disaster management approaches (Meng et al., 2024).

More importantly, the emergence of coalition-based structures reflected a maturing civil society response. Networks like the "Sichuan NGO Disaster Relief Joint Action" (四川民间组织灾后联合行动) coordinated actors not only spatially but thematically. NGOs focusing on children, mental health, or disability services formed issue-based clusters within the platform, enabling specialization and mutual reinforcement (Lu & Li, 2020). These coalitions also served as negotiation spaces for NGOs to articulate concerns, share standards, and develop consensus on ethical practices, such as community entry protocols and data confidentiality. Research shows that these coalitions were instrumental in balancing bottom-up responsiveness with top-down constraints, allowing NGOs to navigate political sensitivities while improving service delivery (Xu et al., 2016; Zhao et al., 2015). Although participation was voluntary and uneven, the very existence of such platforms marked a substantial departure from earlier disaster responses, where coordination was either state-dominated or entirely absent.

While coordination remained limited by resource constraints, information asymmetries, and regulatory uncertainties, these platform mechanisms provided the skeletal infrastructure for government–society interaction in real time. The convergence of shared physical spaces, digital tools, and alliance-based governance structures laid the foundation for emergent collaborative governance. In this sense, the Lushan Earthquake offered not only a test of operational agility but also a blueprint for evolving models of disaster response that center on pluralistic coordination, even in the absence of formal institutional arrangements.

11.2.2 Information Sharing

Information sharing during the Lushan Earthquake response marked a departure from previous disaster events, reflecting gradual progress toward a more open and coordinated operational environment. Despite the persistence of data asymmetries and infrastructural bottlenecks, both governmental and non-governmental actors developed provisional systems for communicating needs, activities, and resource flows. This process, while still fragile and uneven, revealed the emergence of a multidirectional information ecosystem in which situational updates, field assessments, and service coverage reports became critical assets for coordination.

A notable feature of the Lushan response was the growing role of social organizations in conducting and disseminating needs assessments. These organizations collected frontline data on displaced populations, shelter needs, and vulnerable groups, particularly in hard-to-reach areas such as Baoxing County. NGOs also began compiling field-based data and sharing it through coordination meetings or digital platforms. For example, the April 25 briefing featured comprehensive summaries from child protection and psychological support teams, along with disaggregated data on shelter populations and resource gaps (Organisation Z, 2013g). Although initially produced independently, such information was progressively recognized by local authorities and factored into operational decisions. These developments reflected broader trends of NGO professionalization and sectoral specialization in disaster contexts (Xu et al., 2016; Zhao et al., 2015).

In parallel, village-level information loops began to emerge as critical channels for real-time feedback. In several townships, NGOs worked with community leaders to gather updates on population movements, shelter conditions, and urgent medical needs, feeding these insights back to central coordination nodes. On April 26 and 27, township officials and local volunteers transmitted phone-based updates to coordination teams, which facilitated more granular tracking of evolving needs in semi-isolated communities (Organisation Z, 2013h, 2013i). In several areas, NGOs embedded local volunteers to monitor micro-level changes, forming informal "community information relays" that complemented official reporting structures (Chen et al., 2020; Liang et al., 2021). Although the data integration infrastructure remained limited, this active flow of bottom-up information represented a meaningful step toward decentralized situational awareness.

Despite these improvements, structural challenges to information sharing persisted. Government data on casualty figures, shelter capacities, and supply stocks remained unevenly available, and often subject to delays or selective release. Several NGOs reported difficulties in accessing baseline demographic data or official distribution plans, which hindered their ability to avoid service duplication or target underserved populations (Chen et al., 2020). These challenges were particularly acute in remote or ethnic minority areas, where language barriers and local administrative fragmentation further complicated communication (W. Wang et al., 2018). Furthermore, interoperability between government and NGO data systems was limited: while NGOs developed their own mapping tools, spreadsheets, and reporting forms, these formats were rarely compatible with official platforms. As a result, much of the coordination relied on interpersonal networks, WeChat-based communication, and ad hoc reconciliation during joint meetings (Organisation Z, 2013f, 2013g, 2013h).

At the same time, the willingness of local governments to acknowledge and utilize non-governmental data sources marked a subtle but meaningful shift. As noted in empirical studies, the Lushan response witnessed initial steps toward a more inclusive data governance regime, where multiple actors co-produced operational knowledge—even if not yet institutionalized or standardized (Meng et al., 2024). In some instances, township governments used NGO-provided assessments to adjust resource allocation plans or to redirect services to overlooked communities. The integration of information from both formal and informal sources demonstrated not only a shift in operational habits but also the accumulation of mutual trust built since earlier disasters like Wenchuan (Lu & Li, 2020; Rong et al., 2015). This indicates a growing recognition of the epistemic value of social actors, and a move beyond the strictly top-down information monopolies of earlier disasters.

11.2.3 Role Differentiation

The emergency response to the Lushan Earthquake revealed a clearer delineation of roles among key actors—government institutions, non-governmental organizations, and specialized service providers. Compared to the Wenchuan Earthquake, the 2013 response displayed an evolving division of labor, where different entities assumed complementary responsibilities based on their comparative advantages. This operational differentiation was not the product of formalized contracts or mandates but emerged organically through practice, capacity, and situational necessity.

Government agencies retained central authority over coordination, infrastructure repair, and order maintenance. The state managed major logistical tasks such as road clearance, power restoration, and large-scale shelter deployment. The People's Liberation Army and local government departments were responsible for securing key transport routes, deploying heavy machinery, and overseeing centralized shelter sites, particularly in severely affected areas like Lushan County and Baoxing (Organisation Z, 2013e, 2013h). In parallel, local governments handled registration of displaced populations, supervised food and material distribution, and issued safety advisories. These functions reflected the state's continuing dominance over "hard" disaster functions such as infrastructure and security (Lu et al., 2021).

In contrast, NGOs focused on "soft" service provision—targeting vulnerable groups and offering specialized support. Many NGOs, especially those with prior experience from Wenchuan, offered targeted interventions in child protection, psychosocial support, disability services, and education continuity (Zhao et al., 2015). In makeshift shelters, NGOs—including Beijing-based children's foundations—rapidly established child-friendly spaces, trauma counseling services, and safe learning corners to support affected children (Organisation Z, 2013f, 2013i). These services addressed psychosocial and protection gaps that government units were neither mandated nor equipped to handle. Additionally, NGOs brought community-level knowledge and social trust that allowed them to access remote villages and sensitive populations, such as the elderly or those with disabilities, more effectively than official agencies (Bun Ku & Dominelli, 2018; Liang, 2016).

Specialized professional organizations—often operating within or alongside NGOs—provided domain-specific services that bridged operational gaps between state and society. Psychological support teams, including clinical social workers and mental health professionals, were deployed to provide post-trauma care, particularly for children and frontline responders (Chen et al., 2020; Shang et al., 2019). Health-focused NGOs set up mobile clinics, coordinated medicine delivery, and collaborated with local hospitals on hygiene management and maternal care (Xu et al., 2016). Organisation Z records show coordinated health checks, mental health screening, and public health education campaigns being carried out in partnership between NGOs and township clinics, illustrating how these professional actors filled technical voids in local systems (Organisation Z, 2013g).

While role differentiation improved service coverage and reduced redundancy, it also introduced new challenges. Coordination costs increased as actors needed to communicate across sectors with varying operational norms and languages. Moreover, the boundaries between roles were not always stable—some NGOs took on functions traditionally handled by government, while some township authorities began outsourcing certain services to civil society actors. As noted in post-response evaluations, this fluidity occasionally caused jurisdictional ambiguity, especially in overlapping areas such as shelter management or needs assessment (Lu & Li, 2020; Meng et al., 2024).

11.3 Tensions and Constraints: Why Collaboration Remained Limited and Uneven

While the Lushan Earthquake response marked a notable expansion of government–society collaboration, the practice of coordination remained fragmented, unstable, and uneven. This section explores the underlying constraints that limited the quality and durability of collaborative governance. It identifies three interrelated mechanisms: institutional constraints rooted in administrative dominance and regulatory ambiguity; capacity constraints stemming from disparities in NGO professionalism, volunteer management, and sustainability; and coordination costs caused by fragmented information flows, resource mismatches, and spatial imbalances. These constraints highlight why collaboration, though present, remained ad hoc and far from institutionalized.

11.3.1 Institutional Constraints

Despite the emergence of collaborative mechanisms during the Lushan Earthquake response, deeper structural barriers rooted in China's institutional environment continued to constrain the quality, stability, and scope of government–society collaboration. At the heart of these constraints was the persistent dominance of a state-centric logic of emergency governance, which prioritized political control and administrative order over openness and pluralistic coordination.

The administrative logic of command and control was evident in the centralized deployment of rescue forces and decision-making authority. While NGOs were permitted to operate in the affected areas, their activities were often subject to oversight, redirection, or even containment by local authorities. This top-down orientation limited the space for autonomous civil society action, especially in more politically sensitive areas like Baoxing County, where access for NGOs was tightly controlled (Organisation Z, 2013e, 2013f). Empirical studies have shown that despite the growing number of NGOs in China's disaster field, the state remained wary of relinquishing operational or discursive authority to non-state actors (Diakakis, 2016; Sun, 2017).

The legal and institutional status of NGOs further contributed to the ambiguity of their roles. Many participating organizations were not formally registered with the Ministry of Civil Affairs, operating instead as informal networks, university-based teams, or local initiatives without clear legal identity. This lack of official recognition posed challenges in inter-agency coordination and resource allocation. For example, NGOs without proper credentials often struggled to gain access to distribution lists or official coordination meetings, reinforcing a two-tiered system in which "insider" organizations—typically government-affiliated or semi-official entities—were granted privileged access (Wu & Peng, 2017; Zhao et al., 2015).

Moreover, risk control imperatives continued to trump collaborative aspirations. In the wake of the 2008 Wenchuan Earthquake, government authorities had become more cautious of the potential political ramifications of unregulated NGO activity, especially when it came to issues such as media exposure, international funding, or mobilization of large volunteer groups (Xu et al., 2016). During the Lushan response, this translated into tight surveillance of NGO activities and communication channels, particularly digital platforms such as WeChat and blogs, which had become central to real-time coordination and information exchange. This atmosphere of control created a risk-averse environment, where NGOs often self-censored or avoided sensitive areas of work, such as rights-based advocacy or data transparency (Wu & Peng, 2017).

Even where government–NGO collaboration was visible, it remained largely instrumental and temporary. Non-governmental organizations were primarily valued for their capacity to deliver specialized services or access hard-to-reach communities, rather than being recognized as equal partners in governance. Field-based assessments produced by NGOs were sometimes used to complement official data, yet they were seldom incorporated into strategic planning or formal resource allocation processes (Organisation Z, 2013g, 2013h, 2013i). This pattern of selective utilization reinforced a subcontracting logic, limiting opportunities for meaningful co-production and constraining the institutional learning potential of collaborative disaster response.

11.3.2 Capacity Constraints

The capacity of social organizations to engage effectively in post-disaster collaboration during the Lushan Earthquake response was shaped by significant internal limitations. Despite the growing presence of NGOs in China's disaster field, capacity disparities—ranging from professional expertise and volunteer management to project continuity—undermined their ability to operate at scale and engage in equal partnership with government actors.

One of the most persistent challenges was the uneven professionalization among NGOs. While a small cohort of experienced organizations brought specialized skills in fields like psychosocial support, education, and disability services, the broader NGO landscape remained fragmented and varied in quality. Many grassroots NGOs lacked trained personnel, standardized protocols, or operational tools suited for emergency environments. This discrepancy became especially pronounced in high-demand sectors such as mental health or child protection, where only a handful of NGOs possessed the technical expertise to deploy targeted interventions (Lu et al., 2023; Xu, et al., 2018). In several cases, well-intentioned but underprepared teams entered disaster zones without clear mandates or service models, leading to overlap, confusion, or inefficient use of resources (Organisation Z, 2013f, 2013h).

Volunteer management posed another significant constraint. The influx of spontaneous volunteers following the earthquake created both opportunities and operational burdens. Few NGOs had pre-existing frameworks for volunteer screening, training, and deployment. As a result, many teams were overwhelmed by coordination demands, while others failed to integrate volunteers into structured service provision. Ad hoc volunteer mobilization often led to inefficiencies in task allocation and occasionally undermined community trust when expectations were unmet (Lu et al., 2023). Several local authorities expressed concern over volunteer saturation in central shelters, while peripheral areas remained underserved (Organisation Z, 2013g, 2013i). These challenges were compounded by a lack of coordination platforms to manage volunteer flows across organizations.

Project continuity and sustainability also emerged as recurring limitations. A large proportion of NGO efforts remained short-term, often ending with the conclusion of emergency funding cycles. This temporal fragility hindered NGOs from engaging in mid-to-long-term recovery planning or building enduring relationships with local communities and governments. Field data from Baoxing County noted that some NGOs had withdrawn within two weeks of deployment due to budgetary exhaustion or staff turnover (Meng et al., 2024). In the absence of institutional funding or embedded local partnerships, these organizations struggled to maintain operations beyond the immediate relief phase. This impermanence limited their ability to address deeper, structural issues such as trauma recovery, livelihood rebuilding, or educational restoration.

These capacity constraints also shaped the contours of inter-organizational collaboration. Government actors, while increasingly open to civil society participation, tended to gravitate toward NGOs with proven track records, often excluding newer or less professionalized groups. This led to a form of "selective collaboration" that privileged a small circle of institutionalized NGOs, reproducing inequality within the sector and marginalizing emergent actors (Xu, et al., 2018). Consequently, the potential of grassroots innovation and local responsiveness was curtailed, while coordination platforms became skewed toward dominant voices.

11.3.3 Coordination Costs and Inefficiencies

Despite signs of emerging collaborative structures during the Lushan Earthquake response, coordination remained costly and uneven, with several operational inefficiencies hindering effective service delivery. These coordination frictions were not merely logistical but deeply intertwined with the fragmented nature of information flows, resource distribution imbalances, and spatial disparities in attention and access.

A primary source of inefficiency stemmed from the repetition and fragmentation of information. In the absence of standardized data-sharing protocols or unified assessment tools, multiple actors—including NGOs, local governments, and community volunteers—conducted parallel needs assessments. Organization Z's situation briefings from April 25 to 26 reveal that at least four separate organizations visited Shuangshi Township within the same 48-hour period to conduct shelter population surveys and distribute basic relief items, resulting in overlapping data collection and redundant reporting efforts (Organisation Z, 2013f, 2013g). This duplication not only wasted limited human resources but also created confusion for local residents, who were repeatedly asked the same questions without seeing immediate service outcomes. Moreover, some organizations lacked mechanisms for disseminating collected data, further exacerbating siloed decision-making and hampering coordinated planning (Chen et al., 2020).

Resource mismatches were also common. While central townships such as Lushan County seat received a high concentration of supplies and service teams, more peripheral regions, especially Baoxing County and remote villages in Longmen Township, struggled with basic shortages. While some central shelters had accumulated excess donations of bottled water and instant noodles, other villages reported a lack of warm clothing and medical supplies due to logistical delays and poor communication with distribution nodes (Organisation Z, 2013h). This imbalance partly resulted from poor integration between frontline assessment and supply chain coordination. NGOs and donors often directed resources based on media exposure or informal networks, rather than evidence-based needs assessments, leading to supply congestion in well-covered areas and neglect of less visible zones (Liang et al., 2021; Wang et al., 2018).

Another inefficiency was the spatial clustering of NGOs and volunteers in high-visibility zones, a pattern repeatedly observed in disaster contexts. During the initial response in Lushan, over 50 social organizations were observed operating within a 10-kilometer radius of the county center, whereas more remote mountainous villages remained virtually unattended for several days (Organisation Z, 2013e). This concentration led to unnecessary competition for limited space, strained local administrative capacity, and in some cases, even duplicated the same services—such as mobile medical teams operating adjacent to each other without coordination. Meanwhile, the lack of structured mechanisms to redirect surplus capacity to underserved zones resulted in persistent spatial gaps in service coverage (Meng et al., 2024).

These inefficiencies were compounded by the limited scalability of existing coordination platforms. Informal WeChat groups and face-to-face coordination meetings, while useful in small networks, proved insufficient in managing real-time resource tracking, actor mapping, or task allocation across a rapidly evolving operational landscape. The temporary NGO Joint Office lacked technical systems to monitor who was doing what and where, forcing field actors to rely on word-of-mouth updates or outdated spreadsheets (Lu & Li, 2020; Organisation Z, 2013f). As a result, service distribution was reactive rather than strategic, and the overall responsiveness of the collaborative system was constrained.

11.4 Social Outcomes as Evidence: What Collaboration Actually Produced

Beyond institutional arrangements, operational mechanisms, and their inherent constraints, the ultimate significance of collaboration lies in the social outcomes it generated in affected communities. Evidence from the Lushan Earthquake suggests that collaborative practices reshaped patterns of psychosocial support, community recovery, and everyday resilience, while simultaneously reproducing important inequalities in access and benefit distribution. Attention therefore turns to how social support mechanisms influenced mental health trajectories, how community accompaniment, livelihood initiatives, and educational continuity contributed to recovery, and where these gains remained uneven. Examining these outcomes allows the analysis to assess the substantive value of collaboration, not only in terms of coordination efficiency, but in its capacity to produce socially meaningful recovery.

11.4.1 Social Support and Mental Health Outcomes

The Lushan Earthquake response not only mobilized a range of humanitarian and logistical interventions but also revealed the critical role of collaborative action in supporting the psychological recovery of affected populations. Among various outcomes, the strengthening of social support networks—facilitated through NGO involvement—proved essential in mitigating trauma, alleviating mental distress, and promoting resilience. This section explores how such support translated into mental health outcomes, focusing on the interplay between interpersonal support, institutional services, and post-disaster wellbeing.

Empirical research has consistently shown that social support functions as a protective buffer in post-disaster settings, reducing the likelihood of prolonged psychological trauma such as PTSD and depression (Kaniasty, 2020). In the context of Lushan, evidence suggests that individuals who received higher levels of emotional and instrumental support—through community groups, volunteers, or NGO-based services—demonstrated lower prevalence of PTSD symptoms and better psychosocial adjustment over time (Shang et al., 2019; Zhou et al., 2018). One study found that adolescents who had access to both family support and organized psychosocial services experienced significant improvements in perceived recovery and self-efficacy during the six months following the earthquake (Ren et al., 2015). These findings underscore the importance of integrated social support mechanisms that extend beyond material aid.

NGO participation was particularly significant in filling gaps in mental health care, an area traditionally underserved by public disaster response. Immediately following the quake, NGOs specializing in psychological support—including trauma counseling, grief processing, and child protection—were among the earliest actors to set up services in shelters and community centers. Mobile counseling teams were deployed to centralized shelters in Lushan and Baoxing Counties, where they provided one-on-one consultations and facilitated group-based therapy for both children and parents (Organisation Z, 2013g, 2013i). These services were often the only source of psychological care in the early response phase, particularly in remote or marginalized areas.

Importantly, social support was not limited to formal services. Peer support, volunteer accompaniment, and community-based listening spaces also played critical roles. Informal care provided by trained volunteers—including simple check-ins, listening sessions, and activity facilitation for children—substantially improved survivors' emotional expression and stress management (Zhou et al., 2018). In some areas, volunteers embedded in shelters were trained by professional NGOs to identify signs of trauma and refer cases for follow-up care. This bottom-up social infrastructure helped foster a sense of security and continuity during the disorienting aftermath of the disaster.

Nonetheless, limitations in mental health outcomes were also evident. Coverage of psychosocial services remained uneven: urban centers and large shelters received concentrated attention, while more isolated townships reported a lack of consistent follow-up support. While initial outreach efforts were visible, the continuity of care diminished significantly after the first month, due to short funding cycles and personnel turnover (Ren et al., 2015). Gaps in follow-up counseling were also observed in several mountainous villages, where no trained professionals returned after the initial deployment period (Organisation Z, 2013i).

11.4.2 Community Recovery and Livelihood

In the aftermath of the Lushan Earthquake, collaborative governance between government agencies and NGOs played an instrumental role in shaping community recovery processes. Beyond immediate relief, this collaboration contributed to the restoration of daily life through efforts in social reconnection, livelihood support, and education continuity. One notable contribution from NGOs was their ongoing presence in affected areas through approaches centered on community accompaniment. Unlike government units whose presence often diminished after the emergency phase, certain social organizations remained embedded within communities, providing not only logistical aid but also emotional support. Some NGOs established semi-permanent service points in towns such as Lushan and Tianquan, engaging in daily interactions with residents and offering psychosocial assistance to vulnerable groups like elderly individuals and single-parent families (Organisation Z, 2013i). Such embedded engagement enhanced social trust and stabilized communal life during uncertain times. Research suggests that this sustained presence bolstered residents' confidence in long-term recovery and improved subjective perceptions of support (Pu & Chang-Richards, 2022).

Livelihood recovery was another area where collaborative efforts yielded tangible impacts. Many rural households suffered severe economic disruption due to the loss of farmland access, damaged infrastructure, and destruction of productive assets. In response, several NGOs implemented targeted livelihood recovery programs, including small grants for women-led cooperatives, tool replacement schemes, and training in income-generating skills such as mushroom cultivation and tailoring. In Tianquan County, NGO facilitators worked with local farming families to co-develop household-level economic plans, aiming to reduce dependence on short-term aid and rebuild income autonomy within several months (Lu et al., 2021). These micro-level initiatives complemented government subsidies and supported differentiated needs that were not always visible in top-down recovery frameworks. However, the overall scale and sustainability of these efforts remained limited. Many NGO-led programs lacked long-term funding, and organizations without local roots struggled to engage effectively with administrative actors or access relevant land-use and market data (Pu & Chang-Richards, 2024). Several NGOs reported difficulties in aligning livelihood projects with local township development agendas, where infrastructure reconstruction was prioritized over social or economic programming (Organisation Z, 2013j).

Education continuity further highlights the social contributions of post-disaster collaboration. With numerous schools damaged or repurposed as shelters, NGOs played a crucial role in establishing temporary learning spaces to ensure children retained access to educational and psychosocial resources. These spaces, often hosted in tents or community halls, also served as protective environments where trained staff could monitor signs of distress and provide emotional care. By late April, over 2500 children had reportedly participated in structured learning and play sessions, facilitated through coordination between NGOs and local educators (Organisation Z, 2013k). These efforts not only mitigated disruptions to education but also contributed to restoring a sense of normalcy and routine in children's daily lives. Nevertheless, the reach of such interventions was geographically uneven, with more remote areas struggling to access services due to volunteer shortages and weak logistical coordination. Once temporary shelters were dismantled, many of these education-support initiatives also lost continuity due to the lack of institutionalized mechanisms for handover or follow-up. This reflects a broader challenge in sustaining recovery programs that rely heavily on short-term external actors without deeper integration into local governance and community systems.

11.4.3 Uneven Benefits and Social Inequality

Although collaborative efforts during the Lushan Earthquake response produced visible gains in social recovery, these benefits were not equitably distributed. Instead, access to support remained shaped by geographic proximity, information visibility, and institutional embeddedness, reinforcing pre-existing spatial and social inequalities. The urban–rural divide was particularly pronounced. Central zones such as the Lushan County seat and Ya'an city became hubs of intensive resource flows, NGO deployments, and media attention, resulting in faster aid delivery and a wider range of services. In contrast, remote villages in Baoxing County and Longmen Township were significantly underserved, often for extended periods, due to damaged transport links, weak communication infrastructure, and limited organizational presence (Organisation Z, 2013e, 2013g). Despite repeated appeals, some mountain communities still lacked access to safe drinking water, medical support, and adequate shelter more than a week after the quake (Organisation Z, 2013i).

Such spatial imbalances were not merely logistical accidents but were shaped by coordination constraints and visibility-driven deployment patterns. Without centralized mechanisms for data sharing or actor mapping, resource allocation frequently followed informal channels such as interpersonal ties or social media posts, which skewed attention toward already-visible regions. As documented in previous assessments, NGO activities tended to cluster in areas with stronger digital connectivity and clearer information signals, while less visible or harder-to-reach regions were bypassed (Liang, 2016; Wang et al., 2021). This selective concentration created redundancies in well-covered shelters and left peripheral zones with delayed or minimal support.

In addition to spatial disparities, the collaborative system struggled to consistently reach vulnerable groups. While some NGOs designed targeted services for children, the elderly, and persons with disabilities, most interventions were not based on disaggregated population data, limiting the capacity for tailored programming. In several cases, elderly individuals with chronic conditions were housed in general tents without access to specialized care or mobility assistance. Likewise, persons with disabilities were frequently excluded from assessments due to inaccessible terrain or absence from centralized shelters (Organisation Z, 2013h).

The inconsistent presence of NGOs across affected areas further exacerbated inequality. Some communities benefited from recurring psychosocial visits and community-based accompaniment, while others received no external engagement at all. This uneven distribution undermined perceptions of fairness and, in some instances, generated tension among survivor groups. The lack of an equity-sensitive targeting framework contributed to what has been described as "selective recovery," whereby better-connected or better-positioned populations rebounded more quickly, while marginalized groups remained in states of prolonged vulnerability (Pu & Chang-Richards, 2022).

11.5 Conclusion

The Lushan Earthquake response marked a critical turning point in China's disaster governance, revealing both the potential and the limitations of emergent government–society collaboration. Institutional openness, growing NGO capacity, and grassroots coordination platforms enabled more pluralistic forms of engagement than seen in earlier disasters. These dynamics produced meaningful social outcomes—from improved psychosocial care to localized livelihood recovery—while also exposing persistent inequalities in access, data integration, and continuity of support. Collaboration, though increasingly visible, remained informal, uneven, and constrained by structural hierarchies and operational fragmentation. As a transitional moment, Lushan offers a valuable lens into how crisis can catalyze new governance arrangements—while underscoring the need for more institutionalized, inclusive, and equity-sensitive models of partnership in future disaster contexts.

References

Bun, K., & H., & Dominelli, L. (2018). Not only eating together: Space and green social work intervention in a Hazard-affected area in Ya'an, Sichuan of China. *The British Journal of Social Work, 48*(5), 1409–1431. https://doi.org/10.1093/bjsw/bcx071

Chen, W., Zhang, H., Comfort, L. K., & Tao, Z. (2020). Exploring complex adaptive networks in the aftermath of the 2008 Wenchuan earthquake in China. *Safety Science, 125*, 104607. https://doi.org/10.1016/j.ssci.2020.104607

Diakakis, M. (2016). Have flood mortality qualitative characteristics changed during the last decades? The case study of Greece. *Environmental Hazards, 15*(2), 148–159. https://doi.org/10.1080/17477891.2016.1147412

Kaniasty, K. (2020). Social support, interpersonal, and community dynamics following disasters caused by natural hazards. *Current Opinion in Psychology, 32*, 105–109. https://doi.org/10.1016/j.copsyc.2019.07.026

Liang, Y. (2016). Trust in Chinese government and quality of life (QOL) of Sichuan earthquake survivors: Does Trust in Government Help to promote QOL? *Social Indicators Research, 127*(2), 541–564. https://doi.org/10.1007/s11205-015-0967-9

Liang, Y., Zheng, H., Cheng, J., Zhou, Y., & Liu, Z. (2021). Associations between posttraumatic stress symptoms, creative thinking, and trait resilience among Chinese adolescents exposed to the Lushan earthquake. *The Journal of Creative Behavior, 55*(2), 362–373. https://doi.org/10.1002/jocb.460

Lu, Y., & Li, R. (2020). Rebuilding resilient homeland: An NGO-led post-Lushan earthquake experimental reconstruction program. *Natural Hazards, 104*(1), 853–882. https://doi.org/10.1007/s11069-020-04194-3

Lu, Y., Wang, Y., & Zhan, C. (2023). Building rural community disaster resilience in developing countries: Insights from a Chinese NGO's safe rural community programme. *Disasters, 47*(4), 1090–1117. https://doi.org/10.1111/disa.12586

Lu, Y., Wei, L., Cao, B., & Li, J. (2021). Participatory child-centered disaster risk reduction education: An innovative Chinese NGO program. *Disaster Prevention and Management: An International Journal, 30*(3), 293–307. https://doi.org/10.1108/DPM-03-2020-0066

Meng, L., Cheng, W., Deng, Y., & Dai, Z. (2024). Research on the collaborative relationship of task-driven urban earthquake emergency organizations. *International Journal of Disaster Risk Reduction, 113*, 104887. https://doi.org/10.1016/j.ijdrr.2024.104887

Organisation Z. (2013a). Situation report no. 1: Lushan earthquake relief in Ya'an, Sichuan (April 20, 2013a).

Organisation Z.(2013b). Situation report no. 2: Lushan earthquake relief in Ya'an, Sichuan (April 20, 2013b).

Organisation Z. (April 20, 2013c). (2013c). *Situation report no. 3: Lushan earthquake relief in Ya'an. Sichuan.*

Organisation Z. (April 20, 2013d). (2013d). *Situation report no. 4: Lushan earthquake relief in Ya'an. Sichuan.*

Organisation Z. (April 20, 2013e). (2013e). *Situation report no. 5: Lushan earthquake relief in Ya'an. Sichuan.*

Organisation Z. (April 20, 2013f). *(2013f). Situation report no. 6: Lushan earthquake relief in Ya'an. Sichuan.*

Organisation Z. (April 20, 2013g). (2013g). Situation report no. 7: Lushan earthquake relief in Ya'an. *Sichuan.*

Organisation Z. (April 20, 2013h). (2013h). Situation report no. 8: Lushan earthquake relief in Ya'an. *Sichuan.*

Organisation Z. (April 20, 2013i). (2013i). Situation report no. 9: Lushan earthquake relief in Ya'an. *Sichuan.*

Organisation Z. (April 20, 2013j). (2013j). Situation report no. 10: Lushan earthquake relief in Ya'an. *Sichuan.*

Organisation Z. (April 20, 2013k). (2013k). Situation report no. 11: Lushan earthquake relief in Ya'an. *Sichuan.*

Organisation Z. (April 20, 2013l). (2013l). *Situation report no. 12: Lushan earthquake relief in Ya'an. Sichuan.*

Pu, G., & Chang-Richards, A. Y. (2022). Livelihood recovery for sustainable development: A study of the Lushan earthquake. *International Journal of Disaster Risk Reduction, 82*, 103317. https://doi.org/10.1016/j.ijdrr.2022.103317

Pu, G., & Chang-Richards, A. Y. (2024). Identifying critical factors affecting the livelihood recovery following disasters: A comparative analysis of China and New Zealand. *International Journal of Disaster Risk Reduction, 114*, 104958. https://doi.org/10.1016/j.ijdrr.2024.104958

Ren, J., Jiang, X., Yao, J., Li, X., Liu, X., Pang, M., & Chiang, C. L. V. (2015). Depression, social support, and coping styles among pregnant women after the Lushan earthquake in Ya'an. *China. PLOS ONE, 10*(8), e0135809. https://doi.org/10.1371/journal.pone.0135809

Rong, H., Xuedong, L., Guizhi, Z., Yulin, Y., & Da, W. (2015). An evaluation of coordination relationships during earthquake emergency rescue using entropy theory. *Cadernos de Saúde Pública, 31*(5), 947–959. https://doi.org/10.1590/0102-311X00039514

Shang, F., Kaniasty, K., Cowlishaw, S., Wade, D., Ma, H., & Forbes, D. (2019). Social support following a natural disaster: A longitudinal study of survivors of the 2013 Lushan earthquake in China. *Psychiatry Research, 273*, 641–646. https://doi.org/10.1016/j.psychres.2019.01.085

Sun, T. (2017). Earthquakes and the typologies of state–society relations in China. *China Information, 31*(3), 304–326. https://doi.org/10.1177/0920203X17720337

Wang, W., Wu, X., & Tian, Y. (2018). Mediating roles of gratitude and social support in the relation between survivor guilt and posttraumatic stress disorder, posttraumatic growth among adolescents after the Ya'an earthquake. *Frontiers in Psychology, 9*, 2131. https://doi.org/10.3389/fpsyg.2018.02131

Wang, Y., Ge, F., Zhang, J., & Zhang, W. (2021). Heterogeneity in the co-occurrence of depression and anxiety symptoms among youth survivors: A longitudinal study using latent profile analysis. *Early Intervention in Psychiatry, 15*(6), 1612–1625. https://doi.org/10.1111/eip.13101

Wu, F., & Peng, L. (2017). The rise of non-governmental disaster relief in China. *SSRN Electronic Journal.* https://doi.org/10.2139/ssrn.2916221

Xu, J., Wang, Q., Xu, D., & Lu, Y. (2018a). Types of community-focused organisations for disaster risk reduction in the Longmen Shan Fault area. *Environmental Hazards, 17*(3), 181–199. https://doi.org/10.1080/17477891.2017.1383879

Xu, J., Xu, D., & Lu, Y. (2016). Resident participation in post-Lushan earthquake housing reconstruction: A multi-stage field research method-based inquiry. *Environmental Hazards, 15*(2), 128–147. https://doi.org/10.1080/17477891.2016.1142417

Xu, J., Xu, D., Lu, Y., & Wang, Q. (2018). A bridged government–NGOs relationship in post-earthquake reconstruction: The Ya'an service center in Lushan earthquake. *Natural Hazards, 90*(2), 537–562. https://doi.org/10.1007/s11069-017-3056-3

Zhang, H., Zhang, X., Comfort, L., & Chen, M. (2016). The emergence of an adaptive response network: The April 20, 2013 Lushan, China earthquake. *Safety Science, 90*, 14–23. https://doi.org/10.1016/j.ssci.2015.11.012

Zhao, J., Wang, X., Zhu, Y., Zhao, L., & Yu, T. (2015). *Research on the intrinsic mechanism of major disaster and crisis management from the perspective of social governance—Based on the practice of 4.20 Lushan earthquake post-disaster reconstruction* (In J. Xu, S. Nickel, V. C. Machado, & A. Hajiyev (Eds.), *proceedings of the ninth international conference on management science and engineering management* (Vol. 362, pp. 735–741)). Springer. https://doi.org/10.1007/978-3-662-47241-5_62

Zhou, X., Wu, X., & Zhen, R. (2018). Self-esteem and hope mediate the relations between social support and post-traumatic stress disorder and growth in adolescents following the Ya'an earthquake. *Anxiety, Stress, & Coping, 31*(1), 32–45. https://doi.org/10.1080/10615806.2017.1374376

Chapter 12
Ludian Earthquake: Turning Point for Disaster Voluntarism and Social Organizations

Abstract The 2014 Ludian Earthquake marked a critical juncture in the evolution of disaster governance in China, catalyzing a shift from state-dominated relief models to a more pluralistic, networked response system. This chapter examines how civil society actors—spanning national NGOs, local volunteer groups, and spontaneous networks—emerged as key responders, often operating ahead of formal structures. It traces four interconnected trajectories: the rapid mobilization of non-state actors and digital coordination mechanisms; the evolution from fragmented action to structured collaboration and role specialization; the gradual institutional recognition and embedding of civil society through semi-formal platforms; and the lasting governance transformations—both procedural and normative—that emerged from Ludian. These developments redefined the role of civil society in China's disaster response architecture and laid the groundwork for future hybrid governance innovations.

Keywords Spontaneous networks · Digital coordination · Platform governance · Hybrid disaster governance · Institutional transformation

12.1 Rapid Mobilization and the Rise of Spontaneous Networks

The 2014 Ludian Earthquake catalyzed a pivotal moment in China's disaster governance, marked by the rapid emergence of decentralized, self-organized civil society responses. This section traces the unfolding of these dynamics in four parts: the initial "response gap" left by delayed state access, the swift mobilization of diverse volunteer and grassroots actors, the central role of digital tools in enabling informal coordination, and the structural characteristics of spontaneous networks. These developments reveal an evolving pattern of bottom-up disaster engagement—agile, trust-based, and digitally mediated—reshaping the contours of emergency response in China's hybrid governance landscape.

Q. Du and R. Shaw, *Social Organizations in Disaster Response and Recovery*, Disaster Risk Reduction, https://doi.org/10.1007/978-981-92-2846-1_12

12.1.1 Background Context: Disaster Onset and Response Gap

On August 3, 2014, at 16:30 local time, a 6.5-magnitude earthquake struck Longtoushan Township in Ludian County, Yunnan Province. The disaster quickly evolved into one of the most severe seismic events in the region in recent years, with initial reports confirming over 600 deaths and more than 2,400 injuries within the first 48 h (Organisation Z, 2014a). Compounding the devastation was the collapse of basic infrastructure—roads, bridges, telecommunications, and power lines—particularly in remote and mountainous villages, which effectively delayed the entry of formal governmental relief forces.

In the immediate aftermath, a significant "response gap" emerged. Despite rapid national-level mobilization of the People's Liberation Army (PLA), fire services, and emergency medical teams, logistical constraints and blocked access routes limited their timely arrival in severely affected areas. In this temporal vacuum, civil society actors—ranging from national NGOs to local volunteer groups—began to mobilize independently, often reaching the disaster zone before formal command systems were fully activated (Organisation Z, 2014b; Zhang & Yin, 2016).

The earliest waves of response were marked by a notable decentralization of action. Organisations such as Blue Sky Rescue and One Foundation dispatched assessment teams within hours of the quake, while local collectives like "Mountain Action" and volunteer coalitions from Kunming began coordinating on-the-ground efforts without waiting for government clearance (Organisation Z, 2014b, 2014c). The fragmented and rapidly evolving situation meant that traditional state-centric models of command and control could not initially adapt to the speed and complexity of needs on the ground.

What surfaced in these early moments was not simply a set of isolated grassroots reactions, but the embryonic formation of what later literature terms "emergent response networks" or "突生组织网络" (Zhang & Yin, 2016). These networks filled critical voids in information flow, supply delivery, and frontline support, operating in spaces where the state had yet to re-establish operational control.

12.1.2 Spontaneous Mobilization: Civil Society's Rapid Response

The immediate aftermath of the 2014 Ludian Earthquake revealed a significant shift in disaster response dynamics in China. In the critical "golden 72 h," a broad range of civil society actors mobilized rapidly and autonomously, stepping into the response vacuum created by infrastructural collapse, communication disruptions, and the delayed arrival of state-led forces.

Rather than waiting for formal coordination or state approval, non-state actors self-organized with remarkable speed and initiative. By the evening of August 3—the very day the earthquake struck—multiple grassroots and professional volunteer teams had already begun organizing entry into the disaster zone. Within 24 h, over 30 organizations arrived in Zhaotong and Longtoushan, some commencing operations even before any unified official dispatch system had taken shape (Organisation Z, 2014a, 2014b, 2014c).

These early responders represented a diverse constellation of actors. Blue Sky Rescue deployed its frontline team late on August 3 and reached the epicenter the following day, navigating disrupted transport routes to begin search-and-rescue operations in heavily affected areas. The One Foundation set up an on-site command hub in Zhaotong by August 4, mobilizing staff, volunteers, and emergency supplies such as tents, hygiene kits, and clean water. At the local level, smaller community-based organizations, such as Mountain Pathfinder and Dandelion Volunteer Team, responded with psychological support, basic medical assistance, and needs assessments in remote villages unreachable by state actors (Organisation Z, 2014b, 2014c).

This wave of mobilization was notable for its autonomy and flexibility. Volunteers operated using their own resources, relying on previously established digital infrastructures such as WeChat groups, SMS lists, and online coordination forms. These tools facilitated decentralized coordination, enabling fast deployment, real-time information sharing, and basic task allocation even in the absence of hierarchical command. The responsiveness and adaptability of these groups challenged the assumption that only top-down coordination ensures effective disaster relief.

At the same time, the earthquake also triggered the emergence of self-organized aid initiatives within affected communities. Local villagers built temporary shelters, distributed basic supplies, and formed mutual aid groups. These grassroots responses not only demonstrated the latent capacity of communities to act as "first responders," but also complemented the activities of formal civil society organizations, forming a multi-layered, hybrid response network (Wu et al., 2019).

12.1.3 Digital Tools and Informal Coordination Mechanisms

The Ludian Earthquake marked a significant turning point in the digitalization of grassroots disaster response in China. Amid disrupted formal communication channels and limited early governmental coordination, digital platforms—especially social media and mobile tools—became critical enablers of real-time coordination, information exchange, and spontaneous collaboration across civil society actors.

Digital messaging platforms such as WeChat and QQ served as the primary hubs for organizing personnel, relaying needs, and sharing situational updates. Volunteer groups, NGO staff, and spontaneous responders established chat groups within hours after the earthquake, where tasks were distributed, transportation updates were posted, and resources were allocated. In the early days following the earthquake, dozens of WeChat groups were established, often centered around regional hubs (e.g., Kunming-based NGOs) or specific functions such as logistics, psychological support, and medical aid (Organisation Z, 2014b, 2014c). These ad hoc digital communities substituted for formal command chains, allowing geographically dispersed actors to operate with relative cohesion.

A parallel innovation was the use of map-based digital tools to visualize disaster impact zones and direct material flows. Micro-public service platforms enabled the creation of open-source crisis maps, where volunteers marked damaged roads, urgent medical cases, and underserved villages. Frontline responders uploaded real-time geo-tagged images and short descriptions, which were aggregated into living digital dashboards by remote volunteers (Organisation Z, 2014d). This emergent "data stream" architecture significantly improved the targeting of relief efforts and helped prioritize scarce transport and medical resources under chaotic conditions.

Informal coordination centers also began to form, acting as digital "nerve centers" that supported information processing and decision-making. For example, the Zhu Mo Information Service Center quickly became a key hub by compiling organization rosters, distributing coordination tables, and assigning responsibilities based on region and sector (Organisation Z, 2014c). By maintaining shared spreadsheets and updated contact lists, these informal nodes provided the foundation for early horizontal coordination.

The networks emerging from these practices closely resemble what some scholars define as "spontaneous organizational networks," characterized by their non-bureaucratic structure, operational elasticity, and digital-first coordination model (Zhang & Yin, 2016). These networks exhibited notable features: decentralized decision-making, rapid node formation and dissolution, and high dependence on trust-based ties rather than formal agreements. Coordination was not imposed from above but instead evolved through ongoing digital interactions and shared urgency.

12.1.4 Characteristics of Spontaneous Networks

The emergency response to the Ludian Earthquake witnessed the rapid emergence of loosely structured, self-organized networks that operated outside conventional institutional hierarchies. Rather than functioning as ad hoc extensions of state authority or formally registered NGOs, these spontaneous networks reflected an adaptive and fluid form of civic engagement that responded directly to on-the-ground needs.

Decentralization was a defining attribute. In the early phase of the response, no overarching coordination body was in place, and various volunteer groups, rescue teams, and grassroots organizations entered the disaster zone autonomously. Field observations from the first 72 h recorded how these groups self-deployed based on their own needs assessments and resource availability (Organisation Z, 2014a, 2014b, 2014c, 2014d). This enabled swift local interventions but also led to issues such as overlapping services, inconsistent data collection, and fragmented coverage.

These networks also exhibited temporariness, emerging in direct response to the disaster and often dissolving once immediate objectives were met. Many volunteer teams entered the field with short-term goals, ranging from debris clearance to supply distribution. According to records, over 50 civil society actors were active during the first week following the earthquake, though many withdrew as early as the second week, passing responsibilities to more institutionalized responders (Organisation Z, 2014d). This event-specific orientation, while responsive, limited opportunities for sustained capacity building or long-term engagement.

Fluidity further characterized these mobilizations. Personnel, resources, and tasks moved dynamically across groups, with volunteers often shifting roles and affiliations. Cross-group coordination frequently occurred through digital platforms, such as WeChat and shared spreadsheets, allowing for rapid information flow and real-time task allocation. However, this same fluidity obscured lines of accountability and complicated documentation and evaluation processes (Organisation Z, 2014c).

At the foundation of these spontaneous efforts lay trust-based ties. Without the binding force of formal contracts or centralized command, social capital became the primary medium for collaboration. Many of the emergent networks were built on pre-existing personal or organizational relationships—whether through regional NGO alliances, prior disaster collaboration, or shared involvement in online civic communities (Zhang & Yin, 2016). Trust served as both an entry point and an operational currency, facilitating coordination among actors with little to no face-to-face contact.

While these spontaneous networks lacked formal institutional structures, their practices were far from chaotic. Over time, teams developed quasi-institutional mechanisms such as rotating coordination roles, volunteer rosters, daily situation briefs, and shared evaluation templates (Organisation Z, 2014c, 2014d). Some groups even created basic standards for temporary shelter, psychological first aid, and needs assessment—pointing toward an emergent logic of self-regulation and proto-governance within civil society disaster response.

12.2 Collaborative Practices and Functional Differentiation

As the Ludian Earthquake response moved beyond its initial surge of spontaneous volunteerism, civil society organizations began transitioning from dispersed, reactive efforts to more structured and functionally differentiated forms of collaboration. Early operational fragmentation soon revealed the need for coordinated platforms capable of integrating diverse actors while preserving their autonomy. In response, new mechanisms emerged to streamline communication and task distribution across organizations. This shift was accompanied by increasing specialization, as NGOs gravitated toward service areas aligned with their strengths and capacities. Over time, standardized workflows—such as unified assessment forms and daily reporting routines—enhanced procedural consistency across the field. Meanwhile, the focus of civil society action gradually evolved from delivering emergency goods to providing sustained, service-oriented interventions. These changes marked a critical phase of maturation in China's grassroots disaster governance.

12.2.1 From Fragmented Action to Coordinated Platforms

As the number of volunteer teams and social organizations rapidly increased during the first week after the Ludian Earthquake, the limitations of fragmented and loosely connected action became increasingly visible. While early spontaneous mobilization ensured speed and coverage, it also generated operational inefficiencies. Field records documented overlapping service delivery in some villages, uneven geographic coverage in remote areas, inconsistent data reporting, and occasional misallocation of supplies (Organisation Z, 2014d). Information asymmetries and duplicated assessments began to slow down decision-making, signaling the need for a more coordinated mechanism capable of integrating dispersed actors without reverting to rigid hierarchical control.

By the second week of response, a collective effort emerged to stabilize coordination through the establishment of the Yunnan Social Organization Disaster Relief Service Platform. Rather than functioning as a formal government-led command center, the platform operated as a shared coordination infrastructure initiated jointly by multiple civil society organizations. Its core functions included volunteer registration, village-level service check-in, task assignment lists, and consolidated needs mapping (Organisation Z, 2014e). Participating organizations were required to log their locations, service scope, and daily outputs, enabling real-time visibility of who was operating where and doing what.

This platform significantly reduced redundancy and improved horizontal coordination. Villages that had already received assessment visits were marked accordingly, preventing repeated surveys, while unmet needs could be quickly matched with available organizational capacities. Regular updates allowed coordination teams to redirect incoming volunteers toward underserved communities and emerging priorities. In practice, this created a dynamic feedback loop between frontline operations and coordination functions, allowing distributed actors to align their activities without centralized command.

Importantly, the platform represented a form of quasi-institutional coordination rather than formal bureaucratization. Decision-making remained decentralized, but operational transparency increased through shared information standards and reporting routines. This shift gradually transformed the initial "flat" spontaneous network into a more structured, point-to-point distributed system, where coordination was achieved through digital visibility and mutual adjustment rather than top-down authority (Zhang & Yin, 2016).

12.2.2 Emergence of Role Specialization Among NGOs

As collaborative mechanisms matured in the second and third weeks following the Ludian Earthquake, the relief landscape began to exhibit a distinct pattern of role specialization among civil society actors. What began as a broad-based, spontaneous mobilization evolved into a more functionally differentiated system, with NGOs gradually aligning their actions according to domain expertise, resource capacity, and prior disaster experience.

This trend was especially visible in the emergence of thematic service clusters, where organizations focused on specific dimensions of disaster response. For instance, Save the Children led the establishment of Child-Friendly Spaces (CFS), providing psychological support, safe recreational environments, and basic educational activities for affected children. By mid-August, CFS operations had reached multiple village sites and served hundreds of children (Organisation Z, 2014f, 2014h).

Mental health support was led by organizations such as Mountain Pathfinder, which deployed trained counselors and psychosocial support workers to conduct trauma intervention, grief support, and caregiver accompaniment services. These teams worked both independently and in cooperation with community leaders to identify vulnerable groups in need of long-term emotional care (Organisation Z, 2014g).

In the domain of material logistics, organizations such as One Foundation and Amity Foundation took on leadership roles in supply coordination, needs assessment, and hygiene kit distribution. Their operations involved standardized assessment templates and coordinated delivery schedules to avoid service duplication, particularly in the hardest-hit zones of Longtoushan and Huodehong (Organisation Z, 2014f, 2014g, 2014h, 2014i).

Beyond standalone services, collaborative hubs such as the joint coordination site in Longtoushan Township brought together multiple organizations to co-manage daily operations. These joint stations were run on a shared-responsibility model with rotating coordinators and task-based role allocation. Some actors began packaging their interventions into modular service kits, combining shelter, hygiene, and psychological support elements for rapid deployment and replicability across affected villages (Wu et al., 2019).

This emergence of specialization was reinforced by digital filtering mechanisms. Online coordination tools and topic-tagging systems—such as categorized Weibo hashtags and volunteer WeChat groups—enabled organizations to be routed toward tasks aligned with their comparative advantages (Su et al., 2021). Such selective matching facilitated functional clarity and avoided the chaotic overlaps often seen in earlier phases.

12.2.3 Standardized Workflows and Operational Mechanisms

As civil society engagement in the Ludian Earthquake response matured, spontaneous efforts began to adopt increasingly standardized operational processes. This shift marked a transition from ad hoc volunteerism toward a proto-institutional mode of disaster governance, where workflows and coordination mechanisms became progressively codified.

One of the earliest signs of this transformation was the unification of assessment tools. To streamline needs evaluation and resource targeting, participating organizations began using common village-level assessment forms that included data fields on population size, urgent material needs, and the distribution of vulnerable groups such as children, elderly residents, and persons with disabilities (Organisation Z, 2014g). These forms were circulated through WeChat groups and updated in shared cloud drives, allowing organizations to build a coherent picture of service coverage.

In parallel, a service "check-in" mechanism was introduced. Organizations were required to log their presence and service completion status in each village, which served two main purposes: it helped prevent overlapping service delivery, and it enabled coordinators to identify unserved or underserved areas in real time. By August 14, over 70 service entries had been recorded, covering a wide range of tasks from hygiene kit distribution to psychosocial support (Organisation Z, 2014h).

Daily reporting also became a norm among active NGOs. Organizations submitted standardized activity logs, detailing service type, number of beneficiaries, challenges encountered, and adjustments made. These reports were compiled into shared summary tables, which facilitated cross-organizational planning and evidence-based decision-making for resource allocation (Organisation Z, 2014i).

The emergence of service matching lists—dynamic spreadsheets that linked resource offerings with unmet needs—further optimized deployment efficiency. These lists were maintained collaboratively by frontline teams and digital coordination hubs, allowing real-time adjustments based on incoming supply, shifting demand, and road access status (Organisation Z, 2014j).

Physical coordination spaces such as the joint operation camp in Longtoushan and the Aidongli digital platform gradually evolved into semi-formal governance centers. These venues introduced rotating coordination roles, where representatives from participating NGOs took turns managing agenda-setting, task matching, and inter-organizational conflict resolution. This horizontal, distributed leadership model allowed diverse actors to maintain ownership while aligning efforts (Zhou et al., 2024).

12.2.4 From "Delivering Goods" to "Delivering Services"

The Ludian Earthquake response not only tested the capacity of civil society but also catalyzed a transformation in its operational logic. Initially, most non-state actors focused on the delivery of tangible goods—tents, food, blankets, and hygiene supplies—fulfilling urgent material needs. However, as the emergency evolved, these actors gradually shifted toward providing sustained, service-based interventions, marking a significant redefinition of civil society's role in disaster recovery.

By the second week following the earthquake, the focus of NGO activities began to shift. As logistics systems stabilized and material supply chains reached saturation, many organizations pivoted toward providing psychosocial support, child protection, and educational activities in affected communities (Organisation Z, 2014h, 2014i). This evolution reflected not only an operational adjustment but also a conceptual one: disaster-affected populations were increasingly recognized not just as recipients of aid, but as individuals in need of emotional care, safe spaces, and structured engagement.

One illustrative example was the establishment of Child-Friendly Spaces (CFS) by Save the Children and its partners. These spaces offered psychological first aid, recreational activities, and structured supervision for children displaced or traumatized by the earthquake. By August 16, more than 400 children had participated in these activities, with rotating staff and volunteers facilitating games, storytelling, and group therapy sessions (Organisation Z, 2014i). Similar shifts occurred in the deployment of mental health support teams, such as those organized by the Shan Yixing network, which provided trauma counseling, family accompaniment, and grief support to affected households (Organisation Z, 2014j).

This evolution in services demanded a higher degree of professional knowledge, standard operating procedures, and localized integration. Organizations began to train frontline volunteers in psychological first aid, conflict sensitivity, and participatory assessment techniques. Service modules were modularized into "kits"—such as hygiene education kits, emergency schooling packages, or psychological support protocols—that could be rapidly adapted and deployed depending on local needs (Wu et al., 2019).

Moreover, these service-oriented interventions increasingly required integration with local community networks, especially village leaders, schools, and township clinics. Rather than acting as external providers, many organizations adopted a co-implementation model, ensuring cultural appropriateness, trust-building, and long-term acceptance.

12.3 Recognition, Legitimacy, and Institutional Embedding

The Ludian Earthquake response marked a critical juncture in the evolving relationship between civil society and the state in China's disaster governance. Initially peripheral and informal, non-state actors gradually gained legitimacy, procedural access, and strategic relevance. This section traces that transformation across four dimensions: the shift from marginal responders to recognized stakeholders; the emergence of joint governance spaces such as coordination hubs and civil camps; the development of registration and reporting mechanisms that enabled semi-institutional coordination; and, ultimately, the rise of intermediary platforms that facilitated hybrid governance through negotiated state–society collaboration.

12.3.1 From Peripheral Actors to Recognized Stakeholders

In the initial days following the Ludian Earthquake, most civil society organizations operated on the margins of the official emergency response system. Their entry into affected areas was largely self-initiated, relying on informal networks and local knowledge rather than formal invitations or coordination mechanisms. Despite their speed and flexibility, these actors were not formally embedded within the local government's emergency response apparatus and lacked institutional access to critical zones, information, and coordination structures (Organisation Z, 2014i).

However, as the scale of the disaster unfolded and the breadth of civil society engagement became evident, public authorities began to reconsider their stance. The rapid increase in the number of NGOs and volunteer teams—combined with their capacity to mobilize resources, conduct assessments, and deliver services—prompted a gradual shift in how they were perceived by state actors. Coordination challenges, such as overlapping services and unmet needs in peripheral villages, made it increasingly clear that excluding CSOs from the formal governance loop was counterproductive.

By the second week of response, government departments at the county and township levels began taking steps to institutionalize the presence of non-state actors. Local authorities supported the issuance of official volunteer passes, allowing certified personnel to enter restricted zones, thereby formalizing their operational legitimacy (Organisation Z, 2014i). This not only reduced barriers to access but also signaled an important shift from informal tolerance to procedural recognition.

At the same time, a growing number of NGOs were invited to participate in joint coordination meetings and task allocation briefings, often facilitated by platform coordinators. These intermediaries—many of whom were embedded within NGO networks—played a bridging role between local authorities and grassroots responders. Rather than assuming command, they acted as translators across governance logics, allowing civil society actors to align their efforts with evolving government strategies while retaining a degree of operational autonomy (Organisation Z, 2014j).

12.3.2 Formation of Joint Governance Spaces and Civil Camps

As the Ludian Earthquake response progressed beyond its initial emergency phase, a new spatial and organizational infrastructure began to emerge that supported the integration of civil society actors into quasi-institutional governance roles. These structures—most notably field coordination stations and joint civil camps—served as platforms for daily collaboration between NGOs, community leaders, volunteers, and local government representatives, reshaping the geography and dynamics of disaster response.

Among the most prominent of these spaces was the Aidongli Frontline Service Station, which operated as a convergence point for diverse actors in the affected areas. Situated close to priority service zones, Aidongli functioned not only as a logistical base but as a coordination hub for real-time information exchange, task distribution, and stakeholder negotiation. By August 15, the station had registered over 100 active volunteers, coordinated 21 organizational workgroups, and hosted daily briefing sessions that facilitated both vertical (NGO–government) and horizontal (NGO–NGO) interaction (Organisation Z, 2014j).

Complementing these field stations were Joint Civil Camps—semi-permanent cohabitation sites jointly managed by multiple organizations. Unlike temporary shelters or individual NGO camps, these were designed for operational integration. Located in central nodes such as Longtoushan Township, the camps supported a rotating duty system wherein participating organizations alternated responsibilities for coordination, logistics, and inter-agency communication. Each rotation typically lasted 48–72 h and followed a shared agenda that included service updates, needs reassessment, and upcoming task planning (Organisation Z, 2014k).

These camps not only facilitated day-to-day operations but also gradually assumed governance functions, mediating disputes over service boundaries, managing task duplication, and tracking cross-sectoral resource flows. In doing so, they reduced reliance on top-down directives while enhancing mutual accountability. By late August, they had become venues for negotiating resource priorities and community access protocols with growing formality (Organisation Z, 2014l).

The spatialization of coordination through such embedded platforms marked a turning point in the disaster governance landscape. Rather than operating from the periphery, civil society actors occupied central, semi-institutional nodes that enabled sustained, negotiated participation. These developments reflect what scholars have described as a "platform governance shift"—a move from fragmented interventions to spatially anchored, networked coordination infrastructures (Zhang & Yin, 2016). In this context, governance was no longer confined to institutional offices but distributed across hybrid spaces where state and non-state actors co-produced response strategies.

12.3.3 Semi-Institutionalization Through Registration and Reporting

As the Ludian Earthquake response progressed, the integration of civil society actors into disaster governance extended beyond spatial collaboration to encompass emerging procedural norms. This process of semi-institutionalization was characterized by the establishment of standardized registration, reporting, and data feedback mechanisms that aligned non-state operations with broader governance logics, while maintaining their operational autonomy. Rather than becoming fully bureaucratized, NGOs and volunteer groups engaged in coordinated processes that marked a transition toward embedded, yet flexible, participation.

One of the earliest manifestations of this shift was the introduction of a centralized registration system. By the end of the second week, the Aidongli coordination hub and its affiliated platforms had implemented an identity and task registry for all participating organizations. Each NGO was asked to log its name, service domain, target populations, service duration, and geographic coverage (Organisation Z, 2014k). This system enabled both real-time mapping of resource distribution and clearer communication across actors, helping to prevent redundancy. By August 18, the number of registered NGOs had increased from a few dozen to 186—a scale that demanded procedural coordination without hierarchical control (Organisation Z, 2014l).

Complementing registration was the development of a daily reporting routine, which required organizations to submit standardized activity logs summarizing the nature, location, and output of their work. These logs were compiled into aggregated datasets that supported cross-organizational visibility and informed strategic planning by coordination teams. The reports covered key indicators such as number of beneficiaries, type of services delivered, and specific challenges encountered, enabling timely response adjustments and inter-organizational task sharing (Wu et al., 2019).

Further consolidation occurred through the introduction of village coverage mapping and feedback mechanisms. Platforms like Aidongli developed dynamic "service maps" that visualized which communities had received what types of assistance, and which remained underserved. This spatialized accountability tool became central to coordinating subsequent interventions and rebalancing human and material resources. Moreover, informal feedback loops—via daily meetings, volunteer check-in forms, and community liaison notes—contributed to early forms of performance evaluation, guiding later-stage coordination decisions and resource reallocation (Organisation Z, 2014l).

While these processes lacked the full legal authority of state regulation, they served critical governance functions. They offered the government a form of indirect oversight while preserving the agility of non-state actors. In network visualizations of Ludian's post-disaster coordination, these platforms emerged as information-processing nodes, linking decentralized action to systemic responsiveness through soft forms of procedural standardization (Zhou et al., 2024). In this way, semi-institutionalization enabled NGOs to occupy a legitimate and functional position within the evolving disaster governance landscape.

12.3.4 Institutional Hybridization and the Role of Intermediaries

The Ludian Earthquake response illustrates a distinctive path toward institutional hybridization, where disaster governance was neither dominated solely by the state nor entirely driven by autonomous civil society actors. Instead, a new collaborative configuration emerged—anchored by intermediary platforms—that bridged policy frameworks with frontline implementation. These intermediaries played a critical role in translating institutional logic into field-level operations, thus facilitating a triangular model of coordination between government agencies, social organizations, and platform-based governance nodes.

A central feature of this hybrid model was the evolving function of platforms such as the Aidongli Service Station and the Yunnan Social Organization Disaster Relief Service Platform. Initially established as coordination hubs for NGO logistics and information flow, these entities gradually expanded their roles to mediate between state expectations and grassroots practices. Aidongli, for example, transitioned from a task distribution node into a multi-functional governance space, assuming responsibilities such as registration, activity tracking, and service evaluation (Organisation Z, 2014l).

Intermediaries operated as "translators" across institutional boundaries. On one side, they interfaced with township-level officials and local emergency bureaus, relaying policy priorities, safety protocols, and security restrictions. On the other, they supported NGOs in aligning their operational plans with emergent coordination norms, such as unified reporting and community-based targeting. This process was not simply managerial; it redefined the relational structure between state and society in disaster contexts, allowing for non-state actors to exercise influence without formal authority (Zhang & Yin, 2016).

The hybridized model also manifested through the division of governance labor. While the state maintained control over critical functions like emergency declarations and security zoning, platforms took charge of service matching, real-time information exchange, and multi-actor coordination. These arrangements avoided the rigidity of hierarchical control while enhancing collective efficacy. Notably, organizations like One Foundation, as well as platform-based coordinators, took on agenda-setting and convening roles for cross-sectoral meetings, often with local officials participating as observers or partners (Organisation Z, 2014l).

Rather than dissolving traditional boundaries, hybrid governance in Ludian reconstituted them through negotiated practice. This allowed NGOs to retain their identity as independent actors while becoming embedded within a broader, loosely institutionalized system of disaster response. The Ludian case thus provides evidence of a blended governance architecture, where intermediary platforms not only facilitated horizontal coordination but also absorbed partial vertical responsibilities, signaling a shift in China's disaster governance from separation to structured entanglement.

12.4 Limitations, Learning, and Long-Term Transformation

The Ludian Earthquake response revealed both the potential and limitations of civil society participation in disaster governance. While NGOs demonstrated rapid mobilization and growing capacity, they also encountered coordination bottlenecks, uneven service delivery, and institutional ambiguity. These challenges, however, became entry points for organizational learning and adaptive practice, leading to procedural innovation and informal standardization. Over time, key lessons from Ludian were institutionalized into local contingency plans, policy frameworks, and international partnership models. However, persistent structural issues—such as resource inequality, weak post-disaster engagement mechanisms, and legal uncertainty—underscore the need for more durable and inclusive governance architectures.

12.4.1 Operational Bottlenecks and Coordination Gaps

Despite the rapid mobilization of civil society actors in the immediate aftermath of the Ludian Earthquake, the response effort was hindered by persistent operational bottlenecks and fragmented coordination. These limitations revealed underlying structural challenges that shaped—and sometimes constrained—the effectiveness of non-state participation.

One of the most pressing issues was the disorganized distribution of relief supplies. In the absence of unified scheduling or centralized dispatching, relief goods quickly accumulated in urban hubs such as Ludian County, while peripheral areas remained underserved. This imbalance led to repeated cases of material waste, duplicated efforts, and frustration among communities facing delayed or insufficient assistance (Organisation Z, 2014f, 2014g). Volunteer groups, often arriving independently, tended to carry similar supplies—such as bottled water and instant noodles—without access to updated inventories or real-time information on village-level needs.

Geographic and infrastructural barriers further exacerbated these challenges. According to post-disaster assessments, many severely affected villages experienced road blockages and communication outages for multiple days, isolating them from both state-led and civil society responses (Chen et al., 2020). These disruptions hindered coordination among NGO teams, delayed needs assessments, and complicated efforts to match resources to specific community demands. Without access to accurate on-the-ground data, many organizations defaulted to "blind deployment" strategies, which diluted impact and created service redundancies.

The wide variation in organizational capacity highlighted the lack of preparedness among many participating groups. While some NGOs operated with professional specialization, others entered the field with little disaster experience and no clear protocols for safety, logistics, or beneficiary outreach. Service delivery remained uneven, especially in technical areas such as child protection, psychological first aid, and sanitation (An et al., 2015; Organisation Z, 2014h). These inconsistencies made coordination more difficult, particularly when underqualified actors struggled to align with the emerging collaborative framework.

Lastly, structural tensions between local authorities and civil society actors limited cooperation in several areas. While higher-level officials began to recognize the value of NGO involvement, township- and village-level administrators remained cautious. Concerns over liability, reputational risk, and role ambiguity contributed to selective exclusion and sporadic interference (Wong & Li, 2016). In the absence of pre-existing protocols or legal frameworks for collaboration, trust deficits persisted, and some grassroots officials prioritized control over engagement.

12.4.2 Organizational Learning and Adaptive Practices

The operational challenges encountered during the early stages of the Ludian Earthquake response, rather than resulting in systemic failure, became catalysts for rapid organizational learning and procedural adaptation. As civil society organizations confronted coordination inefficiencies, fragmented information, and capacity mismatches, they began to develop context-sensitive tools and workflows to enhance service quality and inter-organizational synergy.

One key area of adaptation involved the standardization of service processes. Tools such as the "village entry assessment sheet" and the "service matching checklist" were developed and disseminated to improve task allocation and enhance accountability across multiple organizations (Organisation Z, 2014i). Resource distribution also shifted from an ad hoc and duplicative mode to a more structured process that incorporated steps like request, approval, dispatch, and receipt confirmation, thereby reducing waste and operational delays (Organisation Z, 2014j).

Cross-organizational information sharing also became more robust. Ad hoc digital tools—including WeChat groups, QQ forums, and shared online spreadsheets—facilitated real-time updates on community needs, resource gaps, and service delivery schedules. These informal mechanisms coalesced into a decentralized but functional information infrastructure that enabled timely adjustments across teams (Su et al., 2021; Wu et al., 2019). The "service mapping" initiative, which visually represented the geographic coverage of different organizations, served both as a coordination guide and as a transparency mechanism for stakeholders (Organisation Z, 2014k).

In parallel, CSOs began experimenting with volunteer training and credentialing as a response to concerns about uneven service capacity. Several platforms piloted basic training modules covering psychological first aid, child protection, and health kit distribution. These initiatives, though limited in scope, laid the groundwork for more structured certification systems in subsequent disasters (Organisation Z, 2014l). Volunteer managers also initiated informal competency tracking, enabling more efficient task assignments based on prior performance and skill sets.

12.4.3 Institutionalization of Lessons in Future Responses

The Ludian Earthquake response not only catalyzed immediate innovations in civil society participation but also laid the foundation for long-term institutional change in China's disaster governance. The mechanisms developed during the Ludian operation—particularly platform-based coordination, unified registration, and civil camp models—were gradually absorbed into local and national emergency planning frameworks, transforming ad hoc experimentation into structured institutional memory.

Following the Ludian experience, multiple local governments revised their emergency response plans to include mechanisms inspired by that operation. These adaptations included the incorporation of NGO registration platforms, such as those modeled after Aidongli, and the formalization of joint civil camps as standard logistical and coordination nodes during major emergencies (Xiao et al., 2016; Zhou et al., 2024). Several provinces initiated cross-sectoral working groups tasked with maintaining updated databases of active NGOs, service domains, and contact points—enabling rapid mobilization and clearer role distribution in future disasters.

Institutional shifts also occurred within the policy domain. Government agencies, particularly those under the Ministry of Civil Affairs and the newly established Ministry of Emergency Management, began to explore formal frameworks for integrating civil society organizations into state-led responses. Policy discussions increasingly referenced concepts such as "community participation" and "coordinated response," suggesting that field-level experimentation in Ludian had filtered upward into regulatory discourse (Xiao et al., 2016). These policy shifts reflected a growing receptiveness to distributed governance models, where non-state actors are not simply tolerated but actively positioned as operational partners.

The Ludian response further influenced the localization strategies of international NGOs operating in China. Organizations such as Save the Children and World Vision adjusted their operational models by strengthening local partnerships and investing in long-term capacity building among domestic NGOs (Hong et al., 2016). This shift toward partnership-based engagement and localized implementation was partially driven by the logistical and legitimacy advantages that local actors demonstrated during the Ludian response, particularly in navigating administrative requirements and cultural contexts.

12.4.4 Lasting Impacts and Ongoing Challenges

The Ludian Earthquake response marked a pivotal shift in China's disaster governance, with enduring impacts on how civil society participation is recognized and operationalized. More than a one-off innovation, Ludian served as a catalyst for redefining the governance architecture of emergency response, particularly in legitimizing the role of NGOs and institutionalizing multi-actor coordination frameworks.

One of the most important legacies was the normalization of civil society organizations as indispensable actors in disaster response. The experience demonstrated that NGOs were not merely auxiliary forces but critical contributors to frontline service delivery, assessment, and logistics management. This recognition contributed to broader shifts in disaster governance discourse, where civil society actors came to be viewed as "necessary collaborators" rather than informal responders (Zhou et al., 2024). At the operational level, the success of coordination platforms and joint civil camps (Organisation Z, 2014k, 2014l) paved the way for further experimentation with standardized workflows, shared data systems, and task-oriented collaboration models.

The Ludian case also opened space for platformization, proceduralization, and localization of emergency response mechanisms. Concepts such as volunteer registration systems, service mapping, and rotating coordination duties were refined through practice and later adapted into formal contingency planning by provincial and municipal authorities (Xiao et al., 2016). Moreover, the integration of domestic and international NGOs through localized partnership models signaled an emergent paradigm of "networked disaster governance" grounded in mutual accountability and capacity sharing (Hong et al., 2016).

However, these achievements coexisted with persistent and emerging challenges. Resource inequality among NGOs remained a fundamental constraint. Smaller, less networked organizations continued to face barriers in accessing coordination platforms or securing stable funding, limiting their long-term viability in complex emergencies (Organisation Z, 2014h). Furthermore, the institutional pathways for civil society participation beyond the initial response phase remained underdeveloped. Few formal mechanisms existed to sustain NGO involvement in long-term recovery, reconstruction, or community resilience-building efforts (Wong & Li, 2016).

Legal and policy ambiguities also persisted. Although coordination improved at the operational level, the regulatory frameworks governing the roles, rights, and responsibilities of social actors in disaster governance remained vague. This ambiguity created uncertainty for both NGOs and local governments, sometimes leading to ad hoc, personality-driven collaboration rather than systemic inclusion (Chen et al., 2020; Zhang & Yin, 2016).

12.5 Conclusion

The Ludian Earthquake catalyzed a foundational transformation in China's disaster governance, revealing both the latent capacity and systemic constraints of civil society participation. From spontaneous mobilization to procedural innovation and semi-institutional integration, the response demonstrated how grassroots agility and digital coordination can complement state-led relief. Through emergent platforms, role specialization, and evolving collaboration models, NGOs shifted from peripheral actors to recognized governance partners. Persistent challenges—resource disparities, weak post-disaster engagement frameworks, and legal ambiguities—continue to limit institutional consolidation. Ludian's legacy lies not in a fully realized governance model, but in the opening of a participatory space where experimentation, negotiation, and adaptation reshaped the interface between state and society. It marks both a turning point and a testing ground for China's evolving hybrid disaster governance.

References

An, J., Bai, X., Xu, J., Nie, G., & Wang, X. (2015). Prediction of highway blockage caused by earthquake-induced landslides for improving earthquake emergency response. *Natural Hazards, 79*(1), 511–536. https://doi.org/10.1007/s11069-015-1859-7

Chen, W., Zhang, H., Comfort, L. K., & Tao, Z. (2020). Exploring complex adaptive networks in the aftermath of the 2008 Wenchuan earthquake in China. *Safety Science, 125*, Article 104607. https://doi.org/10.1016/j.ssci.2020.104607

Hong, H., You, J., & Bi, X. (2016). The Ludian earthquake of 3 August 2014. *Geomatics, Natural Hazards and Risk, 7*(2), 450–457. https://doi.org/10.1080/19475705.2015.1060639

Organisation Z. (2014a). *Situation report No. 1: Ludian earthquake relief in Yunnan (August 3, 2014).*

Organisation Z. (2014b). *Situation report No. 2: Ludian earthquake relief in Yunnan (August 3, 2014).*

Organisation Z. (2014c). *Situation report No. 3: Ludian earthquake relief in Yunnan (August 3, 2014).*

Organisation Z. (2014d). *Situation report No. 4: Ludian earthquake relief in Yunnan (August 3, 2014).*

Organisation Z. (2014e). *Situation report No. 5: Ludian earthquake relief in Yunnan (August 3, 2014).*

Organisation Z. (2014f). *Situation report No. 6: Ludian earthquake relief in Yunnan (August 3, 2014).*

Organisation Z. (2014g). *Situation report No. 7: Ludian earthquake relief in Yunnan (August 3, 2014).*

Organisation Z. (2014h). *Situation report No. 8: Ludian earthquake relief in Yunnan (August 3, 2014).*

Organisation Z. (2014i). *Situation report No. 9: Ludian earthquake relief in Yunnan (August 3, 2014).*

Organisation Z. (2014j). *Situation report No. 10: Ludian earthquake relief in Yunnan (August 3, 2014).*

Organisation Z. (2014k). *Situation report No. 11: Ludian earthquake relief in Yunnan (August 3, 2014).*

Organisation Z. (2014l). *Situation report No. 12: Ludian earthquake relief in Yunnan (August 3, 2014).*

Su, X., Ma, S., Qiu, X., Shi, J., Zhang, X., & Chen, F. (2021). Microblog topic-words detection model for earthquake emergency responses based on information classification hierarchy. *International Journal of Environmental Research and Public Health, 18*(15), 8000. https://doi.org/10.3390/ijerph18158000

Wong, H. T., & Li, S. (2016). Healthcare services demand in post-disaster settings: The 2014 Earthquake in Ludian County, Yunnan Province, China. *International Journal of Disaster Risk Science, 7*(4), 445–449. https://doi.org/10.1007/s13753-016-0107-4

Wu, Y., Wen, J., & Wei, H. (2019). 'Community-centred' as an integrated model for post-disaster social work—The case of earthquake-stricken Ludian, China. *Asia Pacific Journal of Social Work and Development, 29*(2), 163–177. https://doi.org/10.1080/02185385.2019.1567379

Xiao, M.-L., Chen, Y., Yan, M.-J., Ye, L.-Y., & Liu, B.-Y. (2016). Simulation of household evacuation in the 2014 Ludian earthquake. *Bulletin of Earthquake Engineering, 14*(6), 1757–1769. https://doi.org/10.1007/s10518-016-9887-6

Zhang, H., & Yin, M. (2016). Emergent organizational networks in emergency response: A case study of the Ludian earthquake. *Journal of Public Management*, *13*(2). https://kns.cnki.net/kcms2/article/abstract?v=YQFK0XGjx40k4T5rsaWPzAZfVsEb71-KcejxNXzdJdih9UtT8ptHrWxIwtXc0QTKM0dSnqSZ6A0jie66eAaNelQpFVmwQBoDPuJV8Z6LHj5XXJynVr9A8BBamToBj3BaTUOD0BInAuODWPo_wt4hYKdwehF_X1A3hVwgYnAADuA=&uniplatform=NZKPT&language=CHS

Zhou, H., Che, A., Shuai, X., & Cao, Y. (2024). Seismic vulnerability assessment model of civil structure using machine learning algorithms: A case study of the 2014 Ms6.5 Ludian earthquake. *Natural Hazards*, *120*(7), 6481–6508. https://doi.org/10.1007/s11069-024-06465-9

Chapter 13
Jiuzhaigou Earthquake: Institutional Learning and Coordinated Disaster Response

Abstract The 2017 Jiuzhaigou earthquake offered a critical window into the evolving landscape of disaster governance in China, revealing both institutional progress and enduring systemic limits. This chapter explores how the event functioned as a complex governance stress test, exposing vulnerabilities in infrastructure, information systems, and coordination capacity under conditions of high terrain complexity and demographic diversity. It then traces how institutional learning—rooted in post-Wenchuan reforms—shaped legal frameworks, technological integration, and multi-level coordination. Through detailed analysis of intergovernmental collaboration, civil society engagement, and platform-based mechanisms, the chapter highlights a shift from hierarchical command toward polycentric governance. Finally, it evaluates the resilience gains achieved, the structural gaps that remain, and the theoretical implications of an adaptive yet bounded governance model still under construction.

Keywords Polycentric coordination · Institutional learning · Digital governance · Systemic constraints · Adaptive resilience

13.1 Disaster Context and Governance Stress Test

The Jiuzhaigou earthquake of August 2017 functioned as a multidimensional governance stress test, revealing how compound hazards, complex terrain, and socio-demographic diversity can overwhelm existing disaster response systems. This section unpacks four interlinked dimensions of that challenge: the event's compound nature and cascading risks; the vulnerabilities embedded in mobile tourist flows and ethnic minority communities; the failure of critical infrastructure and resulting information asymmetry; and the overextension of institutional coordination under compressed timelines. These elements highlight how Jiuzhaigou exposed not just operational gaps, but systemic governance limits under conditions of emergent complexity and high public visibility.

Q. Du and R. Shaw, *Social Organizations in Disaster Response and Recovery*, Disaster Risk Reduction, https://doi.org/10.1007/978-981-92-2846-1_13

13.1.1 A Compound and High-Complexity Disaster Scenario

The Jiuzhaigou earthquake on August 8, 2017, presented a rare and complex compound disaster, combining seismic rupture, secondary geohazards, infrastructural vulnerability, and high exposure due to tourism concentration. Unlike disasters confined to rural or single-use landscapes, Jiuzhaigou involved a densely visited World Natural Heritage Site, fragile mountain ecology, and high-altitude infrastructure, making it a multidimensional governance challenge.

The earthquake struck at 21:19 local time with a magnitude of 7.0 and a focal depth of approximately 20 km. Its shallow epicenter triggered widespread landslides and slope collapses across the Jiuzhaigou valley, with more than 1800 landslide points identified within a 10 km radius (Zhao et al., 2023). The Environmental Seismic Intensity (ESI) map indicated severe surface ruptures and terrain deformation in areas of steep gradient and soft lithology, exacerbating the spatial distribution of damage (Ni et al., 2019). These secondary hazards not only blocked transportation routes but also impeded emergency access and search-and-rescue operations in the immediate aftermath.

Further compounding the complexity, the earthquake occurred during peak tourist season. Over 38,799 tourists were in the area at the time, many of whom were scattered across scenic zones and had no access to emergency instructions due to power outages and mobile network failure (Organisation Z, 2017). Road blockages and the collapse of provincial highway S301 and multiple mountain passes rendered ground evacuation extremely difficult. Moreover, the sudden nighttime onset of the quake created information asymmetry, disorientation, and panic, especially among tourists unfamiliar with the mountainous terrain.

The location's designation as a World Natural Heritage site added additional policy complexity. Emergency response had to balance life-saving operations with ecological protection mandates. Some landslide areas were designated eco-sensitive zones, making debris clearance and access politically sensitive (Organisation Z, 2017). Moreover, fragile high-altitude infrastructure—including wooden plank roads, heritage hotels, and cableways—sustained disproportionate damage, highlighting the tension between scenic development and disaster resilience.

The Jiuzhaigou event thus went beyond a singular seismic shock; it unfolded as a cascading disaster across ecological, infrastructural, and institutional systems. The interaction of geological risks, tourism intensity, and terrain fragility produced a scenario where governance complexity was both vertical (across scales) and horizontal (across sectors), testing the limits of standard emergency response protocols.

13.1.2 Vulnerable Populations and Governance Blind Spots

The Jiuzhaigou earthquake response was not only shaped by environmental conditions but also deeply affected by the region's demographic and socio-cultural composition. Two population groups—tourists and local ethnic minority residents—posed distinct challenges for emergency coordination due to their mobility, marginality, and informational asymmetry.

The transient nature of the tourist population introduced a severe coordination blind spot. According to official updates, more than 38,799 tourists were present in the region at the time of the earthquake, most of whom were not registered in any local resident databases (Organisation Z, 2017). Without access to localized risk information or evacuation protocols, many visitors remained in high-risk areas or experienced delays in accessing shelter. This mobility-induced vulnerability not only complicated early casualty assessments but also increased the psychological stress experienced during and after the event.

In parallel, the resident population—primarily Tibetan ethnic communities—faced structural vulnerabilities. Many villages were located along steep valleys with limited road access, and traditional homes constructed with wood and stone suffered extensive damage due to low seismic resistance (Jiuzhaigou County Social Organizations and Volunteer Service Coordination Center Workstation, 2017). Some communities were isolated for more than 24 h due to road blockages, delaying aid delivery and medical assistance.

Socially vulnerable subgroups—including children, the elderly, persons with disabilities, and women lacking caregiving support—faced pronounced challenges. Several rural schools were damaged, and elderly evacuees required sustained medical and psychological care during relocation (Jiuzhaigou County Social Organizations and Volunteer Service Coordination Center Workstation, 2017). While age-specific services were attempted, early relief efforts continued to lack key mechanisms such as accessible transportation and gender-sensitive shelter arrangements.

Moreover, post-disaster assessments revealed an increase in psychological trauma among both tourists and local residents. Surveys conducted in the aftermath recorded elevated rates of PTSD and acute stress reactions, especially among children and caregivers (Xi et al., 2020). However, formal psychological support systems were slow to be activated, and coordination between mental health professionals and local responders remained fragmented.

13.1.3 Infrastructure Failure and Information Asymmetry

The Jiuzhaigou earthquake revealed how disaster-induced infrastructural collapse can create a multi-layered disruption of governance functions. As a geographically isolated and tourism-intensive area, Jiuzhaigou was critically dependent on a small number of physical and informational lifelines. Once these failed, the emergency response system entered a state of disarticulation, where decision-making was hampered not only by physical constraints but also by the absence of real-time data and operational clarity.

One of the most immediate governance challenges following the earthquake was the breakdown of telecommunications. Cell towers in the epicentral area sustained physical damage, while backup systems were either absent or non-operational. This left local emergency coordination centers in a state of "informational blackout" during the first 6–12 h, with limited access to external communication or situational updates (Organisation Z, 2017). Without functional radio, mobile, or internet communication, the upstream flow of damage reports and downstream dissemination of orders was largely suspended.

This collapse in connectivity was compounded by physical inaccessibility. Landslides triggered by the quake blocked arterial routes such as the S301 provincial highway, cutting off access to Zhangzha Township and multiple affected villages. Control checkpoints established as part of crowd-management protocols further delayed the entry of emergency personnel and supplies. As a result, multiple convoys carrying tents, medicine, and water were held up en route because of unstable road conditions and the absence of clear, verified clearance pathways, significantly slowing the flow of relief resources into the most affected areas (Jiuzhaigou County Social Organizations and Volunteer Service Coordination Center Workstation, 2017).

At the same time, power outages disrupted basic services in hospitals, temporary shelters, and command posts. The lack of electricity disabled mobile charging, emergency lighting, and water distribution systems in community centers. This not only impaired first-response logistics but also degraded the ability of civil society groups to coordinate shelter provision and deliver supplies (Wang et al., 2018).

The consequences of this multi-system failure were not merely operational, but informational. In the absence of consolidated situation reports, coordination bodies had to make allocation decisions based on fragmented or outdated data. Misestimations of casualty numbers, displacement scale, and road accessibility led to mismatches between supply and need in the early hours. Relief items such as baby formula and medical supplies were unevenly distributed, with some remote communities left without aid for over 24 h (Jiuzhaigou County Social Organizations and Volunteer Service Coordination Center Workstation, 2017).

This convergence of infrastructural failure and information asymmetry reflects a deeper institutional vulnerability: the over-reliance on centralized, infrastructure-dependent coordination models in high-risk geographies. The Jiuzhaigou case demonstrates the need for redundant communication pathways, modular logistics design, and decentralized data platforms capable of functioning under duress. Without such provisions, even highly trained emergency networks may find themselves incapacitated at the moment they are most needed.

13.1.4 Jiuzhaigou as a Systemic Governance Stress Test

The Jiuzhaigou earthquake presented a complex emergency that overwhelmed conventional governance architectures, exposing critical limits in coordination capacity, institutional adaptability, and system integration under high-pressure disaster conditions. The event did not simply test logistical capabilities—it strained the very assumptions underlying disaster governance, including vertical command structures, multi-task handling, and the interface between state and society.

The simultaneity of demands was one defining feature of this stress scenario. Emergency operations were required to manage large-scale evacuation, real-time damage assessment, medical response, traffic control, and heritage protection, all within a compressed time window. For instance, more than 38,000 tourists needed to be safely evacuated from the scenic area within 24 h, while ongoing aftershocks and landslides impeded rescue access and disoriented decision-making (Organisation Z, 2017). Meanwhile, the preservation of Jiuzhaigou's World Heritage landscape further complicated debris clearance and route reconstruction, creating tensions between environmental and operational priorities.

Institutional stress was also visible in the breakdown of coordination between hierarchical levels of government. Although national-level emergency teams arrived swiftly, the initial hours of response revealed disconnects between county-level assessments and provincial-level deployments. Real-time data transmission was hampered by infrastructure outages, leaving upstream authorities with only partial situational awareness. Through intergovernmental network analysis, node centrality and communication density were significantly lower in the immediate aftermath compared to later recovery phases, indicating that vertical coordination was initially fragmented and reactive (Wu et al., 2021).

In parallel, the interface between formal governance and civil society mobilization became a second pressure point. While over 70 social organizations and volunteer teams registered with the local coordination center within 48 h (Jiuzhaigou County Social Organizations and Volunteer Service Coordination Center Workstation, 2017), the lack of a unified platform for volunteer intake, task allocation, and accountability resulted in bottlenecks. In some areas, efforts were duplicated while others were underserved, as humanitarian resources failed to align with assessed needs. Civil society actors faced persistent uncertainty regarding access procedures, division of responsibilities, and communication channels with local officials—an issue further compounded by overlapping mandates and ambiguous protocols.

These patterns suggest that Jiuzhaigou served as a prototypical stress test of a disaster governance system transitioning from hierarchical command to polycentric coordination. Rather than revealing governance failure, the earthquake exposed the limits of an evolving system operating at the edge of its institutional design. The experience underscores the need for anticipatory models that integrate redundancy, decentralization, and adaptive interface mechanisms into emergency governance architecture—especially under complex, compound, and high-density risk scenarios.

13.2 Institutional Learning and Systemic Upgrading

The Jiuzhaigou earthquake response did not emerge spontaneously but reflected a decade of institutional evolution shaped by lessons from past disasters. This section examines how China's disaster governance system has undergone systemic upgrading across three key dimensions: the legal and organizational embedding of emergency protocols, the institutionalization of digital and remote sensing technologies, and the broader governance paradigm shift from centralized control to coordinated, polycentric response. These developments illustrate a maturing disaster governance regime—one increasingly capable of integrating law, technology, and civil society into anticipatory, adaptive, and multi-scalar risk management practices.

13.2.1 Institutional Learning and Legal-Organizational Reforms

The Jiuzhaigou earthquake response in 2017 did not unfold in an institutional vacuum but was shaped by a decade of legal and organizational reforms initiated after the 2008 Wenchuan earthquake. These reforms marked a transition from ad hoc emergency handling to a more codified, multi-layered, and socially integrated governance model. Jiuzhaigou served as an important touchstone to observe how institutional learning was operationalized under real-world disaster conditions.

One of the clearest signs of systemic learning was the emergence of a more standardized incident command system (ICS) embedded within formal administrative hierarchies. By 2017, county-level governments such as Jiuzhaigou had integrated emergency response protocols into standing institutional procedures, including designated leadership groups, cross-sectoral coordination mechanisms, and linkage with higher-level ministries through the Emergency Response Law and its supporting regulations (Hu et al., 2021; Wang et al., 2018). These pre-defined institutional interfaces enabled a relatively swift mobilization of public emergency actors. For example, within hours of the quake, Sichuan Province activated a Level I emergency response, deploying professional rescue teams, while county and township governments began to coordinate sheltering and logistics under a unified command (Organisation Z, 2017).

Another core feature of the evolving governance model was the formal inclusion of civil society actors into the disaster response architecture. The Local Emergency Social Force (LESF) framework—developed after Wenchuan—had begun to take root in multiple counties by the time of the Jiuzhaigou earthquake. Over 70 social organizations and volunteer teams were officially registered with the coordination center in Jiuzhaigou County within 48 h, reflecting an institutional shift toward recognizing non-state actors as structured participants rather than ad hoc contributors (Jiuzhaigou County Social Organizations and Volunteer Service Coordination Center Workstation, 2017). This pre-registration process was supported by provincial-level

LESF policy guidelines, which categorized civil society resources based on functionality (e.g., medical, logistics, psychosocial support) and established contact points between NGOs and government taskforces (Hu et al., 2021).

Moreover, intergovernmental coordination density had improved measurably since previous disasters. Network analysis of the Jiuzhaigou response showed increased node connectivity between different levels of government and across ministries—suggesting that horizontal and vertical integration had become more institutionalized compared to the fragmented coordination seen in Wenchuan (Wu et al., 2021). Nonetheless, gaps remained, especially in integrating real-time field intelligence with command-level decisions. These gaps underscore the ongoing tension between formal planning and the emergent complexity of disaster settings.

Legal codification also played a critical background role. National laws such as the "Emergency Response Law" and provincial implementation measures provided the legal scaffolding for early warning activation, information disclosure, and command hierarchy. These instruments not only enabled faster mobilization but also clarified the legal responsibilities of local governments in coordinating with social organizations and managing information dissemination (Chen et al., 2023).

13.2.2 Embedded Technological Learning and Data-Driven Capacity

The Jiuzhaigou earthquake response highlighted the progressive integration of digital technologies into disaster governance mechanisms, demonstrating a shift from passive data reception to active, real-time data use in operational decision-making. Rather than serving as peripheral tools, technologies such as mobile signaling data, satellite imagery, and UAV-assisted inspection were embedded into institutional routines, forming a nascent infrastructure for anticipatory governance.

One key innovation was the use of mobile phone signaling data to estimate population exposure and displacement in near real-time. Based on location data from telecom providers, emergency agencies were able to reconstruct the spatial distribution of tourists and residents during the event window. This data was crucial given the lack of accurate on-site registries for over 38,000 transient visitors present at the time (Organisation Z, 2017). Spatial estimation models helped inform resource allocation and evacuation planning, particularly in remote valleys where traditional census mechanisms were inapplicable (Guo et al., 2022).

In parallel, geospatial and remote sensing technologies enhanced the accuracy and timeliness of hazard detection. Interferometric Synthetic Aperture Radar (InSAR) and optical imagery were deployed within 48 h to map surface deformation, landslide zones, and potential slope instability hotspots. This allowed emergency teams to prioritize blocked transportation routes and assess secondary risks such as delayed slope failure (Bian et al., 2022; Yin et al., 2024). Notably, the use of InSAR enabled the dynamic updating of risk maps without physical access to dangerous terrain, reducing exposure for first responders.

Technological learning also extended into infrastructure restoration. Unmanned Aerial Vehicles (UAVs) were deployed by power grid companies in coordination with local authorities to assess damage to transmission lines in mountainous terrain where conventional inspection was not feasible. These drone-based missions collected visual and thermal imagery, supporting both diagnostics and rapid decision-making in network recovery (Zheng et al., 2021). The coordinated use of UAVs marked a departure from ad hoc deployment toward systematized human-machine collaboration protocols in lifeline sectors.

Critically, these tools were not only introduced during crisis moments but increasingly pre-integrated into institutional workflows. Data dashboards developed after Wenchuan, now refined, served as platforms for aggregating sensor feeds, crowd-sourced alerts, and satellite layers, enabling local taskforces to engage in scenario planning and resource staging (Hu et al., 2021). Such integration suggests a maturing of China's "digital emergency governance" architecture, where technical tools co-evolve with institutional routines to enable faster, more informed, and more distributed response strategies.

13.2.3 Governance Paradigm Shift: From Control to Coordination

The Jiuzhaigou earthquake marked not only a logistical and institutional test but also a conceptual turning point in China's disaster governance. Rather than adhering strictly to a centralized command-and-control approach, the response revealed a gradual shift toward polycentric coordination, distributed authority, and anticipatory risk governance. This transition was not abrupt but reflected institutional learning accumulated over the preceding decade, especially following the 2008 Wenchuan earthquake.

One of the clearest indications of this shift was the mobilization of multi-actor governance arrangements that moved beyond state-centric intervention. Local Emergency Social Forces—a hybrid mechanism institutionalized after 2008—played a visible role in the Jiuzhaigou response. These actors, composed of registered social organizations, volunteer networks, and local NGOs, were partially embedded within formal emergency plans but operated with considerable autonomy in relief, logistics, and psychosocial support (Hu et al., 2021). Their coordination was supported, though not fully controlled, by local emergency command units, suggesting a structural opening of the governance space to civil society participation.

Furthermore, coordination patterns exhibited signs of decentralization and horizontal expansion. An intergovernmental network analysis showed that in the wake of the Jiuzhaigou earthquake, communication and decision-making nodes were no longer tightly clustered around a single authority node, but were dispersed across local, county, and provincial actors (Wu et al., 2021). This dispersion enabled faster adaptation at the local level, particularly when national teams experienced delays due to infrastructure disruption and informational gaps.

Equally significant was the movement from reactive emergency to anticipatory planning. The incorporation of environmental monitoring, contingency mapping, and early warning systems into routine governance structures—though still incomplete—enabled parts of the response apparatus to engage in scenario-based forecasting and spatial prioritization of rescue (Chen et al., 2023). While this risk-informed orientation was more visible in the technical arms of governance, such as meteorological or geological institutes, it began to influence the broader coordination logic. Local authorities, for instance, relied on pre-mapped hazard zones to guide NGO entry and relief material allocation, even when real-time data was scarce.

The Jiuzhaigou case also revealed evolving interface mechanisms between state and society. More than 70 social organizations registered with the local platform within 48 h (Jiuzhaigou County Social Organizations and Volunteer Service Coordination Center Workstation, 2017), yet gaps in task allocation, information sharing, and legal accountability persisted. These frictions underscored the tension between emergent collaboration and entrenched administrative norms. However, they also reflected a growing institutional acceptance of pluralistic actors as legitimate disaster responders, especially under high-pressure conditions.

13.3 Multi-actor Coordination and Platform Mechanisms

The Jiuzhaigou earthquake response exemplified an evolving model of disaster governance characterized by polycentric coordination, role specialization, and digital integration. Rather than relying solely on top-down command, the response mobilized a constellation of actors—across government tiers, civil society, and data infrastructures—through modular task allocation, formalized NGO engagement, and real-time platform governance. This section unpacks how intra-governmental linkages, institutionalized social force management, and rule-embedded digital platforms collectively enabled multi-actor coordination. It reveals not only the operational mechanisms that structured joint action, but also the institutional logic underpinning this distributed governance architecture.

13.3.1 Intra-Governmental Coordination and Decentralized Task Allocation

The Jiuzhaigou earthquake response showcased a maturing model of intra-governmental coordination, characterized by vertical integration, modular task allocation, and reduced dependence on single-node decision-making. Rather than a top-down command cascade, the governance system activated after the quake exhibited networked features—structured yet flexible linkages among provincial, prefectural, and county-level authorities.

At the core of the response was a vertically integrated three-tier coordination mechanism linking Sichuan Province, Aba Prefecture, and Jiuzhaigou County. This institutional arrangement enabled strategic decision-making, resource mobilization, and operational implementation to be coordinated across administrative levels while allocating responsibilities according to the comparative advantages of each level of government. Within hours of the earthquake, the Sichuan Provincial Earthquake Emergency Command initiated a Level I response, deploying professional rescue units, medical teams, and communication specialists to the affected area (Organisation Z, 2017). Meanwhile, county-level actors assumed immediate operational roles in establishing temporary shelters, mobilizing local transport, and coordinating volunteer services (Jiuzhaigou County Social Organizations and Volunteer Service Coordination Center Workstation, 2017). This structured stratification of responsibilities allowed different governance levels to focus on their comparative strengths—strategic resource deployment at the top, and logistical coordination at the bottom.

Such coordination was underpinned by standardized role distribution across functional domains. Emergency protocols developed after Wenchuan had institutionalized sectoral task groups—including command and planning, logistics, medical, transportation, communication, and psychological support units—each with pre-designated lead departments at county and prefectural levels (Wang et al., 2018). During the Jiuzhaigou response, these task groups operated semi-autonomously, holding regular coordination briefings and submitting situation reports upward. This division reduced the decision-making bottlenecks typical of past disasters.

Network analysis of intergovernmental interactions confirms the emergence of modularity and density in the coordination structure. The Jiuzhaigou response network had multiple active hubs beyond the central command node—especially in logistics and medical care—indicating a transition from centralized control to distributed problem-solving (Wu et al., 2021). These findings reflect a broader governance logic shift from vertical command to node-based operationalism, wherein decision authority is dispersed across functionally specialized clusters that retain the ability to coordinate laterally as needed.

A key enabling condition for this decentralized tasking was the routinization of emergency drills and contingency planning at local levels. Jiuzhaigou County had conducted multiple annual earthquake simulation exercises involving cross-departmental participation prior to 2017, which enhanced operational familiarity with roles and protocols (Cheng et al., 2019). As a result, the earthquake response was not entirely reactive—it activated existing institutional routines, even if under extreme duress.

Nevertheless, friction points remained. The early hours of the response were marked by partial communication blackouts due to infrastructure collapse, which temporarily disrupted the flow of situation reports and slowed higher-level assessments (Organisation Z, 2017). While local actors adapted quickly, inconsistencies in reporting formats and coordination interfaces limited the real-time synthesis of data. These gaps underscore the continuing need for interoperable systems and cross-level redundancy.

13.3.2 Civil Society Integration and Social Force Regulation

The Jiuzhaigou earthquake response highlighted a distinct phase in the institutionalization of civil society participation in China's disaster governance. Unlike earlier post-disaster contexts—where NGO involvement was often ad hoc and loosely coordinated—the 2017 response exhibited signs of structured integration, formal authorization, and regulatory embedding of non-state actors into the broader emergency system. This shift reflected both accumulated policy learning and a maturing apparatus for managing societal forces in high-stakes emergency settings.

A central mechanism for such integration was the pre-established Jiuzhaigou County Social Organizations and Volunteer Service Coordination Center, activated within hours after the earthquake (Jiuzhaigou County Social Organizations and Volunteer Service Coordination Center Workstation, 2017). The Center served as the primary interface for registering, verifying, and coordinating NGO and volunteer participation. Within 48 h, over 70 social organizations had been accredited through this system. These included medical relief teams, logistics providers, psychological support groups, and professional search-and-rescue units, many of whom were pre-registered in LESF databases maintained at the provincial and county levels (Hu et al., 2021).

The registration process was not merely bureaucratic. It included rapid verification of organizational capacity, safety compliance checks, and the issuance of formal permits, which were required to access core affected zones. This system ensured that non-state actors entering the disaster zone were not only legitimate but also accountable. In daily coordination meetings hosted by the Center, task allocation was refined based on real-time needs and performance feedback (Jiuzhaigou County Social Organizations and Volunteer Service Coordination Center Workstation, 2017). These meetings fostered information exchange, matched resources to unmet needs, and minimized redundancy—a notable departure from the fragmented coordination that plagued earlier responses, such as in Yushu or Lushan.

Prevention of unregulated influx was also a key concern. The Coordination Center and associated government actors instituted risk mitigation measures to prevent spontaneous volunteerism and unauthorized NGO operations. These included controlled access checkpoints, media announcements discouraging self-deployment, and public guidance on appropriate donation channels (Organisation Z, 2017). Such regulatory filters not only reduced logistical congestion but also curbed reputational risks from unprofessional actors operating outside formal supervision.

In parallel, the Jiuzhaigou response marked an advance in civil society accountability mechanisms. Insurance coverage, safety protocols, and liability frameworks were introduced into volunteer management systems. Participating organizations were required to carry basic insurance, sign liability agreements, and comply with occupational safety guidelines—reflecting an emerging regulatory culture around social force governance (Hu et al., 2021; Organisation Z, 2017). While enforcement capacity varied across sectors, these institutional safeguards signaled an important shift from permissive inclusion to rule-based participation.

Moreover, evidence suggests that NGOs were not peripheral to the relief effort. In areas where official teams had limited reach—particularly remote valleys cut off by landslides—NGOs provided critical last-mile services, including temporary medical clinics and food distribution (Xi et al., 2020). Their local knowledge, flexibility, and embeddedness enabled them to navigate complex terrains—both physical and social—more nimbly than some official actors.

13.3.3 Platform Governance as Infrastructure for Coordination

In the Jiuzhaigou earthquake response, coordination among multiple actors—government agencies, NGOs, and logistical providers—was not only facilitated by interpersonal networks or hierarchical commands but underpinned by a growing reliance on digital and institutional platforms. These platforms functioned as infrastructures of meta-governance: rule-embedded systems that enabled accountability, synchronized action, and real-time visibility across dispersed operational units.

The central coordinating node was the Jiuzhaigou County Social Organizations and Volunteer Service Coordination Center, which, beyond accrediting civil society actors, also served as a real-time information clearinghouse (Jiuzhaigou County Social Organizations and Volunteer Service Coordination Center Workstation, 2017). All participating organizations were required to submit operational data—team composition, areas of deployment, supply needs—into a structured digital registry. This registry was critical for spatial-temporal planning and minimizing redundancy in field operations. Notably, teams unable to provide such information were not cleared to enter high-risk zones, marking a shift toward information-based access control.

One of the key instruments used to integrate activities was the "4W Table" (Who does What, Where, When), a coordination format adopted from international humanitarian practice. The Coordination Center and its partners updated the table daily, aligning actor roles with geolocated tasks and unmet needs. This tool was not merely descriptive but functioned as a shared protocol for task distribution and resource mobilization (Organisation Z, 2017). For example, when multiple NGOs offered psychological services in the same township, the platform reassigned some to underserved regions to avoid overlap, while linking their activities to broader health system planning via the county's health bureau.

The function of platform governance extended to data dashboards aggregating real-time logistics, traffic status, shelter availability, and environmental risk assessments. These platforms, originally developed as part of the post-Wenchuan emergency modernization initiatives, were now more institutionalized and embedded within official workflows (Hu et al., 2021; Wang et al., 2018). Integration of satellite imagery, mobile signaling data, and relief inventory databases enabled dynamic scenario planning. In some instances, coordination hubs made preemptive decisions on rerouting supply trucks based on anticipated road blockages derived from InSAR data (Bian et al., 2022; Yin et al., 2024).

The platform was not only a technological interface but also a governance mechanism. Participation in the coordination platform was formalized through protocols that outlined reporting duties, safety guarantees, and exit procedures. Organizations were evaluated based on performance metrics derived from their submitted data, such as timeliness, coverage, and compliance with safety standards. These evaluations informed further deployment decisions and post-response assessments (Organisation Z, 2017).

13.4 Governance Outcomes, Resilience Gains, and Remaining Gaps

Building on the preceding analysis, this final section turns to assess the governance outcomes of the Jiuzhaigou earthquake response. It explores how the event revealed not only improvements in institutional readiness and coordination capacity, but also exposed persistent limitations in system integration and equity of access. The discussion unfolds in three parts: First, it highlights key performance gains—such as accelerated response, precision targeting, and platform-enabled coordination; second, it identifies enduring structural gaps, including sectoral silos, data fragmentation, and uneven local capacity; third, it reflects on the broader theoretical implications, viewing Jiuzhaigou as an evolving case of adaptive but bounded disaster governance.

13.4.1 Performance Gains: From Speed to Systemic Integration

The Jiuzhaigou earthquake response demonstrates measurable performance gains that extend beyond rapid emergency control toward a more systemically integrated governance model. These gains are visible in response speed, operational precision enabled by digital tools, transparency in multi-actor coordination, and early coupling between recovery planning and ecological governance.

First, response timeliness and organizational clarity marked a notable improvement over earlier disasters. Emergency shelters were activated within hours, and large-scale evacuation proceeded rapidly despite adverse conditions. More than 38,000 tourists were evacuated from the scenic area in under 24 h, despite facing nighttime darkness, blocked roads, and unstable communication—demonstrating the effectiveness of pre-established mechanisms (Organisation Z, 2017). Casualty management followed structured triage and referral pathways, with temporary medical stations deployed along evacuation routes. These results point not only to mobilization capacity but also to the institutionalization of standardized procedures and inter-agency coordination refined through prior crises such as the Wenchuan earthquake (Hu et al., 2021).

Second, precision in situational awareness and operational targeting improved markedly through the embedded use of data and sensing technologies. Mobile signaling data enabled rapid estimation of population exposure and spatial distribution in areas where tourist density fluctuated dynamically and administrative registries were incomplete. Population flow reconstruction supported prioritization of evacuation routes and shelter placement (Guo et al., 2022). Remote sensing and InSAR mapping provided near-real-time identification of slope deformation and potential secondary landslide risks, enabling safer routing of rescue convoys and early exclusion of unstable zones (Bian et al., 2022; Yin et al., 2024). In lifeline restoration, UAV-based inspections supported rapid diagnosis of transmission line damage in inaccessible mountainous terrain, shortening recovery cycles and reducing personnel exposure to secondary hazards (Zheng et al., 2021). Together, these tools transformed fragmented field signals into actionable intelligence integrated within operational command routines.

Third, transparency and coordination credibility improved through platform-based information disclosure and structured NGO coordination mechanisms. Daily situation briefings, online information updates, and standardized reporting templates enhanced information symmetry between government agencies, social organizations, and the public (Jiuzhaigou County Social Organizations and Volunteer Service Coordination Center Workstation, 2017). The deployment of "4W" tracking ("Who does What, Where, When") reduced duplication in relief delivery and strengthened accountability for task completion. These practices contributed to higher public trust and reduced friction between formal command systems and civil society participation, reinforcing the legitimacy of multi-actor governance arrangements (Hu et al., 2021).

Fourth, recovery planning demonstrated early integration between ecological protection and economic revitalization. As a World Natural Heritage site, Jiuzhaigou faced strict constraints on reconstruction intensity and landscape intervention. Post-disaster planning aligned slope stabilization, scenic restoration, and tourism reopening schedules with environmental carrying capacity assessments, avoiding the rapid but ecologically disruptive reconstruction patterns observed in earlier disasters (Chen et al., 2023). This alignment reflects a maturation of recovery governance, where long-term sustainability objectives are embedded alongside short-term stabilization goals.

Collectively, these performance gains indicate that governance maturity increasingly lies not only in response speed, but in the system's ability to integrate data, actors, and policy objectives into coherent operational feedback loops. Jiuzhaigou thus illustrates a transition from episodic emergency success toward cumulative institutional resilience grounded in coordination capacity, technological embedding, and cross-sectoral alignment.

13.4.2 Remaining Constraints: Structural and Systemic Gaps

Despite the evident advances in coordination and responsiveness during the Jiuzhaigou earthquake response, several structural and systemic constraints persisted. These limitations were not failures of mobilization per se but reflected deeper governance asymmetries, technical dependencies, and unresolved interface challenges among actor groups and governance levels.

One of the most enduring challenges was the lack of effective horizontal coordination across sectors. While vertical alignment between Sichuan province, Aba prefecture, and Jiuzhaigou County was comparatively smooth, interactions between parallel departments—such as emergency management, health, transportation, and public security—were often fragmented. Sectoral silos hindered integrated planning and resource sharing. For instance, although the emergency command center maintained a central information node, cross-departmental data exchange remained inconsistent, especially during the first 48 h when multiple agencies operated on non-interoperable platforms (Wu et al., 2021). This lack of lateral integration constrained situational synthesis, leading to redundant actions and gaps in service continuity.

Data governance itself remained a critical bottleneck. While Jiuzhaigou deployed several digital tools for coordination—including the 4W system and field-level dashboards—many of these platforms were not interoperable across organizations. The absence of standardized data protocols meant that actors often duplicated data entry or failed to access updated situational reports in time-sensitive contexts (Jiuzhaigou County Social Organizations and Volunteer Service Coordination Center Workstation, 2017). In rural townships with limited connectivity, real-time reporting proved even more difficult, weakening the accuracy of deployment planning and resource distribution (Wang et al., 2018). As digital platforms became more central to disaster governance, these fragmentation issues posed increasing operational risks.

Capacity imbalances at the subnational level further reinforced structural inequality in response efficiency. County and township officials, though institutionally mandated to lead localized coordination, often lacked sufficient technical training and digital literacy to fully engage with advanced coordination platforms. In particular, frontline staff in minority villages reported difficulties in understanding or applying data protocols introduced by provincial teams, which resulted in delays in situation reporting and unmet resource requests (Cheng et al., 2019; Wu et al., 2021). These asymmetries produced uneven governance outcomes, especially in less-connected or linguistically marginalized areas.

Psychosocial recovery—especially for vulnerable populations—was another underdeveloped domain. While physical infrastructure repair and emergency shelter provision were rapidly addressed, sustained mental health support was limited in scale and scope. Ethnic minority communities in remote villages faced both cultural and logistical barriers to accessing trauma care, and no dedicated psychological outreach teams were deployed in several high-need areas (Xi et al., 2020). Children, the elderly, and people with disabilities received minimal targeted assistance in this regard, despite evidence of acute post-traumatic symptoms in affected groups.

Finally, the increasing reliance on digital platforms and high-precision tools introduced a new form of exclusion: technology-dependent governance thresholds. Sophisticated systems enhanced coordination for well-equipped actors, but raised barriers for under-resourced organizations, particularly small local NGOs lacking data infrastructure or technical personnel (Hu et al., 2021). This created a paradox wherein technological advancement improved overall system capacity while simultaneously amplifying disparities among participating entities.

13.4.3 Theoretical Implications: Adaptive but Bounded Governance

The Jiuzhaigou earthquake response illustrates a critical juncture in the evolution of disaster governance in China—one that shifts from command-based emergency response to coordination-driven resilience management. This shift is partial and contingent, bounded by persistent structural asymmetries, institutional inertia, and emergent risks from system complexity. Understanding this episode through theoretical lenses reveals how China's disaster governance is developing adaptive features, while still facing hard limits in operational integration and systemic equity.

First, the Jiuzhaigou case reflects a marked movement away from hierarchical, command-centric emergency logic toward a more polycentric, networked form of governance. Institutional routines that previously centered on directive control were gradually replaced by layered coordination across actors and sectors. The deployment of 4W platforms, modular task forces, and decentralized data hubs enabled joint decision-making between government bodies and civil society actors (Jiuzhaigou County Social Organizations and Volunteer Service Coordination Center Workstation, 2017). Coordination was not merely procedural, but embedded in pre-designed mechanisms that structured roles and responsibilities—demonstrating a maturing logic of distributed governance. However, residual command reflexes persisted: critical decisions, such as access authorization to high-risk zones, remained centralized and subject to bureaucratic filtering (Hu et al., 2021). The co-existence of coordination-oriented tools with vertical gatekeeping highlights an ongoing institutional duality.

Second, the Jiuzhaigou experience underscores that institutional maturity does not equate to systemic completeness. Governance frameworks were more robust than in previous disasters, yet remained bounded by platform fragmentation, capacity unevenness, and coordination fatigue. For instance, while remote sensing tools and mobile signaling were used for precision mapping and population estimation (Guo et al., 2022; Zheng et al., 2021), many local actors lacked the technical capabilities to fully utilize such data. Subnational disparities—particularly at township and village levels—led to inconsistent implementation and limited cross-actor interoperability (Cheng et al., 2019; Wu et al., 2021). Moreover, mental health support mechanisms were notably underdeveloped, especially for children and ethnic minorities, revealing cultural and systemic blind spots in trauma governance (Xi et al., 2020). These bounded domains suggest that institutional advancements alone are insufficient without equal emphasis on equitable access, social inclusion, and cross-sectoral capacity building.

Third, the Jiuzhaigou response reinforces the conceptualization of governance not as a fixed design, but as a dynamic capability under construction. What emerged during the response was not a fully institutionalized model, but a resilience-in-progress: an adaptive system incrementally learning through crisis. Feedback mechanisms—such as daily coordination briefings, real-time data dashboards, and operational performance evaluations—enabled continuous recalibration during response phases (Chen et al., 2023; Wang et al., 2018). These mechanisms also revealed systemic friction, where the speed of adaptation was constrained by organizational bandwidth and normative uncertainty. Governance thus appeared as an evolving assemblage—capable of generating resilience through iteration, but still vulnerable to overload and fragmentation under pressure.

In sum, Jiuzhaigou offers a critical case for theorizing disaster governance in transitional systems. It demonstrates the possibility of hybrid models—combining platform-based coordination, data-driven responsiveness, and multi-actor inclusion—while simultaneously exposing the political, institutional, and socio-technical boundaries that shape and constrain governance evolution. The case points not to a resolved architecture of resilience, but to an unfinished institutional project—where learning, failure, and innovation remain tightly interwoven.

13.5 Conclusion

The Jiuzhaigou earthquake response reveals a governance system in active transformation—marked by institutional learning, technological embedding, and expanding multi-actor coordination. The event functioned as both a stress test and a catalyst, exposing structural vulnerabilities in infrastructure, data integration, and psychosocial care, while also demonstrating resilience gains through decentralized tasking, digital tools, and civil society integration. Despite progress, governance remained bounded by capacity asymmetries and coordination friction, particularly at the subnational level. Rather than a finalized model, Jiuzhaigou reflects a resilience regime under construction—adaptive, but uneven; polycentric, yet still anchored in residual hierarchies. The case thus illustrates how disaster governance in China is evolving as a dynamic capability, shaped through iterative practice, contested authority, and a growing imperative for systemic inclusion and anticipatory planning.

References

Bian, Z., Zhang, D., Xu, J., Tang, H., Bai, Z., & Li, Y. (2022). Study on the evolution law of surface landscape pattern in earthquake-stricken areas by remote sensing: A case study of Jiuzhaigou County, Sichuan Province. *Sustainability, 14*(20), 13032. https://doi.org/10.3390/su142013032

Chen, L., Huang, H., Han, D., Wang, X., Xiao, Y., Yang, H., & Du, J. (2023). Investigation on the spatial and temporal patterns of coupling sustainable development posture and economic development in world natural heritage sites: A case study of Jiuzhaigou, China. *Ecological Indicators, 146*, 109920. https://doi.org/10.1016/j.ecolind.2023.109920

Cheng, Y., Chen, Y.-R. R., Jin, Y., & Hung-baesecke, F. (2019). When CSR meets Mobile SNA users in mainland China: An examination of gratifications sought, CSR motives, and relational outcomes in natural disasters. *International Journal of Communication, 13*, 23–23. https://ijoc.org/index.php/ijoc/article/view/10169

Guo, X., Wei, B., Nie, G., & Su, G. (2022). Application of Mobile signaling data in determining the seismic influence field: A case study of the 2017 mw 6.5 Jiuzhaigou earthquake, China. *International Journal of Environmental Research and Public Health, 19*(17), 10697. https://doi.org/10.3390/ijerph191710697

Hu, X., Naim, K., Jia, S., & Zhengwei, Z. (2021). Disaster policy and emergency management reforms in China: From Wenchuan earthquake to Jiuzhaigou earthquake. *International Journal of Disaster Risk Reduction, 52*, 101964. https://doi.org/10.1016/j.ijdrr.2020.101964

Jiuzhaigou County Social Organizations and Volunteer Service Coordination Center Workstation. (2017). *Situation report No. 5: Social organization and volunteer coordination for the Jiuzhaigou "8.8" earthquake (August 15, 2017).*

Ni, H., Ge, H., Gao, Y., Blumetti, A. M., Comerci, V., Di Manna, P., Guerrieri, L., & Vittori, E. (2019). Comparison of earthquake environmental effects and ESI intensities for recent seismic events in different tectonic settings: Sichuan (SW China) and central Apennines (Italy). *Engineering Geology, 258*, 105149. https://doi.org/10.1016/j.enggeo.2019.105149

Organisation Z. (2017). *Analytical situation report: Jiuzhaigou earthquake (August 8, 2017)*

Wang, W., Chen, H., Xu, A., & Qu, M. (2018). Analysis of the disaster characteristics and emergency response of the Jiuzhaigou earthquake. *Natural Hazards and Earth System Sciences, 18*(6), 1771–1783. https://doi.org/10.5194/nhess-18-1771-2018

Wu, M., Gao, X., Cao, M., Papa, E., & Zhu, X. (2021). The changes of intergovernmental collaboration dynamic in post-disaster destination management: Network analysis. *Journal of Hospitality and Tourism Management, 48*, 32–45. https://doi.org/10.1016/j.jhtm.2021.05.006

Xi, Y., Yu, H., Yao, Y., Peng, K., Wang, Y., & Chen, R. (2020). Post-traumatic stress disorder and the role of resilience, social support, anxiety and depression after the Jiuzhaigou earthquake: A structural equation model. *Asian Journal of Psychiatry, 49*, 101958. https://doi.org/10.1016/j.ajp.2020.101958

Yin, S., Dai, Z., & Zeng, Y. (2024). Optimized landslide susceptibility prediction based on SBAS-InSAR: Case study of the Jiuzhaigou Ms7.0 earthquake. *Geomatics, Natural Hazards and Risk, 15*(1), 2366362. https://doi.org/10.1080/19475705.2024.2366362

Zhao, B., Su, L., Xu, Q., Li, W., Xu, C., & Wang, Y. (2023). A review of recent earthquake-induced landslides on the Tibetan plateau. *Earth-Science Reviews, 244*, 104534. https://doi.org/10.1016/j.earscirev.2023.104534

Zheng, Y.-J., Du, Y.-C., Su, Z.-L., Ling, H.-F., Zhang, M.-X., & Chen, S.-Y. (2021). Evolutionary human-UAV cooperation for transmission network restoration. *IEEE Transactions on Industrial Informatics, 17*(3), 1648–1657. https://doi.org/10.1109/TII.2020.3003903

Chapter 14
Changning Earthquake: Social Networks and the New Emergency Management System

Abstract The 2019 Changning Earthquake provides a compelling case through which to observe the practical workings of China's fourth-generation emergency management system. This chapter examines how institutional reform, digital innovation, and societal participation converged to shape a coordinated disaster response. It begins by situating the Changning case within the broader transformation of China's emergency governance landscape, following the establishment of the Ministry of Emergency Management and its key interface institutions and Sichuan Coordination Center for Social Participation in Disaster Prevention, Mitigation and Relief. The analysis then turns to how these entities facilitated real-time, multi-level coordination, enabled by digital platforms and structured through joint situation reports. Attention is also given to the formalized integration of civil society actors, whose roles and responsibilities were increasingly embedded within institutional mechanisms. Finally, the chapter explores how public trust was managed and legitimacy co-produced through digital communication, participatory media, and narrative alignment. These dynamics reveal an emerging model of hybrid governance—one that combines centralized command with collaborative, multi-actor engagement.

Keywords Emergency governance · Hybrid governance · Digital infrastructure · Civil society participation · Narrative strategies

14.1 Institutional Transformation: The New Emergency Management System

The institutional transformation of China's emergency governance system following the 2018 central administrative reform marked a decisive move toward a fourth-generation model of disaster management. This section traces the architecture and operational logic of this transformation across four dimensions: the emergence of the Ministry of Emergency Management (MEM) as a unified authority; its evolving policy and coordination functions; the role of the Emergency Rescue Promotion Center (ERPC–MEM) as a strategic national interface; and the rise of the Sichuan

Q. Du and R. Shaw, *Social Organizations in Disaster Response and Recovery*, Disaster Risk Reduction, https://doi.org/10.1007/978-981-92-2846-1_14

Coordination Center for Social Participation (SCSP-DMR) as a subnational governance platform. These components illustrate a shift toward integrated command, digital coordination, and structured multi-actor engagement.

14.1.1 *The 2018 Reform and the Rise of the Fourth-Generation System*

The establishment of the Ministry of Emergency Management (MEM) in 2018 marked a foundational shift in China's approach to disaster governance. This institutional reform consolidated fragmented emergency functions—previously dispersed across the Ministry of Civil Affairs, Ministry of Public Security, China Earthquake Administration, Ministry of Water Resources, and others—into a unified structure (Zhang, 2022). As a result, China transitioned into what has been described as a "fourth-generation emergency management system," characterized by integrated command, digital coordination, and multi-actor participation.

This new governance model emphasizes centralized leadership at the national level, professionalized response capabilities, and the systematic embedding of social actors into the emergency apparatus. This fourth-generation system differentiates itself from prior models by embracing not only administrative integration but also technological and participatory innovation (Zhang, 2022). The MEM is positioned at the apex of a vertically integrated system that ensures faster resource mobilization, real-time information exchange, and policy enforcement across bureaucratic boundaries.

During the 2019 Changning Earthquake, this institutional redesign was put into practice. The MEM played a critical role in activating national emergency protocols and overseeing the coordination of cross-sectoral response units, including firefighting teams, geological experts, medical rescue teams, and communication restoration crews (ERPC–MEM et al., 2019a). Importantly, the system's capacity to integrate local and non-state actors was also tested.

Two key coordination bodies—the Emergency Rescue Promotion Center of the Ministry of Emergency Management of the People's Republic of China and the Sichuan Coordination Center for Social Participation in Disaster Prevention, Mitigation and Relief—operationalized the logic of the fourth-generation system on the ground. Acting as a strategic interface, ERPC–MEM linked central commands with local action platforms, while SCSP-DMR coordinated grassroots mobilization and civil society engagement (ERPC–MEM et al., 2019a, 2019b, 2019c, 2019d, 2019e).

The SCSP-DMR's activation in Changning was notable as it marked the first time that a provincial coordination platform systematically managed frontline social actors in an earthquake response. Through this institutional arrangement, the fourth-generation system demonstrated its emphasis on "government–society fusion" (政社融合), reflecting a deeper shift from hierarchical control to collaborative governance frameworks.

14.1.2 The Role and Function of the Ministry of Emergency Management

The creation of the Ministry of Emergency Management (MEM) in 2018 redefined the architecture of disaster governance in China. Previously fragmented functions—ranging from earthquake monitoring to flood response and civil relief—were housed in various ministries and agencies. MEM was established to serve as a centralized authority that could bridge institutional divides and lead unified emergency operations (Zhang, 2022). Its mandate includes policy leadership, cross-sectoral coordination, and the integration of both governmental and non-governmental resources.

The Changning Earthquake response in June 2019 offered a critical early test for MEM's operational capacity. Within hours of the M6.0 earthquake, MEM activated the national emergency mechanism and coordinated the dispatch of multi-agency rescue forces, including firefighting units, seismic experts, emergency medical teams, and telecommunications specialists (ERPC–MEM et al., 2019a). This rapid response illustrates the ministry's role as both command center and real-time resource integrator under the fourth-generation system.

Beyond high-level coordination, MEM also worked through its affiliated agency, the Emergency Rescue Promotion Center of the Ministry of Emergency Management, to manage field-level implementation. The situation reports jointly issued by ERPC–MEM and the Sichuan Coordination Center for Social Participation in Disaster Prevention, Mitigation and Relief detail how MEM's policy directives were translated into operational plans on the ground (ERPC–MEM et al., 2019a, 2019b, 2019c, 2019d, 2019e). These reports reveal how MEM leveraged real-time data to prioritize village-level rescue operations, assess damage, and allocate relief materials.

A key innovation in MEM's role post-2018 is its capacity to bridge central-local divides. During the Changning response, vertical integration between MEM, provincial authorities, and county-level governments enabled a coordinated flow of information and resources. This coordination was not limited to administrative actors. Through ERPC–MEM and SCSP-DMR, the ministry also established interfaces with civil society organizations, enabling multi-actor participation while maintaining centralized oversight (Zhang et al., 2021).

14.1.3 The Strategic Interface: ERPC–MEM as a National Coordination hub

A defining feature of the post-2018 emergency governance reform in China is the creation of interface-type institutions that connect central policy mandates with localized and societal implementation mechanisms. The Emergency Rescue Promotion Center of the Ministry of Emergency Management of the People's Republic of China exemplifies such an interface structure. As a functional arm under the Ministry of

Emergency Management (MEM), ERPC–MEM assumes operational responsibilities that bridge national directives and decentralized response systems, particularly in the coordination of social forces.

The 2019 Changning Earthquake response demonstrated how ERPC–MEM operated as a strategic intermediary across vertical and horizontal governance layers. On the vertical axis, ERPC–MEM supported MEM's overarching command system by managing field coordination, real-time reporting, and cross-agency information integration. Its leadership in setting up the information taskforce and compiling the five situation reports—jointly issued with the Sichuan Coordination Center for Social Participation in Disaster Prevention, Mitigation and Relief, with technical support from Organisation Z—served as a foundation for policy-informed resource deployment (ERPC–MEM et al., 2019a, 2019b, 2019c, 2019d, 2019e).

Horizontally, ERPC–MEM functioned as a connector between government and civil society actors. In the aftermath of the earthquake, it worked closely with SCSP-DMR to convene and coordinate over 70 social organizations. These included national-level foundations, local NGOs, volunteer associations, and professional rescue teams. The jointly established Frontline Social Coordination Center, located at Shuanghe Middle School, served as an institutional innovation that enabled ERPC–MEM to jointly manage human resources, service zoning, and daily agenda-setting with SCSP-DMR (ERPC–MEM et al., 2019b, 2019c).

This interface role went beyond logistical functions. It institutionalized the principle of "government–society fusion"—the fusion of governmental and societal response systems. By embedding itself within a hybrid coordination structure, ERPC–MEM enabled a shift from fragmented, ad hoc social participation to structured, protocol-driven engagement. This helped reduce redundancy, avoid inter-organizational conflict, and build mutual trust in rapidly evolving disaster environments (Zhang et al., 2021).

14.1.4 Local Innovation: SCSP-DMR and the Rise of Subnational Governance Platforms

As the Ministry of Emergency Management (MEM) redefined China's top-level command structure after 2018, local institutional innovation emerged as a vital complement to the new centralized framework. The Sichuan Coordination Center for Social Participation in Disaster Prevention, Mitigation and Relief exemplifies such bottom-up governance development. Established as a provincial hub to coordinate non-state actors in disaster response, SCSP-DMR played a pivotal role in institutionalizing civil society participation during the 2019 Changning Earthquake.

SCSP-DMR's involvement in the Changning response was not merely symbolic. From the early hours after the earthquake struck, the center was mobilized to coordinate incoming NGOs, volunteer groups, and foundations (ERPC–MEM et al., 2019a). Its organizational strength was particularly evident in the creation of the Frontline

Social Forces Coordination Center at Shuanghe Middle School, jointly managed with the Emergency Rescue Promotion Center of MEM (ERPC–MEM et al., 2019b). This on-site mechanism facilitated daily planning meetings, task allocations, and sector-specific zoning for more than 70 social organizations operating in the affected area (ERPC–MEM et al., 2019c).

The deployment of SCSP-DMR marked China's first formalized use of a localized coordination platform in an earthquake scenario. Unlike previous disaster responses—characterized by fragmented NGO action and ad hoc local engagement—SCSP-DMR enabled structured mobilization, pre-entry registration, and real-time communication between government and society. By linking remote communities to emergency resources through methods such as "remote interviews + field validation," the platform addressed information blind spots in inaccessible villages (ERPC–MEM et al., 2019d).

Importantly, the Changning case also reflects a shift from informal civil society responses toward embedded institutional arrangements. SCSP-DMR operated not as an outsider to the state system, but as an integrated node within the official emergency framework. This aligns with the broader fourth-generation governance logic of "multi-actor inclusion under unified command" (Zhang, 2022), whereby provincial platforms like SCSP-DMR serve as policy translators, organizational managers, and accountability anchors for non-governmental forces.

In sum, SCSP-DMR demonstrates the potential for subnational platforms to support vertical integration and societal participation in China's evolving emergency governance architecture. Its case offers a replicable model for other provinces seeking to balance central authority with localized responsiveness and societal engagement.

14.2 Governmental Response and Digital Coordination in Action

The 2019 Changning Earthquake became a critical test of China's fourth-generation emergency management system, particularly its capacity for rapid mobilization, cross-sector coordination, and digital integration. In the immediate aftermath, authorities launched a multi-level response combining centralized oversight with localized execution. A key feature was the coordinated action between ERPC–MEM and SCSP-DMR, which enabled inter-agency synchronization alongside structured involvement of civil society groups. Through their jointly issued situation updates, real-time information exchange became institutionalized, while platforms like "One Map for Emergency" supported evidence-based decision-making amid unstable conditions. Together, these developments reflected the emergence of a governance model that blends hierarchical command with adaptive, data-driven collaboration across public and non-governmental actors.

14.2.1 Rapid Response Mechanisms in the Changning Earthquake

The 2019 Changning Earthquake in Sichuan served as an early test of China's fourth-generation emergency management system, particularly its capacity for rapid, multi-agency response under centralized command. Within minutes of the magnitude 6.0 earthquake striking at 22:55 on June 17, 2019, the Ministry of Emergency Management (MEM) activated the national emergency response mechanism and began coordinating a cross-sectoral deployment of rescue forces (ERPC–MEM et al., 2019a). The response was swift and vertically integrated, reflecting the institutional realignment initiated in 2018.

In the first 72 hours, MEM issued urgent directives mobilizing over 3000 personnel from national and provincial firefighting units, earthquake rescue teams, medical emergency responders, and communication restoration crews. Aerial imagery, unmanned aerial vehicles (UAVs), and satellite remote sensing were quickly employed to conduct preliminary damage assessments and identify priority zones (ERPC–MEM et al., 2019a, 2019b). Provincial and county-level emergency command systems were also activated in coordination with MEM, ensuring synchronized local operations within a nationally directed framework.

Inter-agency coordination was a hallmark of the early response. The Earthquake Administration deployed seismic monitoring specialists, while the Ministry of Health organized triage and mobile clinics. The Ministry of Transport and telecommunication enterprises were tasked with clearing roads and restoring signal coverage, especially in severely affected areas such as Shuanghe Town. These efforts were guided by centralized situational awareness facilitated through ERPC–MEM's information taskforce, which fed real-time data back to MEM's central command (ERPC–MEM et al., 2019b).

A key feature of this phase was not only speed, but also coordination across sectors and administrative levels. According to (Zhang et al., 2021), the integration of technical expertise, administrative command, and logistical mobilization in the first three days significantly reduced chaos and duplication of efforts—issues common in earlier disaster responses.

Crucially, the rapid deployment extended beyond state actors. Through its collaboration with the Sichuan Coordination Center for Social Participation in Disaster Prevention, Mitigation and Relief, the government began preparing for the arrival of civil society actors and community-based responders, laying the groundwork for structured multi-actor engagement in subsequent stages (ERPC–MEM et al., 2019b).

14.2.2 The Role of ERPC–MEM and SCSP-DMR in Coordinated Operations

The coordinated operations led by the Emergency Rescue Promotion Center of the Ministry of Emergency Management and the Sichuan Coordination Center for Social Participation in Disaster Prevention, Mitigation and Relief during the 2019 Changning Earthquake reveal a dual-track mechanism of state–society collaboration. This structure not only facilitated vertical integration between central and local levels, but also enabled horizontal coordination between government agencies and social organizations.

Within 24 hours of the earthquake, an integrated coordination initiative was launched through the collaboration of ERPC–MEM and SCSP-DMR, encompassing needs assessments, resource tracking, and stakeholder engagement to address service gaps (ERPC–MEM et al., 2019a). This partnership combined ERPC–MEM's command authority with SCSP-DMR's local reach, operationalizing what was described in coordination updates as a "hybrid coordination model" (ERPC–MEM et al., 2019b). Resource allocation in the early stages was shaped not only by top-down administrative directives, but also by feedback from frontline responders embedded in local communities.

A key innovation was the establishment of the Frontline Social Forces Coordination Center at Shuanghe Middle School—located near the earthquake's epicenter. Co-managed by ERPC–MEM and SCSP-DMR, this on-site hub hosted daily coordination meetings, sector-based task divisions (e.g., medical aid, logistics, children's services), and rotating duty schedules for over 70 registered social organizations (ERPC–MEM et al., 2019c). This model improved operational efficiency and reduced overlapping services in high-need zones such as Shuanghe and Gongxian.

These joint efforts also enabled real-time issue escalation. When supply bottlenecks were identified or village-level access routes were blocked, the coordination center was able to transmit the information to provincial and national command bodies via ERPC–MEM's digital network (ERPC–MEM et al., 2019d). SCSP-DMR, meanwhile, maintained regular contact with local volunteers and grassroots groups, ensuring constant feedback loops between communities and the formal emergency structure.

By embedding coordination at both strategic and operational levels, ERPC–MEM and SCSP-DMR demonstrated a new mode of integrated disaster governance. The Changning experience highlights how institutional collaboration—backed by digital tools, shared protocols, and spatial co-location—can bridge administrative divides and enable responsive, inclusive emergency operations (Guo et al., 2025; Zhang et al., 2021).

14.2.3 Situation Reports as a Tool of Information and Action Synchronization

In the 2019 Changning Earthquake, five jointly produced situation reports marked a key advancement in synchronizing emergency response across government and civil society actors. Rather than serving merely as documentation, these reports actively shaped operational alignment by directing interventions, coordinating information flows, and enhancing transparency. Their publication—co-developed by ERPC–MEM, SCSP-DMR, and supported technically by Organisation Z—signaled a shift toward real-time, actor-inclusive response mechanisms that enabled faster and more structured action during the emergency phase (ERPC–MEM et al., 2019a, 2019b, 2019c, 2019d, 2019e).

The initial report released within 24 hours of the earthquake on June 18, 2019, covered the scope of the disaster, preliminary damage assessment, and the geographic concentration of casualties, recording 12 deaths and over 100 injuries in areas such as Shuanghe and Gongxian (ERPC–MEM et al., 2019a). In subsequent updates, coverage expanded to include more granular operational details—for instance, the third report listed 129 temporary shelters across four counties, documented the activity of 67 social organizations, and flagged urgent shortages in items like infant formula, tents, and portable toilets (ERPC–MEM et al., 2019c).

The integration of data from government teams, NGOs, local authorities, and volunteer groups contributed to the creation of a unified operational picture. This facilitated coordination not only vertically—informing strategic decisions at the national level—but also horizontally, enabling real-time adjustments among provincial departments and civil society actors. Through a structured format incorporating maps, statistical tables, narrative updates, and needs matrices, the reports supported joint action grounded in shared and credible information.

Beyond coordination, the situation reports also contributed to accountability and public trust. Each report contained sections titled "Needs Expressed," "Responses Taken," and "Pending Gaps," thus promoting clarity over who was responsible for what, and where further action was required. According to (Guo et al., 2025), such structured reporting during the Changning response improved public perception of emergency governance by reducing information asymmetry and curbing rumor proliferation on social media platforms.

Ultimately, the situation reports demonstrated that real-time, multi-source documentation—when institutionalized—can serve as a backbone for synchronized action and collective sense-making in disaster governance. The Changning case thus illustrates how knowledge production is not ancillary but central to the operational logic of the fourth-generation emergency management system (Zhang, 2022; Zhang et al., 2021).

14.2.4 Data Collection and Decision Support through Digital Platforms

A core innovation of the fourth-generation emergency management system lies in the institutionalization of real-time, multi-source data to inform rapid decision-making. During the 2019 Changning Earthquake response, this digital turn was operationalized through a suite of platforms and tools that enabled spatial tracking, needs mapping, and inter-agency coordination. The Emergency Rescue Promotion Center of the Ministry of Emergency Management and the Sichuan Coordination Center for Social Participation in Disaster Prevention, Mitigation and Relief played central roles in designing and implementing these information infrastructures.

One of the key tools employed was the "One Map for Emergency" digital dashboard, which served as a centralized platform for compiling, visualizing, and analyzing data from village-level units to national authorities. The Information Taskforce—led by ERPC–MEM in the early hours after the quake—was responsible for uploading real-time data on casualties, building collapses, road conditions, and logistics bottlenecks (ERPC–MEM et al., 2019a). These inputs formed the empirical basis for high-level decision-making by the Ministry of Emergency Management and provincial authorities, helping prioritize rescue operations and allocate critical resources such as tents, sanitation kits, and medical supplies.

While centralized systems enabled top-down coordination, decentralized data collection mechanisms also played a vital role. SCSP-DMR organized a hybrid method of information gathering that combined remote interviews with village leaders and field validation visits to affected sites (ERPC–MEM et al., 2019c, 2019d). This approach addressed the challenge of data asymmetry in remote and mountainous areas—where telecommunications infrastructure had been disrupted—and enabled the identification of "invisible needs" such as psychosocial support, gender-specific aid, and accessibility issues in temporary shelters.

Data collection was not merely technical but deeply embedded in social coordination. Both ERPC–MEM and SCSP-DMR leveraged their roles as interface institutions to standardize data formats across state and non-state actors. Social organizations operating in the field were required to submit daily updates, which were integrated into broader response matrices included in the situation reports (ERPC–MEM et al., 2019b, 2019c, 2019d, 2019e). This integration ensured a shared situational awareness among governmental, social, and technical actors—reducing duplication, enhancing accountability, and reinforcing the evidence base for adaptive decision-making.

14.3 Social Networks and Civil Society in Emergency Response

The 2019 Changning Earthquake highlighted the growing institutionalization of civil society within China's emergency governance framework. This section examines how a diverse array of social actors—ranging from NGOs and volunteer teams to professional responders—contributed to emergency relief through varied roles and interventions. It further explores how coordination mechanisms, particularly the frontline social coordination center, enabled structured integration across sectors. The section also traces the transformation from ad hoc aid to embedded participation, marked by emerging norms of role clarity, reporting, and accountability. Finally, it addresses the challenges of exit procedures and the need for standardization, pointing toward institutional learning for future responses.

14.3.1 The Plurality of Social Actors and Types of Participation

The 2019 Changning Earthquake response illustrates the increasingly diverse ecosystem of social actors engaged in China's evolving emergency governance framework. The coordinated intervention involved not only government agencies but also a wide range of non-state actors, including national foundations, grassroots NGOs, specialized rescue teams, and informal volunteer networks. This plurality of participants—characterized by varied mandates, resources, and operational styles—was crucial in addressing the complex and multidimensional needs of affected communities (Guo et al., 2025; Zhang et al., 2021).

According to the situation reports compiled jointly by the Emergency Rescue Promotion Center of the Ministry of Emergency Management and the Sichuan Coordination Center for Social Participation in Disaster Prevention, Mitigation and Relief, more than 70 social organizations were registered and active in the disaster zone within the first few days following the earthquake (ERPC–MEM et al., 2019c). These organizations operated across sectors, offering services such as emergency rescue, logistical support, medical aid, psychosocial counseling, child protection, and public communication. For example, professional search-and-rescue teams were deployed to collapsed structures, while volunteer groups provided hot meals, mental health support, and assistance to displaced families in temporary shelters.

This diversity was not only functional but strategic. The heterogeneity of social actors allowed for flexible and agile deployment in an uncertain environment. Many organizations, particularly local NGOs and community-based groups, had pre-existing networks in the region and were able to rapidly mobilize volunteers, supplies, and knowledge about local conditions (ERPC–MEM et al., 2019b, 2019d). Meanwhile, larger national foundations contributed logistical capabilities and financial resources, filling critical gaps in procurement and transportation of relief materials.

Importantly, this plurality also required mechanisms for coordination and integration. Without effective alignment, overlapping efforts or inter-organizational friction could emerge. To mitigate this, the Frontline Social Forces Coordination Center—jointly established at Shuanghe Middle School—served as a unifying platform where sectoral responsibilities were allocated, daily tasks assigned, and duplication avoided (ERPC–MEM et al., 2019c). In this way, plural participation did not undermine coherence; rather, it was embedded into a structured framework that allowed distinct actors to contribute while maintaining collective focus.

14.3.2 Coordination Centers and Institutional Mechanisms

The operationalization of civil society participation during the 2019 Changning Earthquake hinged on a robust coordination infrastructure jointly established by governmental and non-governmental actors. This architecture was anchored by the Frontline Social Forces Coordination Center, established at Shuanghe Middle School in Shuanghe Town, near the earthquake's epicenter. This coordination center was co-convened by the Emergency Rescue Promotion Center of the Ministry of Emergency Management and the Sichuan Coordination Center for Social Participation in Disaster Prevention, Mitigation and Relief, reflecting a model of vertically sanctioned yet locally grounded collaboration (ERPC–MEM et al., 2019b, 2019c).

Designed to manage the influx and activities of more than 70 social organizations, the center instituted a daily coordination mechanism that became the cornerstone of multi-actor integration. Organizations arriving on site were required to register and undergo accreditation, a process that ensured accountability and legitimacy while aligning efforts with broader emergency protocols. The Coordination Center used daily morning briefings to assign tasks and divide services across spatial zones and functional sectors—including logistics, shelter management, medical services, psychosocial support, and children's welfare (ERPC–MEM et al., 2019c).

This systematic approach was operationalized through several institutional tools. One was the "Who-What-Where (3W) service map," updated regularly to visualize which organization was operating where, and doing what. This digital matrix enabled real-time resource matching, reduced overlap, and improved information symmetry between actors in the field (ERPC–MEM et al., 2019d). Another innovation was the creation of sector-specific working groups—for example, one dedicated to children and adolescents' needs, another focused on material distribution—which held micro-level coordination meetings in the afternoons, feeding their updates back into the main planning structure.

Importantly, the co-leadership of ERPC–MEM and SCSP-DMR ensured both administrative authority and grassroots responsiveness. While ERPC–MEM provided vertical linkage to national command structures, SCSP-DMR maintained day-to-day engagement with frontline organizations and volunteers. Together, they constituted a dual-anchor mechanism that blended top-down oversight with bottom-up flexibility—an embodiment of the "multi-actor inclusion under unified coordination" logic outlined in China's fourth-generation emergency governance system (Guo et al., 2025; Zhang, 2022).

14.3.3 From Ad Hoc Aid to Embedded Governance: Role Transformation

The 2019 Changning Earthquake marked a significant turning point in the institutional positioning of civil society within China's emergency governance. Historically, social organizations—ranging from grassroots volunteer teams to national foundations—were often mobilized in a reactive and fragmented manner, stepping in to fill service gaps left by formal institutions. In contrast, the Changning response demonstrated an emergent shift: from ad hoc aid to embedded governance, wherein civil actors were not only permitted but structurally integrated into the formal response architecture.

This transformation was enabled by deliberate design. Under the guidance of the Ministry of Emergency Management's affiliated center and the provincial coordination platform, civil society groups were no longer operating on the periphery. Instead, they were incorporated into a platformized governance model that included pre-registration procedures, task allocation protocols, and daily coordination routines (ERPC–MEM et al., 2019b, 2019c). These procedural innovations reduced operational ambiguity and clarified the roles, responsibilities, and reporting duties of participating actors.

Field evidence from Shuanghe and Gongxian shows that social organizations moved beyond emergency relief to offer specialized, institutionalized services. For example, child-focused NGOs were assigned dedicated zones to manage children's emotional support and recreational activities, while professional logistics teams were responsible for warehouse management and supply chain tracking (ERPC–MEM et al., 2019c). Rather than duplicating governmental efforts, these organizations operated within sector-specific workstreams under unified command, suggesting a shift toward functional complementarity rather than informal substitution.

This transition also involved a change in norms. Terms like "coordination discipline," "reporting duty," and "daily rotation" began to structure the conduct of non-state actors (ERPC–MEM et al., 2019d). These procedural norms promoted internal accountability and external legitimacy, allowing civil organizations to occupy stable and recognized roles within a highly formalized emergency apparatus. Furthermore, feedback mechanisms enabled both top-down oversight and bottom-up adaptation, reinforcing iterative learning across institutional boundaries (Guo et al., 2025; Zhang et al., 2021).

14.3.4 Exit, Accountability, and the Challenge of Standardization

While the 2019 Changning Earthquake response marked significant progress in embedding civil society into formal emergency governance, it also exposed institutional gaps in the domains of exit mechanisms, accountability, and operational standardization. These issues emerged most prominently in the later phases of the response, as attention shifted from immediate relief to demobilization, service handover, and post-disaster recovery.

One core challenge lay in the lack of structured exit protocols for participating social organizations. Despite the clear registration and task assignment mechanisms developed at the Frontline Social Forces Coordination Center, few standardized procedures were in place for coordinated withdrawal. Some NGOs departed without formal notice or documentation, leaving ambiguity regarding ongoing responsibilities and community expectations (ERPC–MEM et al., 2019e). This absence of synchronized exit procedures not only hampered continuity of care in psychosocial and educational services but also undermined the capacity for effective post-crisis assessment.

Related to this was the issue of accountability and reporting. Despite the introduction of basic tracking tools—such as daily presence rosters and service zone maps—the response lacked a standardized format for reporting outcomes, evaluating impact, or processing complaints. Final reports or inventories of distributed goods were submitted inconsistently (ERPC–MEM et al., 2019d), leaving coordinating bodies with limited capacity to assess performance, extract lessons, or gauge the quality of services delivered across a landscape of uneven professional standards (Zhang et al., 2021).

The fragmentation of accountability practices highlights the urgent need for standardization in civil participation protocols. This includes not only technical criteria—such as qualifications for medical responders or psychological counselors—but also operational norms, such as conflict resolution procedures, data sharing agreements, and ethical guidelines for engaging with vulnerable populations (Zhang, 2022). Notably, the Changning response revealed a demand for a common accreditation system for volunteer management, which could regulate entry, role allocation, and exit with greater predictability and legitimacy (Guo et al., 2025).

These challenges also provided a foundation for institutional learning. Several follow-up workshops, facilitated by SCSP-DMR and Organisation Z, gathered feedback from frontline actors and proposed revisions to coordination handbooks and exit protocols for future deployments (ERPC–MEM et al., 2019e). These efforts point to an emerging reflexivity within China's fourth-generation emergency system—one that seeks to codify not just the inclusion of civil actors, but the full lifecycle of their engagement, from arrival to withdrawal.

14.4 Communication, Engagement, and Public Trust in the Digital age

The Changning Earthquake response illustrates how digital communication has become central to disaster governance in China's evolving emergency system. This chapter examines four interrelated dimensions: the use of digital platforms as real-time information infrastructure; institutional communication practices that integrate multi-source data and public feedback; video-based engagement that mobilizes emotion and participation; and the coordinated construction of public narratives to manage trust and legitimacy. These practices reflect a shift from one-way broadcasting to hybrid, multi-actor communication ecosystems, where state and civil society co-produce information flows and shape public perception during crisis.

14.4.1 Digital Platforms as Information Infrastructure

The 2019 Changning Earthquake response marked a significant turn in the use of digital platforms as integral components of disaster information infrastructure in China. In the immediate aftermath of the earthquake, government agencies, civil society organizations, and individuals turned to social media and digital communication channels—primarily Weibo, WeChat, and short video platforms—to disseminate updates, signal needs, and coordinate response activities in real time (Guo et al., 2025; Zhang et al., 2021).

These platforms served multiple overlapping functions. First, they became primary vehicles for information dissemination to the public. Local governments and the Emergency Rescue Promotion Center of the Ministry of Emergency Management released verified data about casualties, aftershocks, and safety guidance through official WeChat and Weibo accounts within hours of the initial tremor (ERPC–MEM et al., 2019a). The public's reliance on these digital streams was not only driven by speed but also by their accessibility and perceived legitimacy, particularly in contrast to earlier disasters characterized by fragmented or delayed communication.

Second, civil society organizations used digital platforms for visibility and signaling. Many NGOs posted daily updates on their activities through WeChat public accounts and microblogs, including locations of mobile kitchens, shelter availability, and calls for material donations (ERPC–MEM et al., 2019c, 2019d). This visibility helped reduce information asymmetry between field responders and external supporters, facilitating faster donor responses and volunteer mobilization.

Third, digital platforms enabled a two-way flow of information. Affected residents used comment sections, direct messages, and public posts to report unmet needs, clarify misinformation, or share their own observations on damages. These bottom-up inputs were monitored by the information taskforce of ERPC–MEM and factored into broader situational analysis (ERPC–MEM et al., 2019b). Public digital engagement thus became a de facto participatory monitoring system, enhancing the responsiveness of official interventions.

The convergence of multiple actors on shared digital platforms laid the groundwork for an emergent model of hybrid information governance. Rather than functioning solely as broadcast media, platforms like WeChat became shared coordination interfaces—enabling vertical flows of official messaging, horizontal peer-to-peer updates among NGOs, and grassroots feedback from affected populations. This digitally mediated architecture was not without its limits, particularly in areas with disrupted connectivity, but it nonetheless represented a key evolution in China's fourth-generation emergency response system—embedding public communication within real-time governance logics (Guo et al., 2025; Zhang, 2022).

14.4.2 Institutional Communication and Situation Reports

The 2019 Changning Earthquake underscored the critical role of institutional communication in maintaining clarity, coherence, and public trust in an information-rich yet uncertain emergency environment. In China's fourth-generation emergency governance framework, institutional communication refers not only to the official release of information but also to the design of integrated mechanisms that coordinate data flows, frame authoritative narratives, and synchronize multi-actor response efforts across society. Rather than competing with informal or social media channels, this model seeks to embed communicative legitimacy within the infrastructure of emergency response itself.

In the immediate aftermath of the earthquake, a range of institutional communication tools was activated. Government agencies utilized official Weibo and WeChat accounts, provincial news portals, emergency briefings, and visualized dashboards such as "One Map for Emergency" to disseminate verified updates on casualties, weather risks, road conditions, and ongoing rescue efforts (ERPC–MEM et al., 2019a). These channels allowed the public to access real-time information while also curbing misinformation and rumor circulation. Importantly, these outputs were coordinated under a unified message architecture managed by ERPC–MEM and provincial authorities, ensuring consistency across platforms.

Within this broader system, the series of reports jointly issued by ERPC–MEM and SCSP-DMR exemplified institutionalized information governance. Released five times between June 18 and June 28, 2019, these documents adopted a "fact + needs + response" format, incorporating inputs from frontline organizations, local authorities, and national agencies (ERPC–MEM et al., 2019a, 2019b, 2019c, 2019d, 2019e). Rather than merely describing ongoing developments—such as the setup of shelters or deployment of medical services—they also drew attention to resource shortfalls, volunteer coordination issues, and operational bottlenecks. By facilitating both upward reporting and lateral feedback, the reports functioned as a connective layer between decision-making bodies and frontline realities, reinforcing strategic coordination while promoting public transparency (Guo et al., 2025).

More than a single product, these reports reflected the institutional logic of multi-source information synthesis and procedural accountability. The inclusion of sections like "Pending Gaps" and "Expressed Needs" emphasized not just what the state had done, but what remained unresolved. In doing so, institutional communication moved beyond one-way dissemination to support a model of participatory information governance, wherein credibility was built through openness, responsiveness, and structured interaction among diverse actors (Zhang, 2022; Zhang et al., 2021).

14.4.3 Video-Based Engagement and Participatory Communication

In the aftermath of the 2019 Changning Earthquake, short video platforms emerged as a powerful conduit for participatory communication and public engagement. Unlike traditional broadcast channels or official press releases, video-based media allowed both institutional and civil society actors to construct vivid, emotionally resonant narratives in near real time. This modality offered not only visibility but also legitimacy, particularly in a context where timely, credible, and affective communication was crucial for resource mobilization and trust-building.

Social organizations and individual volunteers rapidly adapted to video-centric platforms such as Douyin, Kuaishou, and WeChat video to document their activities and highlight unmet needs in affected areas. These videos showcased everything from frontline rescue scenes and logistical coordination to the emotional realities of displaced families in shelters. By combining factual content with personal storytelling, video creators were able to foster emotional proximity between disaster zones and distant publics—turning viewers into donors, supporters, or advocates (Guo et al., 2025).

This mode of communication functioned as a two-way bridge. On one hand, it enabled NGOs to communicate operational transparency, showcase their legitimacy, and make targeted appeals for financial or material support. On the other, it allowed online audiences to actively participate through comments, shares, and donations, thereby transforming spectators into distributed stakeholders in the response process. Some organizations even embedded QR codes or donation links directly into their video streams, streamlining the flow from empathy to action (ERPC–MEM et al., 2019c).

Importantly, video was not used solely by civil society. Institutional actors also employed curated video messages to communicate official updates, dispel rumors, and visualize coordination progress. For instance, ERPC–MEM published short video briefings featuring maps, voiceovers, and footage from coordination centers and field hospitals to reinforce their narrative control and operational transparency (ERPC–MEM et al., 2019b, 2019d). This convergence of state and non-state video use suggests a hybrid communication ecology where digital storytelling is institutionalized across sectors.

By foregrounding affective immediacy, visual evidence, and interactive affordances, video-based communication expanded the repertoire of crisis engagement. In the Changning case, it did more than just "inform" or "report"—it activated publics, mediated trust, and shaped how legitimacy was co-produced between institutions and society. As such, short-form video can be seen not merely as a medium, but as a participatory infrastructure for disaster governance in the digital age (Guo et al., 2025; Zhang et al., 2021).

14.4.4 Coordinated Narratives and the Governance of Public Opinion

In the 2019 Changning Earthquake response, the management of public communication extended beyond information dissemination to include the strategic orchestration of narrative authority. The joint coordination between the Emergency Rescue Promotion Center of the Ministry of Emergency Management (ERPC–MEM) and the Sichuan Coordination Center for Social Participation in Disaster Prevention, Mitigation and Relief exemplifies an emerging model of "narrative governance," wherein communication is not only instrumental for transparency, but constitutive of legitimacy and social order (Guo et al., 2025; Zhang, 2022).

From the early stages of the response, ERPC–MEM and SCSP-DMR actively curated public communication channels to ensure consistency in message framing. Official updates, needs assessments, and response actions were integrated into a unified discourse that portrayed state and civil society actors as co-responsible, complementary forces (ERPC–MEM et al., 2019a, 2019b, 2019c, 2019d, 2019e). This coordinated messaging strategy aimed to reinforce institutional credibility while avoiding the appearance of fragmentation or competition among actors.

Crucially, the production of situation reports was complemented by the monitoring of digital platforms such as Weibo and WeChat for rumor control and narrative calibration. When misinformation about casualty numbers or delayed responses circulated online, institutional actors quickly responded with verified updates, often through visually compelling formats such as infographics or video bulletins (ERPC–MEM et al., 2019b, 2019d). This not only curbed rumor proliferation but also asserted communicative sovereignty in an increasingly participatory media environment.

Another key aspect of narrative governance was the framing of civil society roles. Through public statements and media coverage facilitated by coordination bodies, NGOs and volunteer teams were consistently portrayed as disciplined, patriotic, and complementary to the government-led response. This served to validate their participation while maintaining the primacy of state leadership—an approach that balanced inclusion with hierarchy (Zhang et al., 2021). Notably, moments of friction—such as disputes over resource access or duplicative services—were often managed offline, with limited public exposure, thereby preserving the image of collective harmony (ERPC–MEM et al., 2019c).

In sum, the Changning experience reveals a deliberate strategy of message integration, actor alignment, and communicative control. Rather than viewing public opinion as a risk to be managed reactively, the coordination system embedded it into proactive governance design. The result was a communicative architecture that not only informed but also shaped perceptions, regulated emotions, and sustained public trust amid crisis (Guo et al., 2025; Zhang, 2022).

14.5 Conclusion

The 2019 Changning Earthquake response exemplifies China's evolving fourth-generation emergency management system, marked by integrated institutions, digital coordination, and structured civil participation. This chapter has shown how interface bodies like ERPC–MEM and SCSP-DMR operationalized central-local collaboration, how digital platforms and situation reports synchronized multi-actor responses, and how civil society transitioned from ad hoc actors to embedded partners in governance. Importantly, the management of public communication—through participatory media, video-based engagement, and narrative coordination—demonstrates a new model of legitimacy-building in crisis. Together, these developments reflect a hybrid governance architecture, where centralized authority is exercised alongside plural participation, and where digital infrastructures mediate both action and trust. Changning thus offers critical insights into the institutional, technological, and societal dimensions of disaster response in contemporary China.

References

ERPC–MEM, SCSP-DMR, & Organisation Z. (2019a). *Situation report No. 1: Changning earthquake in Yibin, Sichuan (June 17, 2019).* https://shimo.im/docs/kXTWHSG5IKUAhxJD

ERPC–MEM, SCSP-DMR, & Organisation Z. (2019b). *Situation report No. 2: Changning earthquake in Yibin, Sichuan (June 17, 2019).* https://shimo.im/docs/bLGb4vAWJKUkQz4B

ERPC–MEM, SCSP-DMR, & Organisation Z. (2019c). *Situation report No. 3: Changning earthquake in Yibin, Sichuan (June 17, 2019).* https://shimo.im/docs/Xksja0j8hMAn6ezd

ERPC–MEM, SCSP-DMR, & Organisation Z. (2019d). *Situation report No. 4: Changning earthquake in Yibin, Sichuan (June 17, 2019).* https://shimo.im/docs/sWtTKtrQJHIjM9t7

ERPC–MEM, SCSP-DMR, & Organisation Z. (2019e). *Situation report No. 5: Changning earthquake in Yibin, Sichuan (June 17, 2019).* https://shimo.im/docs/93B1bBJZTE8kdBeo

Guo, J., Sun, L., & Fan, C. (2025). Effects of short videos on engaging citizens in disaster communications with the government on social media. *Sustainability Science.* https://doi.org/10.1007/s11625-025-01767-8

Zhang, H. (2022). The fourth generation of China's emergency management: Logic and framework. *Chinese Public Administration.*. https://kns.cnki.net/kcms2/article/abstract?v=YQFK0XGjx413bFdg8I8BedrgTJ1_RFT8vBJcEBadIYl6AHAZj-pQVdVBzuWTCCLDhuBD1yMQnd6WKsH9oxoyQUaiwb7HyWXmG-T2gDdZ0Na8dw_KFCMqgWL3omqiPXFzmhKsAMMGMtsuvtjSEe9kQkcNZ8VxP0S-QdIZaBVgE3ZhrowZal9vnw==&uniplatform=NZKPT&language=CHS

Zhang, M., Qiao, X., Seyler, B. C., Di, B., Wang, Y., & Tang, Y. (2021). Brief communication: Effective earthquake early warning systems: Appropriate messaging and public awareness roles. *Natural Hazards and Earth System Sciences, 21*(10), 3243–3250. https://doi.org/10.5194/nhess-21-3243-2021

Chapter 15
Zhengzhou Extreme Rainfall Disaster: Digital Mobilization and Organizational Transformation

Abstract The Zhengzhou 7.20 extreme rainfall disaster not only exposed systemic weaknesses in urban infrastructure and emergency response, but also triggered a wave of bottom-up mobilization that reconfigured the landscape of disaster governance in China. This chapter examines the unfolding of this transformation across four analytical dimensions. It begins by framing the event as a compound crisis—driven by meteorological extremity, cascading infrastructure failures, and institutional blind spots. It then turns to the emergence of digitally enabled volunteer networks that filled critical gaps in information, logistics, and coordination. The third section analyzes how civil society organizations adapted their roles and entered governance processes through functionality rather than formal invitation. Finally, the chapter assesses the institutional constraints that limited inclusive governance, including legal ambiguity, digital inequality, spatial disparities, and data fragmentation.

Keywords Urban compound risks · Digital response networks · Civil society engagement · Organizational adaptation · Institutional barriers

15.1 The Zhengzhou Extreme Rainfall Event: Systemic Risk and Governance Gaps

The 7.20 Zhengzhou extreme rainfall disaster revealed not only the destructive power of an extreme weather event but also the systemic fragilities embedded in urban infrastructure and governance. This section analyzes the event as a compound disaster rooted in both climatic extremity and institutional failure. It examines four interrelated dimensions: the meteorological drivers and cascading impacts of the rainfall, infrastructure vulnerabilities under stress, disconnects between early warnings and response mechanisms, and the deeper structural governance gaps that hindered timely action. These factors illuminate why traditional emergency systems faltered and why new paradigms of urban risk governance are urgently needed.

Q. Du and R. Shaw, *Social Organizations in Disaster Response and Recovery*, Disaster Risk Reduction, https://doi.org/10.1007/978-981-92-2846-1_15

15.1.1 Extreme Rainfall and the Compound Disaster Scenario

On 20 July 2021, Zhengzhou experienced an unprecedented extreme precipitation event that fundamentally exceeded the design assumptions of the city's hydrometeorological and drainage systems. According to the official investigation, the maximum hourly rainfall reached 201.9 mm, while cumulative precipitation exceeded 624.1 mm within 24 hours, both breaking historical records and approaching what was officially characterized as a "once-in-a-millennium" event (State Council Disaster Investigation Group, 2022). Meteorological analyses further indicate that the event was driven by the combined influence of a stalled subtropical high, moisture transport from Typhoon In-Fa, and strong mesoscale convective systems, producing highly concentrated and persistent rainfall over the urban core (Xiao et al., 2023; Zhang et al., 2023).

The intensity and spatial concentration of precipitation rapidly transformed a meteorological hazard into a compound urban disaster, simultaneously affecting multiple interconnected systems. Urban road networks and underpasses were inundated within hours, paralyzing surface mobility and emergency access. More critically, Metro Line 5 suffered catastrophic water intrusion, trapping passengers in tunnels and resulting in mass casualties, while several river sections and reservoirs experienced overtopping or emergency discharge pressures (State Council Disaster Investigation Group, 2022). Hydrological simulations and post-event assessments confirm that surface runoff volumes exceeded the drainage network's conveyance capacity, generating widespread waterlogging and cascading infrastructure stress (Wang et al., 2024a).

Importantly, the severity of impacts cannot be explained by meteorological extremity alone. The event exposed a tightly coupled socio-technical system in which transportation, power supply, drainage, and communication infrastructures amplified one another's failures. Remote sensing and hydrological observations show that rapid surface ponding coincided with localized power outages and signal disruptions, constraining situational awareness and delaying coordinated evacuation responses (Xiao et al., 2023; Zhang et al., 2023). Urban mobility data further reveal abrupt collapses in travel flows and spatial accessibility during peak rainfall periods, reflecting the systemic nature of functional breakdown rather than isolated infrastructure damage(Wang et al., 2024a).

From a disaster characterization perspective, the Zhengzhou event thus represents a compound risk scenario, where extreme precipitation interacted with urban density, infrastructure interdependence, and governance capacity constraints to generate nonlinear escalation of losses. This framing aligns with emerging disaster research emphasizing that high-impact urban disasters increasingly arise from the coupling of climatic extremes and complex socio-technical systems rather than from singular hazards alone (Zhang et al., 2023). Recognizing this compound nature is essential for understanding why conventional emergency preparedness assumptions failed and why alternative social and digital mobilization mechanisms became critical in subsequent response phases.

15.1.2 Infrastructure Vulnerabilities and Urban Exposure

The extreme rainfall in Zhengzhou on July 20, 2021, did not merely expose meteorological anomalies but critically revealed the fragility and interdependence of urban infrastructure systems under stress. A key structural issue lay in the city's drainage system, which was designed according to a 1-in-20-year rainfall standard. This standard fell drastically short of the once-in-a-millennium intensity recorded during the event (State Council Disaster Investigation Group, 2022), rendering drainage systems overwhelmed and ineffective. Urban runoff rapidly exceeded the system's capacity, resulting in widespread flooding of main roads, residential areas, and commercial zones (Chen et al., 2022).

Transportation infrastructure, particularly the metro and underground tunnels, emerged as critical failure points. The inundation of Metro Line 5, where over 500 passengers were trapped and 14 individuals died, became one of the most tragic outcomes of the event. Investigations revealed that the metro system lacked automatic flood gate mechanisms, and evacuation protocols were either delayed or nonexistent. Floodwater entered the metro system through a 2 km-long tunnel section between Haitansi and Shakou Road stations, rapidly rising to neck-level in some train cars (State Council Disaster Investigation Group, 2022; Yang et al., 2023). At the same time, Zhengzhou's road tunnels—particularly the Jingguang North Road Tunnel—acted as hydraulic traps, with water accumulating to depths of over 4 meters within minutes, severely impeding rescue and drainage efforts (Wang et al., 2024b).

Beyond mobility systems, the cascading impacts extended to power grids and communication networks, which failed in several key areas during peak flooding periods. Power outages affected critical facilities, including hospitals and shelters, while telecommunication disruptions delayed the dissemination of rescue information and coordination across sectors. These systemic interdependencies significantly magnified the disaster's impact, leading to a near-paralysis of emergency functions during the golden hours of response (Gong et al., 2025; Mao et al., 2026).

The nature of these failures points to a cascading disaster model—not a singular infrastructure collapse but a domino-like breakdown across functionally coupled urban systems. Rather than an inevitable consequence of climatic extremity, the disaster can be understood as a failure of urban risk anticipation and design alignment. The inability to simulate and plan for inter-system failure, especially in high-density transit corridors and underground spaces, reflects a broader lack of integrated urban resilience strategy (Chen et al., 2022; Yang et al., 2023).

15.1.3 Gaps in Early Warning and Governmental Response

The 7.20 Zhengzhou extreme rainfall disaster revealed a critical disconnect between early warning systems and effective governmental response, despite multiple meteorological alerts issued in advance. On July 19, Zhengzhou Meteorological Bureau had already released a red warning for heavy rainfall. By 9:00 a.m. on July 20, warnings had been issued five times, escalating from yellow to red within hours (State Council Disaster Investigation Group, 2022). However, administrative response lagged far behind. Key government departments did not take preemptive evacuation measures, and many citizens continued their routine activities unaware of the potential severity (Xu et al., 2022).

This forecast-action gap highlights systemic issues in translating technical knowledge into public safety interventions. Meteorological data were available, but emergency mobilization mechanisms were either too rigid or too slow to adapt. For example, there was no coordinated instruction to suspend public transport or close vulnerable underground facilities like metro systems and tunnels before the flooding occurred (Zhao et al., 2023).

Moreover, public communication remained fragmented and reactive. Alerts were mainly issued through government websites and official social media, with limited outreach to vulnerable or marginalized groups through accessible or trusted channels. Many elderly residents reportedly failed to receive or comprehend evacuation warnings, which were often issued without clear authority or actionable guidance, highlighting persistent gaps in inclusive risk communication (Organisation Z, 2021a, 2021b).

At the institutional level, the lack of interdepartmental coordination further hindered effective response. The emergency command system functioned in silos, with limited real-time data sharing between meteorological, transportation, and urban safety departments. During the critical hours on July 20, delays in activating citywide emergency mechanisms meant that no unified command was in place to prioritize human life, relocate at-risk communities, or deploy rescue teams efficiently (State Council Disaster Investigation Group, 2022; Xu et al., 2022).

Government messaging following the disaster reinforced public perceptions of institutional inertia. The press briefing held on July 20 evening offered limited information and was seen as largely defensive rather than instructive, deferring accountability and offering no clear plan for next steps. Formal acknowledgment of decision-making failures only emerged months later, in the national investigation report, which highlighted systemic deficiencies in risk perception, preparedness, and coordination (State Council Disaster Investigation Group, 2022).

15.1.4 Structural Causes of Governance Gaps

The institutional architecture of emergency management in China heavily reflects a hierarchical governance model, where local officials often rely on top-down instructions before acting. During the Zhengzhou extreme rainfall disaster, this structure became a key bottleneck. Although meteorological alerts had reached critical levels by the morning of July 20, local departments failed to initiate proactive responses, partly due to institutional cultures of "waiting for orders" rather than embracing delegated authority (Feng et al., 2021; State Council Disaster Investigation Group, 2022). This centralized rigidity hindered early decision-making at the grassroots level, especially in fast-evolving urban crises where minutes are critical.

Compounding this issue was the absence of an integrated emergency response system capable of bridging real-time data, inter-agency operations, and public engagement. While meteorological departments, transport bureaus, and local emergency offices held discrete information, there was no platform to synchronize these resources dynamically. As a result, warnings did not translate into operational plans, evacuation orders, or resource allocation in time (Yang & Wang, 2022). Coordination among departments remained fragmented, and institutional routines were not designed for fluid crisis adaptation (Li et al., 2025).

Another structural weakness lies in the over-reliance on formal institutional actors within emergency protocols. Prior to the disaster, civil society organizations and digital volunteer networks were not embedded in preparedness plans or emergency drills. This exclusion limited the state's capacity to leverage non-state actors when formal systems were overwhelmed. The response efforts eventually saw spontaneous social mobilization, but these efforts had to operate outside official frameworks, delaying the creation of effective synergies between the state and society (Feng et al., 2021; Organisation Z, 2021a, 2021b, 2021c, 2021d).

Furthermore, the crisis underscored a governance culture that prioritized stability and reputational control over transparent risk communication and decentralization. Official briefings were delayed, public-facing decisions were reactive, and internal accountability mechanisms remained opaque until external investigations revealed governance lapses. The State Council report ultimately concluded that "emergency preparedness was insufficient, cross-sectoral coordination was weak, and risk perception was lacking", pointing to structural deficiencies that transcended individual errors (State Council Disaster Investigation Group, 2022).

15.2 Digital Mobilization and Emergent Volunteer Networks

The 7.20 Zhengzhou extreme rainfall disaster revealed the transformative role of digital infrastructure and social mobilization in crisis governance. As state-led mechanisms lagged in the early hours, citizens, volunteers, and NGOs activated decentralized, tech-mediated networks to coordinate relief. This section traces four interlinked dynamics: the shift from institutional silence to crowdsourced crisis mapping; the spontaneous rise of decentralized volunteer networks; adaptive decision-making without central command; and the fusion of digital tools with social trust. These dynamics illustrate how digitally enabled self-organization reconfigured disaster response beyond hierarchical control, offering a new paradigm of resilience under extreme urban risk.

15.2.1 From Information Vacuum to Crowdsourced Crisis Mapping

In the early hours of the Zhengzhou extreme rainfall disaster on July 20, 2021, an acute information vacuum emerged as formal government communication mechanisms faltered. Despite the issuance of red alerts by meteorological authorities, critical public information—such as evacuation routes, inundation zones, and rescue coordination—was neither timely nor widely disseminated (State Council Disaster Investigation Group, 2022). As urban infrastructure began to fail, the absence of centralized situational updates led to widespread uncertainty, particularly among trapped residents and responders on the ground.

In response to this vacuum, digital platforms rapidly became alternative infrastructures for information flow and risk mapping. Citizens began using social media tools—including Weibo, WeChat, Douyin, and Baidu Maps—to share real-time updates, distress calls, and location-tagged photos. These fragmented but vital data points laid the foundation for crowdsourced crisis mapping. For example, the Weibo hashtag "#郑州暴雨互助# (Zhengzhou Rainstorm Mutual Aid)" accumulated over 1 billion views within 48 hours, functioning as a participatory information commons where stranded individuals, volunteers, and digital activists converged (Peng & Zhang, 2024).

Private mapping platforms quickly adapted to these citizen signals. Baidu Maps and Amap (Gaode) began integrating user-generated content to mark flooded roads and blocked underpasses, offering dynamic route guidance amid the chaos. On Douyin, short videos not only captured real-time scenes of entrapment but also geo-tagged urgent rescue locations, creating an organic form of crisis visualization (Qu et al., 2023).

NGOs and grassroots organizations, including Organisation Z, played a catalytic role in structuring this bottom-up information. Between July 20 and 21, Organisation

Z collected over 200 geolocated SOS messages from WeChat groups and microblogs, converting them into spreadsheets and digital maps that were shared with volunteer boat teams and local rescue staff (Organisation Z, 2021a, 2021b, 2021c). These decentralized data infrastructures effectively replaced official command channels during critical hours, facilitating improvised coordination among scattered actors.

15.2.2 The Rise of Decentralized Volunteer Networks

The 7.20 Zhengzhou extreme rainfall disaster catalyzed a large-scale grassroots response that fundamentally diverged from China's conventional disaster relief paradigm. Rather than relying solely on hierarchical, state-coordinated mechanisms—such as the Red Cross or official charity federations—emergency mobilization emerged organically through decentralized volunteer networks activated via digital platforms. This horizontal mobilization redefined how social actors self-organized, coordinated, and delivered aid amid infrastructural paralysis and institutional inertia.

Three major categories of actors emerged during the early hours and days of the disaster. First, local volunteers—often operating within pre-existing neighborhood and kinship networks—mobilized boats, organized ad hoc rescues, and distributed essential supplies to flooded communities. Their actions were often based on real-time social media alerts and neighborhood-level situational knowledge. Second, influencers and content creators on platforms like Douyin (TikTok China) and Weibo played pivotal roles in amplifying rescue appeals and resource needs. These Key Opinion Leaders (KOLs) functioned as visibility brokers, enhancing the reach of critical information and facilitating donations and volunteer sign-ups across regions (Peng & Zhang, 2024). For instance, the "Zhouji (舟济) Plan", launched through a Douyin account, coordinated more than 30 inflatable rescue boats and over 200 volunteers to reach severely flooded urban zones.

The third and perhaps most critical actor group comprised cross-regional NGOs. These included experienced disaster relief organizations from cities such as Wuhan, Chengdu, and Guangzhou. Drawing on their past experience from events like the 2020 Yangtze floods, these groups arrived with trained personnel, supply chains, and knowledge of coordination protocols. Their rapid deployment not only brought logistical capacity but also facilitated the transfer of knowledge to emergent local actors (Organisation Z, 2021d, 2021e, 2021f; Wu et al., 2023).

Organisation Z itself offers a telling case of agile volunteer mobilization. Within 48 hours, it had assembled mixed teams of local volunteers and external responders to conduct on-site needs assessments across 21 flooded communities, coordinating evacuation support, food delivery, and communication with emergency services (Organisation Z, 2021d, 2021e, 2021f). This was made possible not through command hierarchies, but through open coordination channels such as WeChat groups, shared digital documents, and community-based mapping tools.

15.2.3 Coordination Without Command: Adaptive Decision-Making in Action

In the absence of centralized command during the Zhengzhou extreme rainfall disaster, emergent volunteer networks demonstrated an adaptive and distributed model of coordination. Confronted with high uncertainty, fragmented information, and limited institutional guidance, these networks employed flexible, bottom-up decision-making mechanisms that responded to real-time needs on the ground. Unlike formal emergency systems, which often require hierarchical approval and follow pre-established routines, volunteer actors adjusted their actions dynamically based on evolving social signals and digital feedback.

One key mechanism was resource reallocation through real-time needs parsing. Volunteer groups tracked help requests—often tagged with geolocated distress calls—in WeChat groups and Douyin videos, then adjusted manpower and supplies accordingly. For instance, rescue boats originally dispatched to the Erqi District were redirected to the submerged Jingguang Tunnel once critical updates on water depth and trapped individuals surfaced on Weibo threads (Peng & Zhang, 2024). This iterative loop of information and action replaced conventional logistics pipelines with fluid responsiveness.

Digital tools facilitated decentralized coordination, enabling geographically dispersed actors to operate with coherence. Shared spreadsheets hosted on Tencent Docs functioned as live operation dashboards, allowing volunteers to assign tasks, track supply status, and log response points. This configuration has been conceptualized as a "WeChat–Map–Form" triad, in which spatial intelligence, communication, and tasking are seamlessly integrated (Peng & Zhang, 2024). Organisation Z, for example, utilized such tools to organize field deployment to Zhengzhou East Station, local hospitals, and school shelters across three districts within 36 hours of initial alerts (Organisation Z, 2021e, 2021f, 2021g, 2021h).

Furthermore, multi-platform synchronization amplified operational efficiency. WeChat provided closed-loop internal coordination; Weibo served as a broadcasting interface for public calls; Douyin channeled donations and visual documentation of field needs. This polycentric architecture enabled vertical and horizontal flows of information, bridging data producers (citizens) with response units (volunteers) in near real-time (Wu et al., 2023). The coordination logic thus resembled a decentralized mesh rather than a command tree—actions were emergent, not ordered; consensus was negotiated, not imposed.

Despite lacking formal mandates, these decentralized systems exhibited notable capacity for collective governance. Their performance hinged on mutual trust, rapid information circulation, and social validation through digital traces. The Zhengzhou case thus challenges the conventional belief that centralized coordination is a prerequisite for effective disaster response, illustrating instead how networked improvisation and digitally mediated trust systems can fill institutional voids during extreme events (Organisation Z, 2021f; Qu et al., 2023).

15.2.4 Digital Infrastructure Meets Social Trust: A New Paradigm of Crisis Response

The Zhengzhou extreme rainfall disaster catalyzed an unprecedented convergence between digital infrastructures and social trust networks, revealing a new governance paradigm that emerged outside traditional state command structures. In the critical window following the collapse of institutional emergency coordination, digital tools provided not only technical functionality but also the scaffolding for distributed trust and collective action. The success of the volunteer-led response was not solely a result of technological availability, but of its embedding in dense social relationships, reputational feedback loops, and mutual accountability among dispersed actors.

Digital platforms enabled real-time visualization, coordination, and communication across actors who had no prior affiliations. Volunteer groups leveraged a suite of tools—such as WeChat for coordination, Tencent Docs for task delegation, and Baidu Maps for geospatial overlays—to synchronize field operations across fragmented zones of the city (Peng & Zhang, 2024; Wu et al., 2023). These tools did not merely support logistics but created a shared informational space that generated trust through transparency and visibility. For instance, Organisation Z reported using real-time "information relay tables" to track completed and pending rescue actions across 21 communities, helping to minimize redundancy and establish operational legitimacy (Organisation Z, 2021e, 2021f, 2021g, 2021h).

Importantly, the infrastructure was sustained by social trust—a decentralized but resilient form of authority derived from interpersonal credibility, peer validation, and digital traceability. Volunteers who shared live videos from flood zones, or who were recommended by trusted contacts, were more likely to receive public donations or on-site coordination support (Chen et al., 2023; Wu et al., 2023). The moral economy of digital trust compensated for the absence of institutional vetting mechanisms, facilitating high-frequency collaboration between previously unconnected actors.

However, this trust-tech nexus also revealed limitations. Crowdsourced data proved difficult to verify, sometimes leading to misallocation of supplies or duplicated rescue efforts (Qu et al., 2023). In the absence of standardized protocols, decisions often relied on informal judgments, which risked inequitable resource distribution, especially for vulnerable populations with limited digital access (Feng et al., 2021; Organisation Z, 2021j). Furthermore, while digital coordination thrived in the response phase, few mechanisms existed to integrate its outputs into state recovery planning or formal policy feedback loops (Gong et al., 2025).

15.3 Civil Society and Organizational Adaptation in Governance

The 7.20 Zhengzhou extreme rainfall disaster revealed not only the limitations of state-centric emergency response but also the adaptive potential of civil society in crisis governance. This chapter traces the evolving role of grassroots organizations from emergent supplements to semi-institutionalized partners. It first examines their informal entry into the response system amid bureaucratic fragmentation, followed by a shift from traditional relief functions to strategic roles as information brokers and needs assessors. The analysis then unpacks the mechanisms of state–society collaboration and concludes with how these dynamics triggered deeper transformations in organizational structures and the governance framework itself.

15.3.1 From Emergency Supplement to Governance Entry: How Civil Society Entered the Response System

The entry of civil society into Zhengzhou's emergency governance system during the extreme rainfall disaster was neither pre-designed nor institutionalized. It emerged instead as a practical response to the crisis-driven breakdown of existing bureaucratic mechanisms. As formal government structures struggled with overwhelmed communication channels, logistical bottlenecks, and fragmented data flows, grassroots organizations stepped in to fill the urgent gaps in frontline delivery, real-time needs mapping, and community-level coordination (State Council Disaster Investigation Group, 2022).

Rather than being invited by mandate, civil society actors entered the scene by performing functions that the state apparatus could not rapidly mobilize. Inundated urban zones required rapid deployment of bottled water, food, hygiene kits, and first aid—tasks that state-led units were too centralized to execute with speed and precision. In this vacuum, a variety of NGOs, volunteer groups, and local networks took on "first-mile" responsibilities, directly responding to citizen-generated SOS messages and neighborhood-level alerts (Chen et al., 2022; Organisation Z, 2021i). Their legitimacy was not bureaucratic, but derived from functionality and responsiveness.

The mode of entry was largely informal and adaptive. Many grassroots actors—including but not limited to Organisation Z—were incorporated into emergency coordination through ad hoc WeChat groups initiated by township and subdistrict officials (Organisation Z, 2021j). These digital channels served as impromptu command centers where civil society organizations could share on-the-ground intelligence, coordinate dispatch, and report delivery updates. In more peripheral townships such as Xingyang and Dengfeng, some NGOs bypassed formal state layers entirely by forming direct links with village committees to expedite aid and conduct localized assessments (Organisation Z, 2021k).

As operations scaled up, several organizations moved from improvised participation to semi-structured integration. For instance, certain NGOs assumed data aggregation and reporting roles, compiling situation updates from dispersed communities and translating them into actionable insights for government use. Organisation Z, among others, curated verified needs data from community volunteers and submitted daily briefs to district-level emergency taskforces between July 24 and 28 (Organisation Z, 2021i, 2021j, 2021k, 2021l, 2021m). This role exemplifies how NGOs were functionally—but not institutionally—recognized as partners in decision-making and logistics allocation.

15.3.2 Role Transformation: From Relief Actors to Information Brokers and Needs Assessors

During the Zhengzhou 7.20 extreme rainfall disaster, civil society organizations evolved from conventional relief providers into multifunctional governance actors. This transformation reflected not only a shift in operational tactics but also a deeper restructuring of their epistemic and infrastructural roles within the disaster response ecosystem. Initially entering as logistical gap-fillers, many organizations quickly assumed more strategic functions, becoming intermediaries for data processing, needs assessment, and service integration.

Traditionally, the roles of grassroots organizations in Chinese disaster relief had centered on donation mobilization, volunteer coordination, and the distribution of supplies (Mao et al., 2026). In Zhengzhou, these functions were evident in the early response: Organisation Z and other groups organized water, food, and emergency medical kits for isolated communities in Jinshui and Huiji districts (Organisation Z, 2021j). However, as the situation evolved, the need for high-resolution situational data, especially from underserved peri-urban and rural areas, exposed the limitations of government-led information channels and elevated the significance of CSOs as information brokers.

In this emergent role, organizations moved beyond delivery to aggregate, verify, and transform scattered field data into actionable intelligence. Organisation Z, for example, built and maintained a dynamic database of local needs reports—collected through WeChat groups, field interviews, and call-in logs—which was regularly updated and shared with other NGOs and, increasingly, with district-level government units (Organisation Z, 2021k, 2021l, 2021m). Their verification of information from remote villages like Gaocun and Songzhuang was especially critical, given the high prevalence of false or duplicate reports during the crisis (Gong et al., 2025).

Parallel to this, many organizations assumed the role of needs assessors, deploying assessment teams to conduct door-to-door visits and collect disaggregated data on vulnerable populations. This included identifying households with elderly residents, left-behind children, and individuals with mobility impairments who were not captured in official registries (Organisation Z, 2021l). Several organizations conducted joint rural assessments across six townships between July 25–29, feeding the findings into a multi-organizational coordination platform (Organisation Z, 2021m).

A third dimension of role expansion involved service integration. Some NGOs began to combine relief delivery with mental health support, epidemic prevention, and even cash aid mechanisms, offering holistic response packages in collaboration with local health workers (Gong et al., 2025; Mao et al., 2026). These actions reflect a shift from task-based execution to systemic integration, requiring new internal capabilities such as digital tracking, cross-sector coordination, and policy interpretation.

15.3.3 Evolving Collaboration Mechanisms Between Government and Civil Society

The Zhengzhou extreme rainfall disaster marked a critical juncture in the evolution of civil society–government collaboration in China's disaster governance. What began as ad hoc and fragmented cooperation gradually transitioned into more structured mechanisms of joint action. This transformation was driven not by institutional mandates, but by operational necessity, shared risk exposure, and emerging mutual dependencies during crisis response.

Four key forms of collaboration emerged during the response phase. First, real-time information sharing proved essential. NGOs compiled dynamic needs assessments through community networks and digital platforms, which were relayed to township-level coordination mechanisms. Between July 24 and 28, several synthesis reports consolidated field data from over 20 communities and were partially referenced in district-level relief planning (Organisation Z, 2021j, 2021k, 2021l, 2021m). Second, joint assessments were conducted in peri-urban and rural areas like Xingyang and Xinmi, where official data were limited. These efforts—co-led by local civil society groups and grassroots cadres—focused on vulnerable populations, including elderly residents, low-income households, and migrant workers (Gong et al., 2025).

Third, cash-based assistance coordination emerged as a novel form of collaborative implementation. Several NGOs, after field verification, proposed direct cash transfers for households that fell outside official registries. In select townships, these proposals were reviewed and approved by local civil affairs bureaus, albeit on a discretionary basis (Li et al., 2025). While not fully institutionalized, this model demonstrated a hybrid interface between public authority and social innovation. Fourth, early-stage discussions began around post-disaster recovery planning, including infrastructure repair, epidemic prevention, and psycho-social support. Though civil society's role remained consultative, their inclusion reflected growing recognition of their field-level knowledge and implementation capacity (Chen et al., 2022).

These forms of collaboration catalyzed the gradual emergence of institutional interfaces. On the one hand, local governments began to depend on NGO-generated data for situational awareness in areas beyond their administrative reach. On the other,

NGOs adapted to public sector norms, adopting government-style documentation and learning compliance procedures for aid distribution. This mutual adjustment established the foundation for longer-term cooperation.

15.3.4 Emergent Organizational Transformation and Governance Reconfiguration

The Zhengzhou extreme rainfall disaster catalyzed not only immediate response innovations but also deeper structural transformations within civil society organizations and the broader governance landscape. These shifts reflected an emergent reconfiguration wherein grassroots actors evolved from operational auxiliaries to semi-institutionalized partners, reshaping both organizational capacities and state-society relations.

At the organizational level, NGOs underwent significant internal adaptation. The crisis exposed the limits of traditional project-based operating models and prompted many groups to upgrade their data capabilities, inter-organizational coordination structures, and cross-regional dispatch systems. Organisation Z, for instance, transitioned from field logistics to managing a dynamic information relay platform, aggregating and updating rescue needs from over 21 communities between July 23 and 30 (Organisation Z, 2021k, 2021l, 2021m). Similar shifts were observed among other national NGOs, which developed modular coordination units capable of operating in parallel across affected districts (Gong et al., 2025; Mao et al., 2026). These changes marked a movement toward a platform organization model, capable of integrating multiple functions—data management, supply chain tracking, and multi-stakeholder collaboration—within a single digital–human architecture.

Concurrently, the governance configuration surrounding disaster response began to evolve. Information no longer flowed unidirectionally from state to society. Instead, civil society actors became institutional knowledge nodes, contributing verified data and local intelligence to official decision-making. In several township governments, situation reports compiled by NGOs were incorporated into daily emergency briefings, influencing deployment priorities (Organisation Z, 2021l). Government bureaus also began to recognize the efficiency of NGO-run needs assessments and included them in post-disaster recovery planning discussions (Chen et al., 2022; Li et al., 2025).

However, these transformations remained contingent on the temporality of crisis. The absence of permanent institutional channels meant that gains in NGO legitimacy and state receptiveness were often fragile. Role boundaries remained blurred, and while civil society assumed more cognitive and coordinative functions, their authority remained informal and conditional. In areas lacking strong local government capacity, NGOs temporarily assumed leadership roles, but these arrangements seldom extended beyond the emergency phase (Mao et al., 2026).

15.4 Institutional Constraints and Challenges of Inclusive Governance

The integration of civil society into disaster governance during the Zhengzhou 7.20 extreme rainfall response exposed multiple structural constraints that limited the scope, equity, and sustainability of inclusive coordination. This chapter examines four interrelated challenges: the ambiguous legitimacy and institutional mistrust that marginalized NGOs despite their critical contributions; the digital divide that systematically excluded vulnerable populations from visibility and assistance; the spatial inequities that left rural and informal settlements underserved; and the data governance gaps that prevented grassroots information from shaping official decision-making. These constraints reveal the institutional fragilities underlying China's evolving model of hybrid emergency response.

15.4.1 Ambiguous Legitimacy and Institutional Mistrust

Despite their operational visibility during the Zhengzhou 7.20 disaster response, civil society organizations remained institutionally peripheral. While grassroots actors played indispensable roles in frontline relief and data relay, their legitimacy was functionally granted rather than formally codified, producing a persistent tension between high participation and low institutional recognition.

A key structural constraint was the ambiguous legal status of many responding organizations. A significant portion of NGOs and volunteer groups lacked local registration and were thus classified as "external collaborators" rather than formal stakeholders within municipal emergency governance (Xie et al., 2023). This ambiguity affected both their access to official coordination platforms and their eligibility to participate in post-disaster policy deliberation. For instance, Organisation Z, though active across multiple districts, was repeatedly referred to in situation reports as "community coordinators" rather than institutional partners (Organisation Z, 2021g, 2021h, 2021i, 2021j, 2021k, 2021l, 2021m).

This limited recognition was compounded by institutional mistrust at the township and subdistrict levels. Local authorities often exhibited caution toward external NGOs, particularly those not embedded in the party-state system. Some township governments declined cooperation requests or restricted access to shelters and aid sites unless organizations could demonstrate prior affiliation with government-endorsed networks (Ye et al., 2025). In the aftermath of the flooding, several local governments explicitly prioritized data collected through official channels, sidelining what they perceived as "unverified" civil society assessments—even when NGO data were more timely and granular (Organisation Z, 2021i, 2021j, 2021k).

The decision-making process during the disaster also reflected this asymmetry. Emergency coordination meetings remained government-dominated, with civil society actors often limited to implementation roles without access to deliberative forums. While NGOs compiled detailed situation reports and needs maps, these were rarely integrated into core resource allocation processes (Ye et al., 2025). Even Organisation Z's multi-day briefs, which documented over 150 urgent needs across 21 neighborhoods, were selectively referenced and not formally adopted into decision systems (Organisation Z, 2021k, 2021l, 2021m).

15.4.2 Digital Divide and the Exclusion of Vulnerable Groups

The digital infrastructure that facilitated rapid civil society mobilization during the Zhengzhou 7.20 disaster also revealed stark exclusionary dynamics. While platforms such as WeChat, Weibo, and online mapping tools enabled the circulation of SOS signals and aid requests, not all residents were equally positioned to participate in or benefit from this digital ecosystem. A digital divide—both technological and socio-demographic—significantly limited the visibility and inclusion of vulnerable groups in the emergency response.

Elderly residents, people with disabilities, and children in caregiver-deficient households were notably disadvantaged. Many lacked access to smartphones, familiarity with social media platforms, or the digital literacy required to post geolocated help requests (Ye et al., 2025). In rural outskirts such as Gaocun and parts of Xinmi, language and literacy barriers further impeded the use of digital tools, with several households unable to complete needs-reporting forms distributed via mobile links (Organisation Z, 2021h, 2021i, 2021j). As a result, these groups were underrepresented in crowdsourced demand maps and online rescue dispatch channels.

Beyond individual-level exclusion, institutional relief mechanisms often favored "digitally visible" populations. Community-level authorities prioritized residents in well-networked housing complexes—typically those managed by property management firms—while migrant workers, informal tenants, and elderly individuals living alone were more likely to be overlooked in aid distribution (Zhao et al., 2024). In several cases, physical site visits revealed clear discrepancies between reported coverage and actual conditions, with some households receiving no assistance several days into the crisis despite being marked as "served" in official records (Organisation Z, 2021i, 2021j, 2021k).

Platform logic further amplified visibility disparities. Algorithms on major social media platforms tended to prioritize posts with high engagement, favoring urban centers and densely connected communities while rendering quieter zones into "silent disaster areas" (Ye et al., 2025). Residents disconnected from digital networks thus became "data blind spots" in both NGO and government response systems.

In response, some organizations reverted to analog modes of assessment. Organisation Z deployed over 40 volunteers for door-to-door checks in non-networked communities between July 23 and 28, confirming and recording unmet needs that were not captured digitally (Organisation Z, 2021h, 2021i, 2021j, 2021k). However, these efforts remained resource-intensive and limited in reach.

15.4.3 Spatial and Territorial Inequities in Disaster Response

Despite the rapid mobilization of digital platforms and social organizations during the Zhengzhou 7.20 extreme rainfall disaster, the reach and responsiveness of aid efforts were spatially uneven. Civil society networks, while agile, were largely concentrated in urban cores, resulting in systematic blind spots across urban-rural fringes, peripheral townships, and informal settlements. These spatial disparities reflected deeper structural limitations in resource distribution, digital visibility, and social connectivity.

Geographically, emergency response activities—including digital SOS coordination and volunteer deployment—were concentrated in central districts such as Jinshui and Erqi, where high population density, digital literacy, and NGO presence enabled quick mobilization (Xie et al., 2023). By contrast, outlying areas such as Xingyang, Xinmi, and parts of Zhongmu experienced significant delays in both public and civil society responses. WeChat rescue groups often lacked active community anchors in these areas, and field coordination suffered from logistical gaps (Organisation Z, 2021i).

Rural townships faced particular disadvantages. Many lacked established NGO networks or local volunteer groups and had limited access to digital coordination infrastructure. Organisation Z noted that in multiple village-level cases, response efforts only became effective after local intermediaries—such as teachers or retired cadres—stepped in to connect affected households with aid organizations (Organisation Z, 2021j, 2021k, 2021l, 2021m). Without these embedded figures, external organizations found it difficult to gain access, verify needs, or distribute resources effectively.

Another layer of spatial inequity emerged in hidden or informal spaces. Flooded underground markets, substandard rental housing, and migrant worker dormitories often lacked formal property management or social representation, making them effectively invisible in both governmental and NGO data systems (Ye et al., 2025). In Organisation Z's post-response summaries, such locations were repeatedly described as "structurally under-represented" in early situational maps and only discovered through manual scouting (Organisation Z, 2021g, 2021h, 2021i, 2021j, 2021k).

The cumulative result of these spatial disparities was a disproportionate allocation of resources to areas with stronger digital presence and louder online signals. This "signal-based triage" prioritized visible crises and reinforced existing urban biases in governance structures. Communities that failed to generate digital visibility—whether due to remoteness, infrastructure, or demography—were slower to receive aid, despite potentially higher vulnerability.

15.4.4 Data Governance Gaps and Institutional Absorption Failure

Despite the abundance of grassroots-generated data during the Zhengzhou 7.20 extreme rainfall disaster, significant structural barriers prevented this information from being absorbed into formal government decision-making systems. Civil society actors—particularly NGOs such as Organisation Z—accumulated large volumes of real-time, community-sourced data through digital platforms like WeChat, spreadsheets, and field-based reporting systems. However, these data streams largely remained outside the institutional boundaries of governmental disaster governance.

A primary obstacle was the ambiguity of data ownership. Much of the crisis intelligence originated from commercial platforms such as Weibo, WeChat, and Douyin, whose data are governed by proprietary terms and not directly accessible to public authorities. This limitation restricted the state's ability to verify or incorporate citizen-generated alerts into official dispatch systems (Ye et al., 2025). Even when civil society compiled secondary datasets—such as Organisation Z's "needs map" developed from field interviews and digital SOS signals—there was no clear protocol for integrating this information into government-led coordination (Organisation Z, 2021k).

Equally critical was the incompatibility between grassroots data structures and bureaucratic information systems. NGOs often submitted updates in semi-structured formats: annotated images, conversational texts, and geolocated social media posts. In contrast, public sector workflows were designed for standardized Excel sheets, situation reports, and top-down statistical summaries (Xie et al., 2023). As a result, volunteer-generated insights were difficult to process using government-approved templates and reporting chains.

Ethical and regulatory gaps further complicated integration. During the emergency, some volunteer networks uploaded identifiable images and location data of stranded individuals without informed consent. In the absence of clear data protection standards for disaster scenarios, many grassroots platforms operated in a legal grey zone, deterring government actors from formally engaging with or adopting these datasets (Ye et al., 2025).

Ultimately, the data collected by social organizations failed to transform into what could be termed "governance assets." Although field actors like Organisation Z played crucial roles in identifying isolated households and priority zones (Organisation Z, 2021i, 2021j, 2021k, 2021l, 2021m), these efforts were seldom referenced in official resource allocation or recovery planning. Each round of grassroots mobilization had to rebuild information infrastructures from scratch, with limited institutional memory or procedural continuity.

15.5 Conclusion

The Zhengzhou 7.20 extreme rainfall disaster revealed both the vulnerabilities of state-centric emergency governance and the emergent capacities of digitally mobilized civil society. As infrastructure collapsed and formal responses faltered, grassroots actors filled critical operational and informational gaps, leveraging digital tools to coordinate relief, map needs, and build trust networks. Their entry into governance processes was improvised yet impactful, illustrating new possibilities for hybrid crisis response. However, structural constraints—including legal ambiguity, spatial and digital exclusion, and data absorption failures—limited the institutionalization of these contributions. The Zhengzhou case thus underscores both the promise and precarity of digitally enabled, society-state co-governance under conditions of extreme risk. Moving forward, integrating these emergent practices into formal systems remains a key challenge for building inclusive and resilient urban governance.

References

Chen, Y., Hu, M., Chen, X., Wang, F., Liu, B., & Huo, Z. (2023). An approach of using social media data to detect the real time spatio-temporal variations of urban waterlogging. *Journal of Hydrology, 625*, 130128. https://doi.org/10.1016/j.jhydrol.2023.130128

Chen, Z., Kong, F., & Zhang, M. (2022). A case study of the "7-20" extreme rainfall and flooding event in Zhengzhou, Henan Province, China from the perspective of fragmentation. *Water, 14*(19), 2970. https://doi.org/10.3390/w14192970

Feng, W., Zhang, Y., Wang, S., Xu, M., Li, Y., Wang, P., Zhu, C., Han, S., Shi, L., & Guo, Y. (2021). Integrated risk assessment of multiple air pollutants and influence factors in an urban agglomeration of China. *Polish Journal of Environmental Studies, 30*(5), 4521–4529. https://doi.org/10.15244/pjoes/133718

Gong, Z., Deng, Z., Tang, J., Zhao, H., Liu, Z., & Zhao, P. (2025). Uncovering human behavioral heterogeneity in urban mobility under the impacts of disruptive weather events. *International Journal of Geographical Information Science, 39*(5), 951–974. https://doi.org/10.1080/13658816.2024.2372783

Li, P., Zhuang, L., Lin, K., She, D., Chen, Q., Wang, Q., & Xia, J. (2025). New perspectives on urban stormwater management in China, with a focus on extreme rainfall events. *Natural Hazards, 121*(4), 3745–3774. https://doi.org/10.1007/s11069-024-06994-3

Mao, J., Cao, X., Zhang, R., Tang, J., Zhao, P., & Qin, H. (2026). Characterising the high spatial heterogeneity of urban flood resilience under extreme rainfall using mobile signal data. *Geomatics, Natural Hazards and Risk, 17*(1), 2605116. https://doi.org/10.1080/19475705.2025.2605116

Organisation Z (2021a). *Situation report no. 1: "7.20" extreme rainstorm disaster in Zhengzhou, Henan* (July 21, 2021). https://shimo.im/docs/1lq7MXmMomuJQgAe

Organisation Z. (2021b). *Situation report no. 2: "7.20" extreme rainstorm disaster in Zhengzhou, Henan* (July 21, 2021b). https://shimo.im/docs/tjw8WqrHRGkKggxc

Organisation Z. (2021c). *Situation report no. 3: "7.20" extreme rainstorm disaster in Zhengzhou, Henan* (July 21, 2021c). https://shimo.im/docs/WXQYt9Gc3QHVxwq8

Organisation Z. (2021d). Situation report no. 4: "7.20" extreme rainstorm disaster in Zhengzhou, Henan (July 21, 2021d). https://shimo.im/docs/NJkbEvxXpRCNLlqR

Organisation Z. (2021e). *Situation report no. 5: "7.20" extreme rainstorm disaster in Zhengzhou, Henan* (July 21, 2021e). https://shimo.im/docs/loqeW9nNBeh5LyAn
Organisation Z. (2021f). *Situation report no. 6: "7.20" extreme rainstorm disaster in Zhengzhou, Henan* (July 21, 2021f). https://shimo.im/docs/KrkEVngG9BhyWyAJ
Organisation Z. (2021g). *Situation report no. 7: "7.20" extreme rainstorm disaster in Zhengzhou, Henan* (July 21, 2021g). https://shimo.im/docs/5rk9drMKLBubm5qx
Organisation Z. (2021h). *Situation report no. 8: "7.20" extreme rainstorm disaster in Zhengzhou, Henan* (July 21, 2021h). https://shimo.im/docs/m8AZVe8E0NcVL9Ab
Organisation Z. (2021i). Situation report no. 9: "7.20" extreme rainstorm disaster in Zhengzhou, Henan (July 21, 2021i). https://shimo.im/docs/thgVwRDXRCWpjHQH
Organisation Z. (2021j). Situation report no. 10: "7.20" extreme rainstorm disaster in Zhengzhou, Henan (July 21, 2021j). https://shimo.im/docs/RKAWVbWZlNhYdvk8
Organisation Z. (2021k). *Situation report no. 11: "7.20" extreme rainstorm disaster in Zhengzhou, Henan* (July 21, 2021k). https://shimo.im/docs/e1Az4MoQoXcyVxqW
Organisation Z. (2021l). *Situation report no. 12: "7.20" extreme rainstorm disaster in Zhengzhou, Henan* (July 21, 2021l). https://shimo.im/docs/m4kMLbY8Bnu5ywqD
Organisation Z. (2021m). *Situation report no. 13: "7.20" extreme rainstorm disaster in Zhengzhou, Henan* (July 21, 2021m). https://shimo.im/docs/B1Aw1MDR78IbaEqm
Peng, J., & Zhang, J. (2024). Spatiotemporal assessment of urban flooding hazard using social media: A case study of Zhengzhou '7·20.'. *Environmental Modelling and Software, 176*, 106021. https://doi.org/10.1016/j.envsoft.2024.106021
Qu, Z., Wang, J., & Zhang, M. (2023). Mining and analysis of public sentiment during disaster events: The extreme rainstorm disaster in megacities of China in 2021. *Heliyon, 9*(7), e18272. https://doi.org/10.1016/j.heliyon.2023.e18272
State Council Disaster Investigation Group. (2022). *Investigation report on the "7.20" extraordinarily heavy rain disaster in Zhengzhou, Henan*. https://www.mem.gov.cn/gk/sgcc/tbzdsgdcbg/202201/P020220121639049697767.pdf.
Wang, J., Feng, J., Sun, B., Guo, J., Hu, S., & Zhang, M. (2024a). Characterization of dam failure flooding in an urban reservoir under different rainfall conditions. *Scientific Reports, 14*(1), 27560. https://doi.org/10.1038/s41598-024-78968-1
Wang, Y., Li, S., Hu, C., Ren, J., Liu, P., Zhao, C., & Zhu, M. (2024b). Analysis of surface runoff characteristics in Zhengzhou City under extreme rainfall conditions. *Sustainability, 16*(16), 6980. https://doi.org/10.3390/su16166980
Wu, Y., Chen, Y., Zhang, R., Cui, Z., Liu, X., Zhang, J., Wang, M., & Wu, Y. (2023). A spatial information extraction Method based on multi-modal social media data: A case study on urban inundation. *ISPRS International Journal of Geo-Information, 12*(9), 368. https://doi.org/10.3390/ijgi12090368
Xiao, C., Zhong, Y., Feng, W., Gao, W., Wang, Z., Zhong, M., & Ji, B. (2023). Monitoring the catastrophic flood with GRACE-FO and near-real-time precipitation data in northern Henan Province of China in July 2021. *IEEE Journal of Selected Topics in Applied Earth Observations and Remote Sensing, 16*, 89–101. https://doi.org/10.1109/JSTARS.2022.3223790
Xie, X., Huang, L., Marson, S. M., & Wei, G. (2023). Emergency response process for sudden rainstorm and flooding: Scenario deduction and Bayesian network analysis using evidence theory and knowledge meta-theory. *Natural Hazards, 117*(3), 3307–3329. https://doi.org/10.1007/s11069-023-05988-x
Xu, Q., Han, L., & Xu, K. (2022). Causal analysis and prevention measures for extreme heavy rainstorms in Zhengzhou to protect human health. *Behavioral Sciences, 12*(6), 176. https://doi.org/10.3390/bs12060176
Yang, Z., Dong, X., & Guo, L. (2023). Scenario inference model of urban metro system cascading failure under extreme rainfall conditions. *Reliability Engineering and System Safety, 229*, 108888. https://doi.org/10.1016/j.ress.2022.108888
Yang, Z., & Wang, Q. (2022). Mining method of traffic impact areas of rainstorm event based on social media in Zhengzhou city The International Archives of the Photogrammetry, Remote

Sensing and Spatial, Information Sciences, XLVIII-3/W1-2022, 79–86. doi:https://doi.org/10.5194/isprs-archives-XLVIII-3-W1-2022-79-2022.

Ye, Q., Ma, Y., Xu, S., Sotelo, M. Á., & Li, Z. (2025). Understanding urban mobility responses to extreme precipitation events: A case study of Zhengzhou, China. *IET Intelligent Transport Systems, 19*(1), e70045. https://doi.org/10.1049/itr2.70045

Zhang, Q., Li, R., Sun, J., Lu, F., Xu, J., & Zhang, F. (2023). A review of research on the record-breaking precipitation event in Henan Province, China, July 2021. *Advances in Atmospheric Sciences, 40*(8), 1485–1500. https://doi.org/10.1007/s00376-023-2360-y

Zhao, H., Liu, Y., Yue, L., Gu, T., Tang, J., & Wang, Z. (2024). Unraveling the factors behind self-reported trapped incidents in the extraordinary urban flood disaster: A case study of Zhengzhou City, China. *Cities, 155*, 105444. https://doi.org/10.1016/j.cities.2024.105444

Zhao, X., Li, H., Cai, Q., Pan, Y., & Qi, Y. (2023). Managing extreme rainfall and flooding events: A case study of the 20 July 2021 Zhengzhou flood in China. *Climate, 11*(11), 228. https://doi.org/10.3390/cli11110228

Part IV
Practices of Social Organizations in Emergency Management: Overseas Chinese Experiences

Chapter 16
Nepal Earthquake: Cross-Border Action and Local Integration

Abstract The 2015 Nepal Earthquake marked a significant early moment of international engagement for Chinese social organizations, offering both a humanitarian imperative and a testing ground for nascent cross-border capacities. This chapter traces their response across four interrelated dimensions: the motivations and mobilization of diverse actors entering Nepal's humanitarian landscape; the improvisational field operations conducted under institutional uncertainty; the evolving efforts at community engagement and cultural adaptation; and the subsequent institutional learning that catalyzed an emergent global orientation. Through detailed analysis, the chapter reveals how these organizations operated within—and against—the constraints of limited experience, legal ambiguity, and coordination gaps. Their actions in Nepal not only addressed urgent needs but also reshaped organizational identity, setting in motion a shift from domestic relief providers to aspiring transnational actors.

Keywords Cross-border response · Operational improvisation · Local integration · Organizational learning · Global orientation

16.1 Early Cross-Border Mobilization: Actors, Motivations, and Organizational Capacity

The 2015 Nepal Earthquake marked a pivotal moment in the early international engagement of Chinese social organizations. This section examines how the disaster created a rare cross-border entry point for civil actors, analyzing the geopolitical and cultural conditions that facilitated their involvement. It outlines the diversity of participating organizations and their decentralized mobilization strategies, explores the humanitarian, strategic, and institutional motivations behind their actions, and critically assesses the operational constraints they faced. These dynamics reveal the experimental nature of China's nascent global civil society and its emergent role in transnational humanitarian response.

Q. Du and R. Shaw, *Social Organizations in Disaster Response and Recovery*, Disaster Risk Reduction, https://doi.org/10.1007/978-981-92-2846-1_16

16.1.1 Contextualizing the Earthquake and the Cross-Border Opportunity

The 2015 Nepal Earthquake, registering a magnitude of 7.8, struck on April 25 and rapidly escalated into a large-scale humanitarian emergency. The epicenter near Barpak in the Gorkha District triggered widespread destruction across central Nepal, severely affecting the Kathmandu Valley and neighboring rural regions. The disaster caused more than 8800 deaths and injured over 22,000 people (Dubey et al., 2015; Kargel et al., 2016). Infrastructural collapse included the destruction of over 600,000 homes and critical damage to roads, healthcare facilities, and UNESCO World Heritage sites (Jones et al., 2016; Wang et al., 2016). Compounding the crisis was the region's mountainous geography, which significantly hindered logistics and communication, creating urgent gaps in immediate disaster response capacity (He et al., 2018; Organisation Z, 2015a).

In this context, Nepal became a focal point for regional and international humanitarian mobilization. The Chinese government reacted swiftly, dispatching search-and-rescue teams, military aircraft, and emergency medical units within 24 hours of the earthquake (China News Service, 2015; Xinhua News Agency, 2015a). However, what distinguished this response from previous episodes was the simultaneous emergence of civil society actors as cross-border responders. For Chinese social organizations—ranging from established foundations to informal volunteer collectives—the Nepal disaster presented a rare and accessible opportunity for overseas engagement. Several contextual factors made this possible.

Geopolitically, Nepal's status as a close neighbor with long-standing diplomatic and cultural ties to China reduced the perceived risks often associated with overseas interventions. Unlike complex conflict zones, Nepal was seen as politically neutral, socially proximate, and logistically reachable—especially for NGOs based in southwest China and the Tibetan plateau (Liang & Zhou, 2016; Seale-Feldman, 2020). Culturally, shared Buddhist heritage and people-to-people exchanges along border regions fostered a sense of transnational empathy and solidarity, which further lowered the psychological barrier for civil society entry (Bhattarai et al., 2023; Zhao, 2016).

Domestically, the rapid state-led deployment created both infrastructural support and political endorsement for NGO involvement. Early media reports signaled a positive framing of non-governmental participation (China Youth Daily, 2015), and several organizations cited this state response as a "green light" for action (Organisation Z, 2015c). In this sense, the Nepal Earthquake not only exposed the humanitarian need but also generated a rare alignment of logistical, political, and cultural conditions that enabled Chinese NGOs to experiment with international engagement—albeit in a nascent and improvised form.

16.1.2 Key Actors and Forms of Mobilization

The humanitarian response to the 2015 Nepal Earthquake witnessed the rapid mobilization of a diverse array of Chinese social organizations. These actors represented a broad spectrum of civil society, including philanthropic foundations, university-based initiatives, faith-based groups, private-sector volunteers, and semi-formal professional associations. Notably, organizations such as the One Foundation and Amity Foundation deployed rapid assessment teams and relief supplies, while smaller units—ranging from university rescue squads to Buddhist volunteer networks—contributed medical aid, food distribution, and situational mapping (Organisation Z, 2015a, 2015d; Xinhua News Agency, 2015b).

A defining characteristic of this mobilization was its decentralized and non-institutionalized structure. Rather than operating through pre-existing international coordination mechanisms, many Chinese groups relied on digital platforms, such as WeChat and QQ, to form ad hoc teams, coordinate logistics, and share real-time field intelligence (China Youth Daily, 2015). Drone-assisted reconnaissance and crowdsourced geographic mapping, for example, were facilitated through grassroots volunteer collaborations, indicating a high level of technological adaptability despite the absence of formal training (Zhang & Pan, 2022).

Forms of mobilization exhibited hybrid features that blurred the lines between professionalized humanitarianism and spontaneous volunteerism. For instance, volunteer groups from Shenzhen and Chengdu self-organized into rescue units equipped with basic search-and-rescue gear and partnered with local intermediaries upon arrival in Kathmandu (Organisation Z, 2015b, 2015g). Meanwhile, medical teams affiliated with university hospitals initiated temporary clinics in rural areas, often operating without formal host-country accreditation (Organisation Z, 2015h).

These efforts co-existed with China's state-led assistance, producing a dual-track model of engagement. While official aid focused on material supply—such as the dispatch of over 10,000 emergency kits and multiple rounds of shelter equipment (China News Service, 2015)—social organizations filled operational gaps in remote villages, often navigating terrain and language barriers without institutional support (Organisation Z, 2015e, 2015m).

16.1.3 Motivations and Strategic Rationales

The mobilization of Chinese social organizations during the 2015 Nepal Earthquake was underpinned by a blend of humanitarian concern, strategic positioning, and the desire for institutional evolution. These motivations were shaped both by the specificities of the Nepal crisis and by broader structural transformations within China's civil society landscape.

Humanitarian impulse was a central motivator. The severe devastation caused by the 7.8-magnitude earthquake on April 25, 2015—resulting in over 9000 deaths, more than 22,000 injuries, and large-scale displacement—triggered a strong sense of urgency and empathy among Chinese civil actors (Kargel et al., 2016; Xinhua News Agency, 2015a). For many organizations, such as those documented in field reports by Organisation Z, the decision to mobilize was framed explicitly in moral terms: a duty to provide life-saving support and alleviate suffering beyond national borders (Organisation Z, 2015e, 2015m).

Beyond moral imperatives, the Nepal response was also seen as a reputational and strategic opportunity. Entering the international humanitarian field allowed Chinese NGOs to raise their visibility, both domestically and globally. For emerging NGOs, participation offered a chance to rebrand themselves as internationally relevant actors, while for more established foundations, such as One Foundation and Amity Foundation, it provided an opportunity to consolidate their credibility in international civil society (Li et al., 2025; Organisation Z, 2015a, 2015b, 2015c, 2015d, 2015e, 2015f, 2015g, 2015h, 2015i, 2015j, 2015k, 2015l, 2015m, 2015n, 2015o). This pursuit of visibility was not merely symbolic—it was linked to long-term goals of resource diversification, international partnerships, and legitimacy within a competitive NGO landscape.

Political context also played an enabling role. The broader "Going Out" strategy promoted under the Belt and Road Initiative (BRI) emphasized the internationalization of Chinese civil actors as part of soft power projection (Lin, 2021). While the Nepal case predated some of the more formal BRI mechanisms, it was already operating within a discursive environment that encouraged Chinese social organizations to seek cross-border engagement. The earthquake thus functioned as a relatively "safe" testing ground—close to China's borders, with cultural affinity and limited geopolitical sensitivity (Fehr, 2022).

Cultural and emotional proximity further strengthened mobilization. Shared religious heritage, particularly through Buddhism, and long-standing people-to-people links in the border regions enhanced the moral and emotional appeal of aiding Nepal (Seale-Feldman, 2020). In some cases, existing volunteer networks or faith-based affiliations facilitated rapid deployment. Some volunteers had previously worked in Tibetan communities or Buddhist institutions, easing the process of cultural adaptation and reducing psychological distance between actors and affected communities (Organisation Z, 2015d; Zhang & Pan, 2022).

For many organizations, the Nepal response was not only an emergency deployment but also a valuable opportunity for institutional learning. Engagement with international standards, cluster coordination, and multilingual environments served as a platform to build experience for future participation in more complex humanitarian scenarios. Internally, the mission was often framed as a preparatory step toward deeper integration into global disaster governance systems (He et al., 2018; Organisation Z, 2015f).

16.1.4 Constraints, Gaps, and the Nature of "Early" Mobilization

Despite the rapid and committed cross-border mobilization of Chinese social organizations following the 2015 Nepal Earthquake, their engagement revealed significant limitations indicative of an early-stage phase of international humanitarian participation. These constraints manifested across institutional, operational, and logistical dimensions.

One of the most pervasive challenges was the lack of prior international experience. For many participating groups—particularly volunteer collectives and newly formed networks—this was their first cross-border humanitarian mission, with little precedent to draw upon for planning, coordination, or implementation (Yu et al., 2018; Zhao, 2016). The absence of standardized operational procedures or prior international partnerships resulted in a largely ad hoc mode of operation that, while flexible, lacked institutional robustness.

Legal and institutional gaps further constrained Chinese NGOs on the ground. Most organizations lacked legal registration or formal recognition in Nepal, complicating customs clearance, local coordination, and field deployment. Delays in transporting relief materials were frequently caused by incomplete documentation or unclear customs protocols, as reflected in several coordination accounts. Additionally, few organizations held international insurance coverage or had access to legal advisors familiar with cross-border regulatory regimes.

Communication was another key barrier. Language limitations hampered coordination with local authorities and international NGOs. The lack of Nepali and English-speaking staff frequently delayed information verification and needs assessment (Organisation Z, 2015e). Most coordination instead relied on internal Chinese communication tools like WeChat, reflecting a reliance on domestic digital ecosystems rather than integration into international humanitarian platforms.

Operationally, organizations demonstrated strong motivation but relied heavily on improvisational logistics. For example, delivery of medical supplies and rescue equipment was largely dependent on commercial carriers arranged informally in China, without standardized procurement or warehousing systems (Organisation Z, 2015d, 2015i). Volunteers and technical teams were often assembled on short notice, drawing on social media networks and personal contacts.

These features underscore the "experimental" nature of Chinese NGOs' early cross-border engagement: fast-moving but fragile, enthusiastic but institutionally thin. Mobilization efforts were often described as "learning-by-doing" (Zhao, 2016), highlighting both the opportunities and vulnerabilities inherent in China's emerging global civil society presence.

16.2 Early-Stage Field Operations: Humanitarian Practice and Coordination Under Institutional Uncertainty

The early-stage deployment of Chinese social organizations to Nepal following the 2015 earthquake revealed the operational realities of cross-border humanitarian action under institutional uncertainty. Lacking prior exposure to global coordination systems, these actors faced challenges in integrating into formal humanitarian frameworks, navigating legal ambiguities, and managing transnational logistics. In response, they relied on adaptive field strategies, informal partnerships, and digital tools to sustain operations. This section analyzes four interrelated dimensions of their early engagement: institutional unfamiliarity, legal-operational constraints, improvisational field practices, and the enabling role of technology and informal networks in shaping emergent humanitarian capacities.

16.2.1 Operating in an Unfamiliar Institutional Environment

The deployment of Chinese social organizations to Nepal in the aftermath of the 2015 earthquake marked one of their earliest collective attempts to operate within an international humanitarian setting. However, these actors entered a field dominated by pre-existing global coordination systems—particularly the United Nations-led Cluster system—without prior exposure to its operational norms, terminologies, or bureaucratic pathways. The result was a pronounced sense of institutional unfamiliarity, which limited their integration into formal response frameworks.

Unlike many Western humanitarian NGOs accustomed to the Office for the Coordination of Humanitarian Affairs (OCHA) and its cluster-based coordination model, Chinese NGOs were largely unacquainted with procedures such as humanitarian registration, sectoral coordination, or inter-agency task allocation. Multiple Chinese teams arrived without understanding how to attend coordination meetings, submit needs assessments, or report activities to central response bodies (Organisation Z, 2015b, 2015c, 2015d, 2015e, 2015f). This disconnect created an operational vacuum, where Chinese responders functioned parallel to—but not within—the main international coordination grid.

Furthermore, interviews and field documentation showed that many organizations were unaware of the mechanisms through which NGOs access humanitarian information, such as ReliefWeb or HumanitarianResponse.info. Instead, coordination tended to rely on personal networks and informal exchanges. For instance, one team collaborated with a local monastery to establish a medical station, circumventing formal government or UN channels entirely (Organisation Z, 2015d). Another partnered directly with a village school to distribute supplies, illustrating a preference—or necessity—for "single-point" grassroots coordination over multilateral institutional engagement (Organisation Z, 2015h).

The fragmented nature of Chinese participation was compounded by linguistic and procedural barriers. Few organizations had English- or Nepali-speaking staff, which severely hampered their ability to communicate with international counterparts or interpret shared coordination documents. Delays in needs assessment and duplication of effort were common due to poor information flow and minimal data integration with other actors on the ground (Organisation Z, 2015e). This led to operational inefficiencies, including overlapping distributions and under-served communities, particularly in remote districts such as Sindhupalchowk and Gorkha.

Although some Chinese NGOs attempted to bridge these gaps—by sending staff to observe cluster meetings or relying on local intermediaries—the lack of prior experience created a steep learning curve. In comparison to long-standing international NGOs, Chinese organizations lacked not only institutional access but also familiarity with the soft norms of humanitarian diplomacy, such as accountability, coordination etiquette, and standard reporting procedures (Cook et al., 2018; Dubey et al., 2015).

16.2.2 Coping with Legal Ambiguities and Operational Constraints

During the 2015 Nepal Earthquake response, Chinese social organizations confronted significant legal, financial, and operational constraints that underscored the institutional limitations of early-stage cross-border humanitarian engagement. These constraints not only complicated on-the-ground implementation but also exposed the absence of a legal-institutional architecture to support international civil society action from China.

A major structural barrier was the lack of legal registration in Nepal. Most participating Chinese NGOs and volunteer teams entered the country without formal recognition under Nepalese law. This led to repeated complications in customs clearance, bank transfers, facility rentals, and inter-agency coordination. During the response, numerous Chinese teams faced delays or obstacles in transporting relief supplies due to missing documentation or ambiguities surrounding their legal status. Lacking local legal identity, these organizations were often unable to open official bank accounts or sign formal agreements, which restricted their ability to function as recognized operational actors. These constraints significantly limited cross-border coordination and formal engagement in host-country mechanisms (Organisation Z, 2015c, 2015e).

The financial operations of many organizations were similarly constrained. Few had transnational financial mechanisms in place, resulting in an overreliance on domestic bank accounts, individual remittances, or informal cash-based arrangements (Liang & Zhou, 2016). For example, several teams coordinated fundraising in China via WeChat or Alipay, then physically carried cash across borders or wired money to personal accounts in Kathmandu to circumvent institutional bottlenecks. This made transactions more opaque and exposed organizations to regulatory risk, particularly in the absence of official auditing or compliance systems.

Logistical challenges further compounded operational difficulties. The lack of pre-existing partnerships with local customs brokers or legal intermediaries meant that most Chinese teams were unfamiliar with Nepal's import procedures. Relief materials—such as tents, medical kits, and food packages—were often detained at the border or at Tribhuvan International Airport for extended periods due to improper manifests or insufficient clearance support (China Youth Daily, 2015; Organisation Z, 2015d). These setbacks hindered the timely delivery of aid to affected communities and strained volunteer capacity in the early days of the crisis.

Equally concerning was the lack of insurance and security protocols. The majority of volunteer teams, especially youth-led and faith-based groups, deployed without international travel insurance, accident coverage, or field security training (Organisation Z, 2015e, 2015g). Many first responders were unaware of basic risk mitigation practices, such as safety briefings or emergency evacuation procedures. In a highly unstable post-disaster setting—with aftershocks, damaged infrastructure, and limited medical services—this exposed both aid workers and beneficiaries to potential harm.

Despite their strong moral commitment and operational enthusiasm, Chinese social organizations thus found themselves operating within a "gray zone"—present in the field, yet outside the formal humanitarian framework. Their ambiguous status constrained access to international coordination platforms, complicated relationships with local authorities, and undermined long-term institutional learning. As early entrants in transnational disaster response, these organizations encountered not just the challenges of physical relief delivery, but the deeper constraints of legal-institutional absence that continue to shape the contours of China's emerging global civil society.

16.2.3 Adaptive Field Practices Under Uncertainty

In the absence of stable institutional infrastructures and formal coordination mechanisms, Chinese social organizations operating in post-earthquake Nepal adopted a range of adaptive practices that reflect both improvisation and resilience. These practices emerged as functional responses to institutional uncertainty, legal ambiguity, and logistical fragmentation—challenges characteristic of early-stage cross-border engagement.

Rather than relying on structured humanitarian protocols, many Chinese NGOs and volunteer teams established informal collaborations with local actors, including schools, temples, community-based organizations, and informal volunteer groups. These partnerships enabled rapid deployment of resources, particularly in rural areas beyond the reach of large international agencies. For example, Chinese medical teams cooperated with local Buddhist monasteries to set up temporary clinics, while volunteer teams used temple courtyards and school grounds as distribution hubs for relief goods such as tents, medicine, and food (Organisation Z, 2015f, 2015h). In the absence of pre-arranged facilities, some organizations even constructed makeshift shelters or repurposed local infrastructure to support their operations.

Coordination within and across Chinese organizations relied heavily on digital platforms. WeChat, QQ groups, and SMS messaging served as key operational tools for real-time decision-making, personnel dispatch, and intelligence sharing (Zhang & Pan, 2022). This decentralized use of mobile platforms allowed dispersed teams to communicate needs, avoid duplication, and reroute supplies dynamically. However, the same reliance also led to fragmented coordination: the lack of centralized data-sharing or a unified logistical framework resulted in repeated deployments to the same villages, overlapping efforts, and occasional misallocation of resources (Organisation Z, 2015i).

Despite limited field experience, teams demonstrated notable flexibility. Many volunteers adapted their roles based on evolving needs—from first aid providers to supply transporters, interpreters, and community liaisons. Action plans were frequently revised in response to local feedback, weather shifts, and transportation delays (Organisation Z, 2015g, 2015l). This adaptability, while often necessitated by weak institutional anchoring, also fostered localized responsiveness and rapid learning cycles.

16.2.4 Digital Tools and Informal Networks in Action

In the context of limited institutional support and coordination barriers, Chinese social organizations involved in the 2015 Nepal Earthquake response leaned heavily on digital tools and informal networks to sustain and operationalize their humanitarian missions. These technological and social infrastructures functioned as "temporary scaffolding," enabling Chinese actors to navigate uncertainties and implement field operations despite the absence of formal mechanisms.

Mobile communication platforms such as WeChat and QQ played a critical role in real-time coordination, functioning as the backbone of intra- and inter-organizational communication. Through chat groups, organizations were able to dispatch volunteer teams, report security risks, share supply updates, and circulate rapidly evolving needs from field sites (Zhang & Pan, 2022). In some cases, WeChat groups with over 500 members became de facto command centers, substituting for more formal coordination bodies which Chinese teams found inaccessible or opaque (Organisation Z, 2015f, 2015h).

To compensate for the lack of standardized data collection systems, Chinese NGOs implemented flexible mobile-based reporting systems. Volunteers used tools like Google Sheets, mobile phone cameras, and social media posts to document distributions and register affected populations (Organisation Z, 2015k). These methods—though often lacking methodological consistency—offered real-time visibility and adaptability, especially in remote zones where bandwidth and official communication channels were limited.

Informal translation networks also emerged as a critical enabler. Given the scarcity of Nepali- or English-speaking personnel in many Chinese teams, ad hoc language support was crowdsourced through volunteer interpreters recruited from overseas

Chinese communities, local students, and diaspora networks. These translators mediated between Chinese responders and local residents, police, and health officials, ensuring basic communicative functioning in otherwise linguistically fragmented environments (Organisation Z, 2015j, 2015l).

Moreover, digital mapping technologies, including unmanned aerial vehicles (UAVs), were used by some tech-savvy teams to generate visual damage assessments and prioritize target areas for distribution. While these practices remained relatively limited in scale, they marked a significant shift toward data-informed operations among emerging humanitarian actors (Zhang & Pan, 2022).

16.3 Early Attempts at Local Integration: Community Engagement, Cultural Adaptation, and Trust-Building

In the early phase of the 2015 Nepal Earthquake response, Chinese social organizations faced the dual challenge of delivering urgent aid while navigating an unfamiliar social and cultural terrain. This section examines their initial efforts at local integration, tracing how operational priorities evolved from improvised entry points to deeper community engagement. It focuses on four key aspects: early contact mechanisms and the use of community intermediaries; barriers stemming from language and cultural distance; the transition from transactional relief to relational trust-building; and the emerging, though uneven, awareness of vulnerable groups. These themes highlight the experimental nature of early cross-border engagement.

16.3.1 First Encounters: Entry Points and Initial Community Contacts

The initial entry of Chinese social organizations into post-earthquake Nepal in 2015 was marked by a combination of improvisation and pragmatism. Lacking prior experience with Nepalese civil society or humanitarian logistics in the region, most Chinese NGOs sought immediate and visible access points into local communities. In practice, this meant establishing operations through neutral and accessible community sites such as temples, schools, and local health outposts (Organisation Z, 2015c, 2015e). These spaces not only offered logistical convenience but were often viewed as culturally respected and politically non-contentious venues for relief activities.

Early community engagement was largely facilitated through informal intermediaries. Many Chinese teams worked with local religious leaders, schoolteachers, and village volunteers to gain initial trust and distribute supplies. In Sindhupalchowk and Dhading districts, Chinese medical teams collaborated with Buddhist monasteries to set up mobile clinics and temporary shelters. At the same time, volunteer groups repurposed schoolyards and public squares for food and tent distribution (Organisation Z, 2015f). These arrangements, while effective in rapid mobilization, reflected the absence of formalized local partnerships or Memoranda of Understanding with host communities.

In several cases, partnerships were ad hoc and highly situational. When medical supplies were delayed or routes to remote villages were blocked, organizations adjusted by embedding with Nepali volunteer networks or using Chinese-speaking local residents as cultural and logistical brokers (Jones et al., 2016; Organisation Z, 2015h). These intermediaries played a crucial role in facilitating introductions to village heads, coordinating entry routes, and managing expectations on both sides.

Despite the ingenuity of these entry strategies, most Chinese NGOs operated without a clear framework for long-term community engagement. Engagements were typically structured around immediate delivery rather than sustained presence, and decision-making remained centralized within the Chinese teams (Organisation Z, 2015g). The absence of locally embedded staff or institutional memory further limited continuity and learning between sites of intervention.

Notably, a few mixed Chinese-Nepali volunteer teams succeeded in building more sustainable connections by staying longer in affected villages and participating in daily routines. Even in these instances, the lack of prior cultural exposure and institutional preparedness constrained deeper integration (Organisation Z, 2015f; Seale-Feldman, 2020).

16.3.2 Cultural Distance and Communication Barriers

Language gaps and cultural unfamiliarity posed serious challenges for Chinese social organizations attempting to engage local communities in Nepal during the early stages of the 2015 earthquake response. Without prior cross-cultural training or embedded field experience, many teams found themselves navigating environments where both linguistic and socio-cultural misalignment complicated effective coordination and trust-building.

Language barriers significantly constrained effective engagement. Without adequate English or Nepali proficiency, Chinese volunteer teams often depended on informal interpreters recruited from local communities or the Chinese-speaking Nepalese population. However, these interpreters were frequently too few in number and lacked formal training, resulting in repeated communication failures with residents, officials, and international organizations (Organisation Z, 2015g, 2015i). One incident during a medical consultation illustrated this gap: Patients misunderstood dosage instructions due to inconsistent translation, leading to treatment errors and reduced trust among local communities (Organisation Z, 2015i).

Beyond language, cultural distance became a recurring source of friction. Several Chinese teams struggled to navigate local religious practices, dietary customs, and gender norms—factors essential for respectful community engagement in Nepal's diverse social landscape. In some cases, aid distributions inadvertently clashed with local customs—for example, the provision of non-vegetarian food in predominantly Hindu villages or the deployment of male volunteers to distribute supplies to female-headed households, which led to discomfort and refusal of aid (Organisation Z, 2015j). These incidents were not isolated; similar mismatches also occurred in relation to public space use and religious festivals (Fehr, 2022; Seale-Feldman, 2020).

Moreover, the lack of cultural mediation reinforced a perception of "externality" among some Nepali communities. In areas such as Nuwakot and Gorkha, residents reported hesitancy in engaging with Chinese teams due to limited understanding of their identity, purpose, or affiliations (Organisation Z, 2015j). This "information gap" often produced what responders themselves described as "social isolation zones," in which their efforts remained peripheral to local relief dynamics, even when material assistance was successfully delivered.

16.3.3 From Material Delivery to Relationship Building

In the early weeks of the 2015 Nepal Earthquake response, Chinese social organizations primarily engaged in direct material delivery. Relief operations were initially centered on the distribution of tents, medicine, food, and other essentials, often in a "drop-and-go" format with limited community consultation or participatory planning (Organisation Z, 2015d, 2015e). This approach reflected both the urgency of immediate humanitarian needs and the unfamiliarity of Chinese teams with participatory relief practices common in international settings (Dubey et al., 2015; Jones et al., 2016). While these early interventions addressed acute shortages, they often bypassed local governance structures and did not foster sustained community relationships.

As operations progressed, however, several organizations recognized the limitations of transactional aid delivery. This awareness prompted a gradual shift toward more relational and embedded approaches. In areas such as Bhaktapur and Sindhupalchowk, Chinese teams began engaging local stakeholders—including village committees, school principals, and Buddhist monks—to co-manage distribution efforts and coordinate follow-up visits (Organisation Z, 2015k). In some cases, local leaders were involved in beneficiary registration and site selection, which improved the perceived legitimacy and effectiveness of aid delivery. These interactions marked early steps toward collaborative humanitarian practice, albeit still constrained by linguistic and institutional barriers.

Medical teams were among the first to cultivate deeper ties with host communities. Some health teams established semi-permanent service points within affected villages, providing repeated check-ups for women and children, health education sessions, and psychological support. These continued presences enabled the development of trust-based relationships, as residents came to associate Chinese responders not only with supplies but also with accessible and caring services (Organisation Z, 2015l).

In addition, a few organizations experimented with community-based needs assessments and participatory rural appraisal (PRA) tools to better understand local priorities. For instance, one volunteer team in Gorkha piloted a village mapping exercise in collaboration with local youth, aiming to identify high-need households and overlooked terrain (Organisation Z, 2015m). While such practices remained sporadic and often informal due to limited technical capacity, they demonstrated an evolving recognition of the value of local knowledge and co-produced relief strategies (He et al., 2018; Zhang & Pan, 2022).

16.3.4 Identifying and Responding to Vulnerable Groups

In the immediate aftermath of the 2015 Nepal Earthquake, Chinese social organizations primarily approached humanitarian aid through general distribution models. Relief items such as tents, rice, medicine, and hygiene kits were allocated without systematic differentiation between households or population groups (Organisation Z, 2015d, 2015e). This "one-size-fits-all" approach stemmed in part from logistical urgency and limited field knowledge, but it also reflected the initial absence of vulnerability assessment mechanisms within many Chinese teams.

However, as the humanitarian response unfolded and local engagements deepened, some organizations began to recognize the heterogeneity of needs across affected populations. By the third week of operations, several teams had started to identify vulnerable sub-groups, particularly women, children, persons with disabilities, and the elderly (Organisation Z, 2015m). For instance, in remote villages of Nuwakot and Dhading, volunteers reported cases of elderly survivors who were unable to access distribution points, prompting teams to establish mobile delivery units. Similarly, pregnant women and lactating mothers were identified as requiring specialized nutritional and medical support, which led to targeted distributions of maternal care kits in selected zones (Organisation Z, 2015m).

Gender-related vulnerabilities emerged as a particularly overlooked area in the early phase. According to situational observations, some female recipients expressed discomfort at mixed-gender shelter arrangements and lack of privacy. In response, a number of teams began to designate "women-only tents" and to create child-safe spaces in cooperation with local women's groups and school teachers (Organisation Z, 2015n). These ad hoc adjustments marked an important step toward incorporating gender sensitivity into operational design, though often without the support of formal tools or frameworks.

Despite these emerging adaptations, comprehensive vulnerability mapping remained limited. Many teams continued to rely on visible indicators of distress or ad hoc local referrals rather than standardized assessment tools. The absence of prior training in international protection standards—such as Sphere guidelines or Do No Harm principles—also constrained the depth and consistency of vulnerability-focused programming (Fehr, 2022; Seale-Feldman, 2020). In several documented cases, attention to marginal groups only followed incidents of exclusion or complaints from affected communities, reflecting a reactive rather than anticipatory orientation (He et al., 2018; Organisation Z, 2015j).

16.4 Institutional Learning and the Formation of an Early Global Orientation

The 2015 Nepal Earthquake offered a rare and catalytic opportunity for Chinese social organizations to engage in cross-border humanitarian action at an unprecedented scale. This section examines how this experience triggered a process of institutional learning, exposing the structural limits of domestically-oriented capacities, prompting internal reflection on operational gaps, and initiating early encounters with international norms and ethical frameworks. Through these challenges, Chinese NGOs began to redefine their roles within global civil society. The Nepal mission, though improvised and exploratory, laid the groundwork for a more deliberate global orientation, marking a pivotal stage in the international evolution of China's nonprofit sector.

16.4.1 Revealing the Boundaries of Domestic-Centered Capacity

The 2015 Nepal Earthquake marked one of the earliest large-scale overseas deployments of Chinese social organizations in an international humanitarian context. This engagement revealed substantial mismatches between domestically-formed capacities and the institutional demands of cross-border disaster relief. Most Chinese NGOs were accustomed to operating within a state-led framework in China, where coordination with local authorities, predictable regulatory environments, and nationally-integrated logistics chains provided structural support for project implementation (Liang & Zhou, 2016; Lin, 2021). Once in Nepal, however, these assumptions no longer held.

From the outset, organizations encountered legal and logistical hurdles that they were ill-prepared to navigate. Many teams lacked overseas bank accounts or currency exchange mechanisms, making fund transfers and procurement extremely difficult. Delays in accessing funds due to the absence of foreign exchange-ready financial infrastructure, and several teams relied on individual volunteers carrying cash or using personal overseas cards to purchase supplies (Organisation Z, 2015f).

Further compounding operational inefficiencies were customs clearance issues and poor coordination across the supply chain. Relief supplies sent from China faced hold-ups at the border due to a lack of familiarity with Nepali import protocols, insufficient documentation, or delays in securing temporary import exemptions (Organisation Z, 2015i, 2015j). In one case, a batch of urgently needed medical equipment remained in transit for several days, resulting in gaps in mobile clinic service delivery in Sindhupalchowk.

Administrative and institutional gaps also emerged in relation to personnel deployment. Most NGOs did not have formalized processes for obtaining Nepali visas, registering teams, or insuring staff for overseas humanitarian work. The lack of standardized deployment protocols exposed volunteer teams to potential legal and safety risks (Fehr, 2022). For example, Organisation Z's field notes highlighted confusion around local NGO registration requirements, which in some districts resulted in Chinese teams being temporarily restricted from operating until proper documentation was arranged (Organisation Z, 2015j).

These operational frictions were not merely procedural. They exposed the deeply embedded assumptions of domestic NGO practices—namely, reliance on hierarchical coordination, familiarity with the linguistic and cultural environment, and proximity to supportive government agencies. In Nepal, where governance structures were decentralized and the humanitarian space involved a diversity of local and international actors, Chinese organizations struggled to adapt. This gap limited their ability to engage in joint coordination meetings, participate in UN-led Cluster Systems, or integrate into established relief pipelines (Cook et al., 2018; Jones et al., 2016).

16.4.2 Institutional Reflection and the Discovery of Gaps

The conclusion of the 2015 Nepal earthquake response marked a critical moment of institutional reflection for many participating Chinese social organizations. Confronted with the operational dissonance between domestic practices and international field realities, NGOs began to systematically assess internal shortcomings and identify areas in urgent need of reform. Several post-mission reviews highlighted a growing awareness that China's emergent humanitarian actors were institutionally underprepared for cross-border engagement, especially in contexts shaped by transnational norms, multilingual coordination, and complex risk environments (Organisation Z, 2015o).

Language-related challenges were widely recognized as a key source of miscommunication and operational inefficiency during the response. The lack of English or Nepali proficiency among many Chinese teams obstructed engagement with local and international actors, while also limiting access to key coordination systems such as OCHA's cluster meetings and ReliefWeb (He et al., 2018; Organisation Z, 2015o). In light of these constraints, some organizations began discussing the need to recruit bilingual personnel or implement language training as part of future preparedness planning.

In addition to language barriers, many field teams lacked exposure to international humanitarian principles and operational standards. Terms such as "beneficiary targeting" or "accountability to affected populations" were frequently encountered but seldom understood in practice. Without prior engagement in formal coordination settings, numerous organizations resorted to improvisational methods, resulting in duplicated efforts, procedural gaps, and limited collaboration opportunities (Cook et al., 2018; Organisation Z, 2015n).

Safety concerns related to untrained personnel operating in hazardous environments also became a focal point of institutional critique. Without standardized pre-departure training, personal insurance coverage, or risk assessment protocols, many volunteers were placed in high-risk situations without adequate protection (Organisation Z, 2015g; Yu et al., 2018). This prompted a growing consensus around the need to develop formal safety mechanisms for overseas deployment, particularly for youth volunteers and first-time responders.

Furthermore, the experience in Nepal sparked broader conversations about internal standardization. Multiple organizations noted the absence of cross-border financial management systems, leading to ad hoc cash handling, informal transfers, and auditing gaps (Organisation Z, 2015f). Calls were also made for standardized documentation protocols, such as needs assessment templates and post-distribution monitoring forms, which were largely missing during the Nepal operation (Zhang & Pan, 2022).

16.4.3 *Normative Awakening: Ethics, Standards, and Coordination*

The Nepal earthquake response marked a pivotal moment in the normative learning trajectory of Chinese social organizations. Confronted with an unfamiliar global humanitarian system, many actors began to recognize the importance of international standards, ethical principles, and coordination frameworks—despite entering the field with limited prior exposure.

Chinese NGOs encountered international coordination infrastructures such as the UN Cluster System and the Sphere Standards for the first time, often with limited preparation. Many teams struggled to navigate unfamiliar protocols and terminologies, yet gradually adapted through participation in cluster meetings and field-level coordination. Volunteers attended sessions in Kathmandu as observers, gaining familiarity with reporting procedures and multilateral expectations (Cook et al., 2018; Organisation Z, 2015g). Some organizations—particularly medical teams—attempted to align service delivery with Sphere guidelines for health and shelter, though practical implementation remained uneven due to limited training.

Normative engagement extended into practical learning about beneficiary identification, needs assessment, and equitable distribution. Early on, the use of "first come, first served" models at distribution sites often resulted in inequities and tensions (Organisation Z, 2015j). In Bhaktapur and Gorkha districts, some organizations responded by introducing simple registration systems and community-generated beneficiary lists (Karna et al., 2017; Organisation Z, 2015k). Though lacking formal standardization, these practices reflected an emerging awareness of transparency, inclusion, and accountability in aid delivery.

Equally significant was the nascent discussion of humanitarian ethics. Field notes and post-mission reflections described increasing concern with equitable access, gender sensitivity, and the principle of "Do No Harm." For example, several teams adjusted distribution protocols after realizing that male-only volunteer groups created discomfort among women beneficiaries in Hindu-majority villages (Fehr, 2023; Organisation Z, 2015n). Others responded to community feedback by modifying food items incompatible with local religious practices—demonstrating a shift toward cultural sensitivity (Zhao, 2016). While these adaptations were often reactive rather than guided by formal codes, they marked the beginning of ethical awareness as an integral part of field operations.

16.4.4 Nepal as a Reference Point for Emerging Global Orientation

The 2015 Nepal Earthquake not only marked a moment of emergency response but also emerged as a critical reference point for Chinese social organizations beginning to engage in transnational humanitarian action. As their early encounter with large-scale international disaster relief, the Nepal mission functioned as both a testing ground and a turning point—prompting these actors to rethink institutional identity, strategic orientation, and global responsibility.

During the mission, several Chinese NGOs recognized the gap between domestic success and international relevance. In internal reviews, organizations described the Nepal response as a "live simulation" that revealed essential weaknesses in their cross-border operating capacities, while also serving as a formative learning context (Organisation Z, 2015o). This event catalyzed their efforts to document operational experience, assess procedural shortcomings, and revise future engagement plans. For example, multiple post-action evaluations explicitly positioned Nepal as a prototype case for future outbound disaster interventions, with recommendations to standardize overseas deployment protocols and improve foreign coordination mechanisms (Organisation Z, 2015o).

Beyond internal documentation, the Nepal case became a symbolic benchmark in organizational planning. In the months following the mission, a number of NGOs began to formalize their global ambitions. Some initiated negotiations for Memoranda of Understanding (MoU) with international partners (Xinhua News Agency, 2015b), while others applied to join cross-border platforms for disaster coordination. Notably, organizations such as China Foundation for Poverty Alleviation (CFPA) and Amity Foundation subsequently developed independent international departments to support long-term overseas programs (Chen et al., 2018; Wang et al., 2016).

Strategically, the Nepal mission also shaped how Chinese NGOs positioned themselves within global civil society. Prior to this event, international presence was often reactive and symbolic. Afterward, however, participating organizations began to frame their work with a stronger awareness of global standards and geopolitical relevance. This was evident in revised strategic reports that included language about "international humanitarian mandates," "Belt and Road civic engagement," and "community-based cross-border resilience" (China Youth Daily, 2015; Organisation Z, 2015o). These narratives signaled a shift from incidental participation to proactive international orientation.

In essence, the Nepal earthquake experience marked the emergence of a transnational learning loop. Chinese NGOs not only responded to a neighboring disaster but also redefined their own trajectories through the act of participation. This institutional pivot—from domestic comfort zones to global experimentation—laid the groundwork for a new chapter in China's civil society development.

16.5 Conclusion

The 2015 Nepal Earthquake served as a formative turning point for Chinese social organizations, marking one of their earliest large-scale forays into cross-border humanitarian action. Their improvised yet determined response highlighted both the promise and the precarity of China's emerging global civil society. From diverse motivations and decentralized mobilization to adaptive field practices and early community engagement, Chinese actors operated under significant institutional uncertainty. Through these challenges, they began to confront the limits of domestically-oriented models, internalize international norms, and experiment with global identity formation. The Nepal mission thus not only addressed urgent needs but also catalyzed a longer process of institutional learning—laying critical groundwork for more structured, normative, and globally oriented humanitarian engagement in the years to follow.

References

Bhattarai, H. K., Hung, K. K. C., MacDermot, M. K., Hubloue, I., Barone-Adesi, F., Ragazzoni, L., Della Corte, F., Acharya, R., & Graham, C. A. (2023). Role of community health volunteers since the 2015 Nepal earthquakes: A qualitative study. *Disaster Medicine and Public Health Preparedness, 17*, e138. https://doi.org/10.1017/dmp.2022.47

Chen, H., Xie, Q., Feng, B., Liu, J., Huang, Y., & Chen, H. (2018). Seismic performance to emergency centers, communication and hospital facilities subjected to Nepal earthquakes, 2015. *Journal of Earthquake Engineering, 22*(9), 1537–1568. https://doi.org/10.1080/13632469.2017.1286623

China News Service. (2015). *China's earthquake relief assistance to Nepal reaches RMB 140 million*. https://www.xinhuanet.com/politics/2015-05/15/c_127805336.htm.

China Youth Daily. (2015). *Why have Chinese civil rescue organizations fallen into "islands" of isolation?*. https://www.xinhuanet.com//politics/2015-05/27/c_127846106.htm.

Cook, A. D. B., Shrestha, M., & Htet, Z. B. (2018). An assessment of international emergency disaster response to the 2015 Nepal earthquakes. *International Journal of Disaster Risk Reduction, 31*, 535–547. https://doi.org/10.1016/j.ijdrr.2018.05.014

Dubey, R., Luo, Z., Xu, M., & Wamba, S. F. (2015). Developing an integration framework for crowdsourcing and internet of things with applications for disaster response. In *2015 IEEE International Conference on Data Science and Data Intensive Systems* (pp. 520–524). https://doi.org/10.1109/DSDIS.2015.50

Fehr, T. (2022). Nepal's post-earthquake development surge. *Sociology of Development, 8*(3), 272–293. https://doi.org/10.1525/sod.2021.0021

Fehr, T. (2023). The multiplicity, contradictions, and shifting status of widowhood in post-earthquake Nepal. *Gender, Technology and Development, 27*(3), 366–386. https://doi.org/10.1080/09718524.2023.2231791

He, L., Aitchison, J. C., Hussey, K., Wei, Y., & Lo, A. (2018). Accumulation of vulnerabilities in the aftermath of the 2015 Nepal earthquake: Household displacement, livelihood changes and recovery challenges. *International Journal of Disaster Risk Reduction, 31*, 68–75. https://doi.org/10.1016/j.ijdrr.2018.04.017

Jones, S., Oven, K. J., & Wisner, B. (2016). A comparison of the governance landscape of earthquake risk reduction in Nepal and the Indian state of Bihar. *International Journal of Disaster Risk Reduction, 15*, 29–42. https://doi.org/10.1016/j.ijdrr.2015.10.011

Kargel, J. S., Leonard, G. J., Shugar, D. H., Haritashya, U. K., Bevington, A., Fielding, E. J., Fujita, K., Geertsema, M., Miles, E. S., Steiner, J., Anderson, E., Bajracharya, S., Bawden, G. W., Breashears, D. F., Byers, A., Collins, B., Dhital, M. R., Donnellan, A., Evans, T. L., … Young, N. (2016). Geomorphic and geologic controls of geohazards induced by Nepal's 2015 Gorkha earthquake. *Science, 351*(6269), aac8353. https://doi.org/10.1126/science.aac8353

Karna, R. K., Ye, Q., & Karn, A. L. (2017). Super-connected: How techno tourism redesigns our lives through meaningful travel. In *International Conference on Transformations and Innovations in Management (ictim-17), Shanghai, China*. https://doi.org/10.2991/ictim-17.2017.43

Li, S., Liu, Q., & Sun, X. (2025). HTCNN-Attn: A fine-grained hierarchical multi-label deep learning model for disaster emergency information intelligent extraction from social media. *PeerJ Computer Science, 11*, e2992. https://doi.org/10.7717/peerj-cs.2992

Liang, G., & Zhou, N. (2016). Background and reflections on Gorkha earthquake of April 25, 2015. *Natural Hazards, 81*(2), 1385–1392. https://doi.org/10.1007/s11069-015-2127-6

Lin, P. (2021). China's evolving humanitarian diplomacy: Evidence from China's disaster-related aid to Nepal. *Asian Journal of Comparative Politics, 6*(3), 221–237. https://doi.org/10.1177/20578911211019255

Organisation Z. (2015a). *Analytical situation report: Nepal earthquake* (April 25, 2015).

Organisation Z. (2015b). *Situation report no. 1: Nepal earthquake relief* (April 25, 2015).

Organisation Z. (2015c). *Situation report no. 2: Nepal earthquake relief (April 25, 2015)*.

Organisation Z. (2015d). *Situation report no. 3: Nepal earthquake relief (April 25, 2015)*.

Organisation Z. (2015e). *Situation report no. 4: Nepal earthquake relief (April 25, 2015)*.

Organisation Z. (2015f). *Situation report no. 5: Nepal earthquake relief* (April 25, 2015).

Organisation Z. (2015g). *Situation report no. 6: Nepal earthquake relief (April 25, 2015)*.

Organisation Z. (2015h). *Situation report no. 7: Nepal earthquake relief (April 25, 2015)*.

Organisation Z. (2015i). *Situation report no. 8: Nepal earthquake relief (April 25, 2015)*.

Organisation Z. (2015j). *Situation report no. 9: Nepal earthquake relief (April 25, 2015)*.

Organisation Z. (2015k). *Situation report no. 10: Nepal earthquake relief (April 25, 2015)*.

Organisation Z. (2015l). *Situation report no. 11: Nepal earthquake relief (April 25, 2015)*.

Organisation Z. (2015m). *Situation report no. 12: Nepal earthquake relief (April 25, 2015)*.

Organisation Z. (2015n). *Situation report no. 13: Nepal earthquake relief (April 25, 2015)*.

Organisation Z. (2015o). *Situation report no. 14: Nepal earthquake relief (April 25, 2015)*.

Seale-Feldman, A. (2020). The work of disaster: Building Back otherwise in post-earthquake Nepal. *Cultural Anthropology, 35*(2). https://doi.org/10.14506/ca35.2.07

Wang, X., Lv, J., Ding, X., & Dou, A. (2016). Digital disaster evaluation and its application to 2015 M_S 8.1 Nepal earthquake. *IOP Conference Series: Earth and Environmental Science, 46*, 012056. https://doi.org/10.1088/1755-1315/46/1/012056

Xinhua News Agency. (2015a). *All for our Nepalese brothers: A chronicle of China's major rescue operation after the Nepal earthquake.* https://www.xinhuanet.com//world/2015-05/04/c_1115174371.htm.

Xinhua News Agency. (2015b). *Nepal earthquake relief: China's practice in building a community of shared destiny with neighboring countries.* https://www.xinhuanet.com//world/2015-05/08/c_1115224648.htm.

Yu, L., Peng, C., Regmi, A. D., Murray, V., Pasuto, A., Titti, G., Shafique, M., & Priyadarshana, D. G. T. (2018). An international program on silk road disaster risk Reduction–A Belt and road initiative (2016–2020). *Journal of Mountain Science, 15*(7), 1383–1396. https://doi.org/10.1007/s11629-018-4842-4

Zhang, L., & Pan, Y. (2022). Information fusion for automated post-disaster building damage evaluation using deep neural network. *Sustainable Cities and Society, 77*, 103574. https://doi.org/10.1016/j.scs.2021.103574

Zhao, B. (2016). April 2015 Nepal earthquake: Observations and reflections. *Natural Hazards, 80*(2), 1405–1410. https://doi.org/10.1007/s11069-015-2001-6

Chapter 17
Ecuador Earthquake: Collaboration with Local Chinese Communities and International Partners

Abstract Rather than approaching the 2016 Ecuador Earthquake as a standalone humanitarian episode, this chapter uses it to explore how collaboration was formed, enabled, and constrained within a multi-layered disaster governance environment. It begins by outlining the structural conditions under which coordination became a practical necessity, as governance capacity coexisted with persistent spatial and operational gaps. The analysis then foregrounds the role of local Chinese communities as intermediary infrastructures that connected transnational actors with host-state and international systems. Building on this foundation, the chapter examines how Chinese social organizations operated as embedded participants across health, shelter, and relief practices, before situating Ecuador within a broader trajectory of overseas engagement marked by a shift from emergency mobilization toward more structured—but not fully institutionalized—participation.

Keywords Disaster governance · Intermediary roles · Local collaboration · Transitional engagement · Practice-based learning

17.1 Disaster Governance and the Conditions for Collaboration

The response to the 2016 Ecuador Earthquake unfolded within a complex governance environment in which collaboration emerged as a structural feature rather than an incidental choice. This section traces how a multi-layered disaster governance framework—combining state authority, international humanitarian coordination, and socially embedded actors—shaped both the possibilities and limits of response. It shows that while governance capacity was institutionally robust, its implementation was constrained by spatial, institutional, and operational tensions, giving rise to persistent coordination gaps. These conditions generated a structural demand for complementary forms of action, through which differentiated capacities could be aligned. By conceptualizing collaboration as a product of governance design

Q. Du and R. Shaw, *Social Organizations in Disaster Response and Recovery*, Disaster Risk Reduction, https://doi.org/10.1007/978-981-92-2846-1_17

and constraint, this section establishes the analytical foundation for examining how collaborative spaces were subsequently accessed and operationalized.

17.1.1 A Structured Disaster Governance Framework

The 2016 Ecuador Earthquake did not unfold within a governance vacuum. On the contrary, the response was embedded in a clearly articulated and multi-layered disaster governance framework, in which state authority, international humanitarian coordination, and social and community actors coexisted and interacted. Understanding this structured setting is essential for explaining why collaboration emerged as a systemic feature of the response rather than an ad hoc arrangement.

At the national level, the Ecuadorian government rapidly asserted a state-led emergency governance structure following the 7.8-magnitude earthquake on April 16, 2016. A nationwide state of emergency was declared, enabling the mobilization of the armed forces, police, and emergency management agencies to secure affected areas, manage logistics, and coordinate rescue operations (Organisation Z, 2016a, 2016b). Central authorities assumed responsibility for damage assessment, fiscal allocation, and the regulation of humanitarian access, reinforcing the primacy of state sovereignty in disaster response (Córdova et al., 2023; Guerrero-Miranda & Luque González, 2021). This centralized framework provided legal authority and organizational capacity, but it did not imply uniform reach across all affected territories.

Alongside state leadership, the international humanitarian system constituted a second, clearly defined governance layer. United Nations agencies and international organizations activated established coordination mechanisms, including sector-based clusters, to organize relief efforts in health, shelter, water and sanitation, and food assistance (Furtado et al., 2021). These mechanisms emphasized standardized procedures, reporting requirements, and inter-agency coordination, aligning the Ecuador response with broader humanitarian governance norms. International actors operated with formal mandates and technical expertise, yet remained dependent on host-state authorization and local interfaces (Organisation Z, 2016b).

A third layer consisted of social and community actors, including local NGOs, volunteer groups, faith-based organizations, and diaspora networks. These actors were neither marginal nor informal in a purely residual sense; rather, they represented a persistent and socially embedded component of Ecuador's disaster governance landscape (Lai & Hsu, 2019; Norris, 2017). Community-based initiatives and social organizations played visible roles in needs identification, local logistics, and grassroots mobilization, particularly in areas where institutional reach was uneven (Organisation Z, 2016a).

17.1.2 Governance Capacity and Its Structural Limits

The existence of a structured disaster governance framework did not automatically translate into comprehensive coverage or uniform implementation across the affected areas. In the Ecuador Earthquake response, governance capacity was evident in centralized coordination, legal authority, and resource mobilization, yet these capacities encountered structural limits when translated into territorially and socially uneven contexts.

A fundamental distinction emerged between institutional capacity and operational coverage. While national authorities exercised centralized control over emergency declarations, security deployment, and fiscal mechanisms, the reach of these measures varied significantly across regions. Damage was concentrated in coastal provinces such as Manabí and Esmeraldas, where pre-existing infrastructural disparities and dispersed rural settlements complicated uniform implementation (Organisation Z, 2016b, 2016c). This spatial tension highlighted the limits of centralized coordination in addressing localized and heterogeneous needs, particularly outside major urban centers (Guerrero-Miranda & Luque González, 2021; Rodríguez-Ríos, 2024).

Institutional tensions further shaped governance outcomes. Formal emergency systems, including national agencies and internationally coordinated humanitarian actors, operated through standardized procedures designed to ensure accountability and efficiency. However, these formal mechanisms often intersected with dense networks of informal actors—community groups, volunteers, and local associations—whose modes of operation were relational rather than procedural (Testori et al., 2021). The coexistence of formal and informal systems did not indicate institutional breakdown, but rather reflected the layered nature of disaster governance, in which standardized coordination frameworks struggled to fully integrate socially embedded practices.

Operational tensions also emerged between standardized humanitarian responses and context-specific demands. International and national relief efforts prioritized sectoral standards in areas such as shelter, health, and water provision, yet these standards were not always readily adaptable to local conditions or rapidly evolving needs (Organisation Z, 2016c, 2016d). The emphasis on uniform procedures, while essential for coordination, limited flexibility at the margins and contributed to uneven service delivery, particularly among rural populations and vulnerable groups (Rodríguez-Ríos, 2024).

17.1.3 Coordination Gaps and the Need for Complementary Actors

The structural tensions identified in the Ecuador Earthquake response translated into concrete coordination gaps that could not be resolved through further centralization alone. These gaps emerged not from the absence of authority or resources, but from mismatches between governance design and the heterogeneous conditions of disaster-affected communities.

One prominent gap concerned information flows. While national authorities and international agencies relied on formal reporting channels and standardized assessment tools, information from remote and rural areas often circulated through informal networks, community leaders, and volunteer groups. This asymmetry produced delays in situational awareness and hindered the timely identification of localized needs, particularly in peripheral settlements and smaller towns (Lozano & Vaca, 2017; Organisation Z, 2016c). The challenge was not a lack of data per se, but the difficulty of translating fragmented, multilingual, and context-specific information into centralized coordination systems (Lai & Hsu, 2019).

Coverage gaps further reflected uneven access to relief across social and spatial lines. Despite large-scale mobilization, rural populations and vulnerable groups—including informal workers, women-headed households, and elderly residents—were less consistently reached by standardized relief mechanisms (Organisation Z, 2016d; Valderrama-Martínez et al., 2022). These disparities were exacerbated by infrastructural damage and logistical bottlenecks, which limited the reach of formal distribution channels beyond easily accessible urban centers.

Cultural and relational barriers also shaped coordination outcomes. Disaster response operated across linguistic, cultural, and institutional boundaries, increasing the transaction costs of coordination among state agencies, international actors, and local communities. Trust deficits, unfamiliarity with local norms, and limited community engagement mechanisms constrained the effectiveness of top-down interventions, even when material resources were available (Lai & Hsu, 2019; Norris, 2017).

Finally, temporal mismatches between resource allocation and evolving needs contributed to coordination challenges. Relief supplies and financial assistance often arrived in phases aligned with institutional planning cycles, while community needs shifted rapidly during the response and early recovery periods (Organisation Z, 2016e). This misalignment underscored the limits of standardized humanitarian workflows in dynamic post-disaster environments.

17.1.4 Collaboration as a Structural Condition of the Response

Taken together, the preceding analysis demonstrates that collaboration in the Ecuador Earthquake response did not emerge as an incidental or voluntary choice, but as a structural condition shaped by the interaction of governance design, operational capacity, and persistent coordination gaps. The response unfolded within a system that was simultaneously structured and constrained: state authority was present, international humanitarian mechanisms were activated, yet neither level could fully address the diversity nor dynamism of needs across affected territories.

The coexistence of centralized capacity and localized limitations created a governance environment in which no single actor or institutional layer could operate autonomously. Central authorities retained legal and organizational control, while international humanitarian agencies contributed technical expertise and standardized coordination frameworks (Córdova et al., 2023). At the same time, spatial disparities, informational asymmetries, and operational rigidities limited the reach and adaptability of these systems, particularly in rural and socially heterogeneous contexts (Organisation Z, 2016e, 2016f). Under such conditions, coordination could not be achieved solely through hierarchical command or procedural standardization.

Collaboration thus emerged as a structurally conditioned response to these systemic constraints. It was neither a substitute for state authority nor a sign of institutional insufficiency. Instead, collaborative arrangements functioned as mechanisms through which differentiated capacities could be aligned, allowing formal governance structures to interact with socially embedded forms of action. This complementarity was especially evident in areas where information needed translation across linguistic and institutional boundaries, where logistical access depended on locally embedded networks, and where rapidly shifting needs exceeded the responsiveness of standardized planning cycles (Pacheco-Miranda et al., 2025).

Importantly, the structural nature of collaboration does not imply uniformity or stability. Collaborative practices were contingent, context-specific, and unevenly distributed across sectors and locations. They operated within existing governance boundaries rather than replacing them, reinforcing the central role of the state while enabling adaptive responses at the margins. The Ecuador case therefore illustrates how collaboration can be simultaneously indispensable and constrained, embedded within governance structures rather than external to them.

17.2 Local Chinese Communities as Intermediary Infrastructures

Rather than treating diaspora communities as contextual background or ad hoc sources of support, this chapter approaches local Chinese communities in Ecuador as intermediary infrastructures that made transnational humanitarian engagement operationally possible. Building on this perspective, the section first redefines diaspora presence as a durable and relational configuration embedded in local governance environments. It then traces how informational mediation and translational capacity enabled alignment across linguistic and institutional boundaries, before examining resource mobilization and operational brokerage as mechanisms through which dispersed inputs were converted into coordinated humanitarian capacity. Finally, a comparative discussion highlights the conditional nature of these intermediary roles and clarifies how they created enabling conditions for the embedded collaborative practices analyzed in the next section.

17.2.1 From Diaspora Presence to Intermediary Infrastructure

In analyses of disaster response, diaspora communities are often treated as contextual background variables—cultural groups located abroad whose relevance lies primarily in identity, solidarity, or ad hoc charitable contributions. Such a perspective, however, is insufficient for explaining how transnational social organizations are able to enter and operate within complex humanitarian governance environments. In the context of the 2016 Ecuador Earthquake, local Chinese communities cannot be adequately understood as passive social groups or residual resource pools. Instead, they functioned as intermediary infrastructures that enabled cross-border engagement to become operationally viable.

Conceptually, an intermediary infrastructure refers not to a single organization or event-driven initiative, but to a relatively stable configuration of social relations, organizational presence, and institutional connectivity. Unlike temporary volunteer mobilization, such infrastructures are characterized by durability, repeated usability, and the capacity to link otherwise segmented systems. In Ecuador, Chinese communities had been established well before the earthquake through business associations, hometown networks, and community organizations, forming embedded social structures with regular interaction with local authorities and economic actors (Torres & Vaca, 2019). This pre-existing presence meant that when the earthquake occurred, Chinese social organizations from outside Ecuador did not encounter a social vacuum, but rather a functioning relational landscape.

The infrastructural quality of these communities becomes evident when considering how disaster governance operates. As discussed in Sect. 17.1, the Ecuador response was structured but incomplete, relying on multiple governance layers that did not automatically interface with one another. Within such a setting, intermediary infrastructures perform connective functions that are neither purely cultural nor purely institutional. They reduce transaction costs across linguistic, administrative, and social boundaries, while remaining embedded in local contexts (Lai & Hsu, 2019). This distinguishes them from both informal volunteerism and formal humanitarian agencies.

Chinese community organizations were present from the earliest stages of response, actively participating in information circulation, coordination meetings, and local relief operations—not as external actors, but as embedded nodes within the affected social landscape. Their significance lay not in the volume of material resources they managed, but in their capacity to connect Chinese civil society, Ecuadorian authorities, and international stakeholders (Organisation Z, 2016a, 2016b). This connective role echoes broader findings on the importance of socially embedded networks in disaster resilience and coordination, particularly in contexts where formal institutional systems face constraints in reach and adaptability (Norris, 2017).

17.2.2 Informational Mediation and Translational Capacity

Disaster response is often framed as a problem of logistics and material supply, yet information constitutes an equally critical dimension of humanitarian governance. Information does not circulate automatically across disaster settings; rather, it is mediated by language, trust, and network access. In the aftermath of the 2016 Ecuador Earthquake, coordination challenges were shaped not only by physical damage and institutional capacity, but also by the uneven flow, translation, and interpretation of information across governance levels.

Formal disaster governance relied heavily on standardized reporting systems, situation updates, and sectoral coordination mechanisms. National authorities and international humanitarian actors depended on structured channels for needs assessment, damage reporting, and operational planning (Organisation Z, 2016b). However, as documented in multiple assessments, these channels struggled to capture real-time conditions in dispersed rural areas and socially heterogeneous communities, where information circulated primarily through informal and community-based networks (Lozano & Vaca, 2017). This produced informational asymmetries that delayed response alignment and constrained situational awareness.

Within this context, local Chinese communities performed a distinctive informational mediation function. Their role was not limited to ad hoc translation but involved sustained translational capacity across linguistic, institutional, and social boundaries. Linguistically, community members operated across Chinese, Spanish, and English, enabling the conversion of locally grounded observations into forms intelligible to external social organizations and international actors. This multilingual capacity reduced misinterpretation and accelerated the transmission of situational updates during the early and volatile phases of the response (Organisation Z, 2016c).

Beyond language, informational mediation also involved real-time situational relay. Community networks were embedded in local commercial districts, residential areas, and social associations, allowing rapid circulation of updates on damage, accessibility, and emerging needs. Such information often preceded formal assessments and complemented official data streams, particularly in areas where institutional presence was limited (Organisation Z, 2016d). Importantly, this was not an episodic function but a repeatable one, drawing on pre-existing communication routines and trust-based relationships (Lai & Hsu, 2019).

A third dimension of informational mediation concerned governance interface alignment. Disaster response required constant interaction among local governments, national coordination platforms, and international humanitarian systems. Chinese community actors frequently served as connectors between these domains, facilitating communication with municipal authorities, emergency response platforms, and international organizations operating under cluster coordination frameworks (Valderrama-Martínez et al., 2022). This interface function reduced coordination costs and mitigated friction arising from institutional unfamiliarity.

17.2.3 Resource Mobilization and Operational Brokerage

Beyond information, effective humanitarian engagement depends on the ability to mobilize, coordinate, and deploy resources across institutional and territorial boundaries. In the Ecuador Earthquake response, resource-related challenges were not defined solely by scarcity, but by fragmentation: material donations, financial contributions, and operational capacities were dispersed across actors operating at different scales. Within this fragmented landscape, local Chinese communities functioned as intermediaries that brokered resources and operational access, transforming dispersed inputs into coordinated humanitarian capacity.

Material resources constituted the most visible dimension of this brokerage role. In the immediate aftermath of the earthquake, donations from Chinese enterprises, associations, and individuals in Ecuador were mobilized through community-based networks rather than through isolated acts of giving. These networks enabled the aggregation of funds and relief items, facilitating alignment with identified needs and existing distribution channels (Organisation Z, 2016c). Rather than operating as independent donors, community actors coordinated contributions to avoid duplication and to synchronize with broader relief efforts, reflecting patterns observed in post-disaster social mobilization more broadly (Furtado et al., 2021).

Equally important were operational resources, including transportation, storage, access permits, and personnel coordination. Damage to infrastructure and the imposition of emergency regulations created bottlenecks that limited the autonomous movement of humanitarian actors, particularly those lacking established local connections (Organisation Z, 2016d). Chinese community organizations and business networks played a brokerage role by leveraging pre-existing relationships with local authorities, logistics providers, and commercial actors. This facilitated access to transport routes, warehousing, and administrative procedures, enabling relief materials and personnel to move within constrained operational environments (Rodríguez-Ríos, 2024).

This brokerage function operated in two directions. Internally, community leaders coordinated among enterprises, associations, and individuals to consolidate resources and align expectations. Externally, they interfaced with municipal governments, emergency authorities, and other humanitarian actors to ensure that mobilized resources could be integrated into ongoing response structures (Organisation Z, 2016e). Such dual positioning distinguished intermediary brokerage from simple charitable donation, emphasizing coordination and interface management over unilateral action.

Crucially, this form of resource mobilization was embedded in existing social and organizational infrastructures rather than improvised for the crisis alone. Community associations, business chambers, and social organizations provided stable platforms through which resources could be repeatedly mobilized and operationalized (Cordero-Reyes et al., 2017). This stability reduced transaction costs and enhanced reliability, particularly during the transition from immediate relief to early recovery phases.

17.2.4 Intermediary Infrastructures and Comparative Implications

Viewed in comparative perspective, the Ecuador Earthquake illustrates a distinct configuration of overseas humanitarian engagement in which local communities functioned not primarily as cultural mediators or adaptive recipients, but as operational infrastructures enabling collaboration. This configuration differs from cases in which diaspora communities are mobilized mainly to facilitate cultural understanding or social acceptance. In Ecuador, the local Chinese presence assumed a more infrastructural role, providing stable interfaces through which transnational social organizations could connect with host-state institutions and international humanitarian systems.

This distinction becomes clearer when contrasted with disaster contexts characterized by weaker institutional environments or higher levels of political volatility. In such settings, community engagement often centers on cultural translation and social adaptation, with local actors helping external responders navigate unfamiliar norms and sensitivities. By contrast, the Ecuador response unfolded within a relatively stable governance framework, where state authority, emergency regulations, and international coordination mechanisms were firmly in place (Córdova et al., 2023). Under these conditions, the primary challenge was not cultural access, but operational integration. Local Chinese communities were therefore positioned to function as infrastructures that linked systems rather than as buffers between them.

The effectiveness of this intermediary role was contingent upon several structural conditions. First, the presence of a mature and organized diaspora network provided continuity and organizational density. Community associations, business networks, and social organizations constituted durable platforms that could be activated repeatedly rather than assembled ad hoc during the crisis (Organisation Z, 2016e). Second, host-country governance capacity mattered. The existence of functioning emergency institutions, clear regulatory frameworks, and identifiable coordination points enabled intermediary actors to operate within, rather than outside, formal governance arrangements (Pacheco-Miranda et al., 2025). Third, the accessibility of the international humanitarian system—through cluster mechanisms, standardized procedures, and recognized interfaces—created opportunities for brokerage rather than competition (Organisation Z, 2016f).

These conditions also highlight the limits of transferability. Intermediary infrastructures are not universally available, nor do they emerge automatically from diaspora presence alone. Where community networks are fragmented, where host governance is unstable, or where international systems are inaccessible, similar mechanisms may not materialize or may take different forms. The Ecuador case therefore underscores the conditional nature of diaspora-mediated collaboration, cautioning against generalization without attention to institutional context.

17.3 Embedded Multi-actor Collaboration in Humanitarian Practice

Building on the intermediary infrastructures outlined in the previous section, this section examines how collaboration was enacted in practice within the Ecuador Earthquake response. Rather than treating humanitarian action as a collection of parallel interventions, it analyzes how Chinese social organizations participated as embedded actors operating across community, state, and international systems. Focusing on health and medical response, shelter and WASH provision, and relief distribution, the section shows how collaboration functioned as a mechanism for coordination and gap-filling within a structured governance environment. It further delineates the relational, context-specific, and case-based nature of these practices, clarifying both their effectiveness and their structural limits.

17.3.1 Embedded Participation in a Multi-actor Humanitarian System

Humanitarian engagement by Chinese social organizations in the 2016 Ecuador Earthquake cannot be accurately characterized as parallel response or independent humanitarian provision. Contrary to depictions that frame overseas actors as self-contained responders operating alongside state and international systems, the Ecuador case demonstrates a pattern of participation that was fundamentally embedded within an existing multi-actor humanitarian governance structure. Chinese social organizations entered the response space not by establishing autonomous operational tracks, but by situating their actions within community networks, host-state coordination mechanisms, and internationally coordinated humanitarian arrangements.

This embedded positioning reflected the institutional configuration of disaster governance in Ecuador. The response environment was shaped by strong state authority, formal emergency regulations, and internationally recognized coordination platforms. Within this setting, autonomous action was neither feasible nor desirable. Relief activities involving Chinese social organizations were carried out under host-country authorization, aligned with local emergency priorities, and integrated into existing coordination channels rather than managed through independent command structures (Organisation Z, 2016c, 2016d). Participation was thus conditioned by compliance with regulatory frameworks and adherence to shared coordination norms, rather than driven by unilateral agenda setting.

Embeddedness also manifested through relational integration with local communities and intermediary actors. Chinese social organizations did not operate directly at scale upon arrival; instead, their engagement was mediated through locally embedded Chinese community networks that had established relationships with municipal authorities, emergency platforms, and civil society actors. This mode of entry enabled alignment with local needs assessments and reduced friction associated with institutional unfamiliarity. As a result, participation unfolded through negotiated access and relational trust rather than through formal delegation of authority or independent deployment (Guerrero-Miranda & Luque González, 2021).

At the international level, embedded participation entailed operating within, rather than outside, humanitarian coordination mechanisms. Although Chinese social organizations were not central actors in cluster leadership, their activities intersected with internationally coordinated sectors through information sharing, referral mechanisms, and complementary task allocation. This form of engagement reflected broader patterns in contemporary humanitarian governance, where diverse actors contribute differentiated capacities without displacing established coordination hierarchies (Córdova et al., 2023). Importantly, embeddedness did not imply subordination or passivity; rather, it denoted a mode of action shaped by institutional constraints and relational interfaces.

17.3.2 Health and Medical Collaboration Across Governance Levels

Health and medical response constituted one of the most institutionalized yet operationally demanding domains in the Ecuador Earthquake response. Medical intervention unfolded within a tightly regulated governance environment shaped by national health authorities, emergency regulations, and internationally coordinated medical assistance. At the same time, the scale of injuries, damage to local health facilities, and uneven access to care—particularly in coastal and rural areas—required flexible coordination across governance levels rather than rigid sectoral implementation (Organisation Z, 2016c, 2016d).

At the national level, the Ecuadorian Ministry of Public Health maintained formal authority over medical services, hospital coordination, and the deployment of emergency health personnel. International medical teams operated under host-state authorization and within established humanitarian coordination frameworks, aligning their activities with national priorities and sectoral standards (Furtado et al., 2021). This institutional structure ensured regulatory oversight and professional accountability, but it also limited the ability of any single actor to respond autonomously to rapidly evolving local conditions.

Within this constrained yet structured environment, collaboration emerged through embedded coordination rather than parallel provision. Chinese social organizations did not operate independent medical systems or deploy standalone clinical services. Instead, their participation was integrated into joint operations that connected local health departments, international medical teams, and community-level needs. Medical outreach and consultation activities were carried out in alignment with existing health facilities, with clear role differentiation that avoided duplication and supported coordinated delivery efforts (Organisation Z, 2016e).

A critical enabling factor in this collaboration was the intermediary role of local Chinese communities, particularly in translational and logistical support. Health interventions required constant communication across Spanish-, Chinese-, and English-speaking actors, as well as alignment with local administrative procedures. Community-based actors facilitated real-time translation, assisted with patient communication, and supported coordination between visiting medical personnel and local health authorities. These functions did not substitute for professional medical expertise but enabled it to be operationally effective within local contexts (Organisation Z, 2016f).

Collaboration in the medical sector was therefore characterized by joint operations and task differentiation. International teams contributed specialized expertise and emergency capacity, national institutions retained regulatory control and continuity of care, and Chinese social organizations—working through community networks—supported interface functions that reduced coordination friction. This complementarity allowed medical responses to adapt to spatial and institutional constraints without undermining existing health governance arrangements (Vega et al., 2021).

17.3.3 Shelter, WASH, and Relief Distribution as Collaborative Practice

Shelter provision, WASH services, and relief distribution formed a core operational domain of the Ecuador Earthquake response, one characterized by heavy logistical demands and pronounced spatial unevenness. Unlike health interventions, which were anchored in relatively centralized institutional systems, these sectors depended on extended supply chains, access to damaged infrastructure, and localized knowledge of settlement patterns. As a result, they exposed the limits of centralized humanitarian distribution models, particularly in reaching rural, peri-urban, and geographically fragmented communities (Organisation Z, 2016d; Testori et al., 2021).

International humanitarian actors contributed standardized relief items, technical guidelines, and sectoral coordination mechanisms for shelter and WASH, ensuring alignment with global humanitarian norms. State authorities, in turn, exercised control over access, security, and prioritization, regulating the movement of goods and personnel under emergency conditions (Organisation Z, 2016e). While these arrangements were essential for order and accountability, they tended to privilege urban centers and accessible locations, where logistical costs were lower and coordination more straightforward. Consequently, significant coverage gaps emerged in outlying and less visible areas, despite the overall scale of mobilized resources (Valderrama-Martínez et al., 2022).

Within this context, collaboration functioned less as a means of increasing efficiency than as a mechanism for correcting spatial and operational imbalances. Chinese social organizations, operating as embedded actors, did not introduce parallel distribution systems. Instead, they worked within existing relief architectures to extend reach toward peripheral communities. Distribution activities were carried out in coordination with local authorities and aligned with international supply streams, while implementation drew on flexible, locally embedded capacities (Organisation Z, 2016f, 2016g).

A key feature of this collaboration was task differentiation. International organizations focused on procurement, standards, and bulk logistics; government agencies ensured regulatory compliance and access; and Chinese social organizations, often through diaspora-mediated networks, facilitated last-mile delivery, local coordination, and sequencing of distribution in areas where formal systems faced bottlenecks. This arrangement did not accelerate distribution uniformly, but it reduced exclusion by enabling relief to reach settlements that were otherwise deprioritized due to logistical complexity or limited visibility within centralized planning frameworks (Testori et al., 2021).

Importantly, collaboration in shelter and WASH did not imply a redistribution of authority. Decisions regarding standards, eligibility, and overall allocation remained anchored in state-led and internationally coordinated systems. What collaboration altered was the operational interface between centralized planning and localized delivery. By bridging institutional systems and community-level realities, embedded collaboration transformed relief distribution from a purely hierarchical process into a more adaptive one, capable of addressing uneven coverage without undermining governance structures (Organisation Z, 2016d, 2016e, 2016f, 2016g).

17.3.4 Coordination, Gap-Filling, and the Boundaries of Collaboration

The collaborative practices observed across health, shelter, WASH, and relief distribution in the Ecuador Earthquake response shared a set of common characteristics that define both their effectiveness and their limits. Rather than constituting a unified operational model or an institutionalized response framework, collaboration functioned as a relational and case-based mechanism for addressing coordination gaps that emerged within an otherwise structured humanitarian governance system.

First, collaboration in Ecuador was fundamentally relational. Coordination did not rely primarily on standardized platforms or formalized inter-organizational protocols, but on trust-based ties linking community networks, social organizations, and institutional actors. These relationships enabled rapid alignment in contexts where formal coordination mechanisms were either overstretched or insufficiently granular, particularly at the local level (Organisation Z, 2016g). Such relational coordination allowed actors to navigate administrative complexity and logistical constraints without altering the formal authority of state or international systems (Córdova et al., 2023).

Second, collaborative practices were highly context-specific. The effectiveness of embedded coordination depended on local social structures, governance stability, and the availability of intermediary infrastructures. As earlier sections have shown, collaboration was most pronounced in settings characterized by spatial fragmentation, informational asymmetry, or operational bottlenecks, rather than uniformly across all sectors or regions. This situational dependence meant that collaboration could not be abstracted into a generalized template; instead, it emerged in response to concrete mismatches between centralized planning and localized needs (Organisation Z, 2016h).

Third, collaboration remained case-based rather than programmatic. Joint actions were organized around specific tasks—medical outreach, last-mile distribution, or access facilitation—without evolving into standing coordination bodies or permanent platforms. Learning occurred primarily through practice rather than through codified procedures, reinforcing adaptability while limiting scalability. This case-based orientation proved effective for immediate gap-filling but constrained the ability to replicate collaboration across different disaster contexts without renewed relational investment (Pacheco-Miranda et al., 2025).

These characteristics also delineate the boundaries of collaboration in Ecuador. Embedded coordination depended heavily on intermediary networks and personal ties, lacked modular components that could be easily transferred, and did not result in durable institutional arrangements. Importantly, collaboration neither replaced state authority nor reconfigured international humanitarian governance. Instead, it operated within existing boundaries, compensating for structural limits without challenging the overarching governance architecture.

Recognizing these boundaries is essential for situating the Ecuador case within a broader comparative trajectory. The practices documented here differ fundamentally from regional coordination mechanisms and from later forms of institutionalized international response. Ecuador represents neither a regionalized coordination model nor a fully institutionalized platform, but an intermediate stage in which collaboration was effective precisely because it remained embedded, relational, and constrained.

17.4 From Emergency Mobilization to Structured Overseas Engagement

Rather than treating the Ecuador Earthquake as a discrete humanitarian episode, this section situates it within a longer temporal and analytical arc of overseas disaster engagement. Shifting attention away from immediate outcomes toward evolving modes of participation, the analysis traces how emergency involvement gradually gave rise to more structured forms of overseas engagement while stopping short of full institutionalization. Ecuador neither represents a culmination of international humanitarian capacity nor a consolidated response model; instead, it marks a transitional moment in which rapid mobilization was reshaped into patterned, procedurally aligned participation, grounded in relational and case-based coordination rather than permanent institutional platforms.

17.4.1 Beyond Outcomes: Ecuador as a Turning Point in Overseas Engagement

Evaluating the 2016 Ecuador Earthquake response primarily through humanitarian outcomes risks narrowing its analytical significance. While relief delivery, coordination efficiency, and operational contributions remain important, focusing exclusively on performance obscures the broader transformation that unfolded through the response process. The Ecuador case invites a shift in perspective—from assessing what was accomplished to examining how modes of participation evolved over time, and how this evolution reconfigured patterns of overseas engagement.

Such a shift is necessary because Ecuador does not represent a culmination of Chinese social organizations' international humanitarian capacity. Nor does it constitute a fully consolidated model of overseas disaster response. Instead, it marks a transitional moment in which earlier practices of rapid, situation-driven mobilization began to give way to more structured and procedurally aligned forms of engagement.

Participation increasingly unfolded within defined governance parameters shaped by host-state authority, regulatory frameworks, and internationally recognized coordination arrangements (Córdova et al., 2023). This environment constrained unilateral action while simultaneously encouraging alignment with existing systems, signaling a qualitative change in how overseas involvement was conceived and enacted.

Evidence from later phases of the response illustrates this transition. As emergency operations progressed, coordination routines became more regularized, reporting practices more explicit, and interfaces with government agencies and international actors more clearly delineated (Organisation Z, 2016h). These developments point toward an emerging emphasis on structured engagement—characterized by rule alignment, role differentiation, and repeated interaction—rather than episodic intervention. At the same time, engagement remained heavily reliant on intermediary networks, relational trust, and case-specific coordination practices. The absence of permanent platforms, standardized replication mechanisms, or formalized institutional mandates underscores that this process had not yet matured into full institutionalization.

From a trajectory perspective, Ecuador occupies an intermediate position. It reflects neither the early stage of overseas engagement dominated by ad hoc mobilization nor the later stage associated with institutionalized international response capacity. Instead, it represents a critical inflection point at which accumulated experience, governance exposure, and collaborative practice began to reshape the orientation of Chinese social organizations abroad. This pattern is consistent with broader analyses of humanitarian governance that emphasize incremental transition through practice-based learning rather than abrupt organizational transformation (Pacheco-Miranda et al., 2025).

17.4.2 A Three-Phase Trajectory of Overseas Engagement

When situated along a temporal axis, the Ecuador Earthquake response reveals a discernible trajectory in the overseas engagement of Chinese social organizations. Rather than unfolding as a single, homogeneous mode of action, participation evolved through successive phases that reflected changing conditions, learning processes, and degrees of institutional alignment. This trajectory can be analytically described through three overlapping phases: emergency mobilization, embedded collaboration, and the emergence of structured engagement.

The initial phase was characterized by rapid emergency mobilization. In the immediate aftermath of the April 16 earthquake, responses were driven by urgency, uncertainty, and the imperative to act quickly. Swift fundraising, ad hoc deployment of volunteers, and spontaneous coordination emerged to meet immediate relief needs (Organisation Z, 2016a, 2016b). During this stage, engagement relied heavily on informal networks and personal initiative, with limited differentiation of roles and minimal procedural standardization. Actions were reactive and situation-driven, reflecting patterns commonly observed in early disaster response when information is fragmented and institutional coordination mechanisms are still consolidating (Guerrero-Miranda & Luque González, 2021).

As the response progressed, a second phase of embedded collaboration became more pronounced. From late April into May 2016, participation increasingly unfolded within existing governance arrangements rather than alongside them. Alignment with host-country coordination structures became more evident, with communication channels to local authorities clarified and engagement with international humanitarian actors growing more consistent (Organisation Z, 2016c, 2016d, 2016e, 2016f). During this phase, Chinese social organizations did not abandon flexibility, but their actions were increasingly shaped by negotiated access, task differentiation, and reliance on intermediary infrastructures embedded in local communities. Collaboration functioned as a means of integrating diverse capacities into ongoing response systems, rather than as an assertion of autonomous operational control (Rodríguez-Ríos, 2024).

In the later stage of the Ecuador response, coordination routines became more regularized, reporting practices more explicit, and interfaces with state agencies and international systems more clearly defined (Organisation Z, 2016g, 2016h). This period saw clearer role boundaries and greater procedural awareness, signaling a shift from ad hoc to more structured modes of overseas engagement. Nevertheless, institutionalization remained partial. Many activities still relied on diaspora-mediated networks, case-specific negotiation, and trust-based relationships rather than formalized mandates or standardized mechanisms.

17.4.3 Structured Engagement Without Full Institutionalization

The Ecuador Earthquake response highlights the importance of distinguishing between structured overseas engagement and institutionalized international response. While these two modes are often conflated in discussions of humanitarian professionalization, the Ecuador case demonstrates that structured participation can emerge without crossing the threshold into full institutionalization. Making this distinction is essential for accurately situating Ecuador within the broader trajectory of Chinese social organizations' overseas engagement.

Structured engagement refers to the presence of recognizable coordination patterns, clearer role differentiation, and increasing alignment with formal governance frameworks. In Ecuador, such structuring became visible during the later stages of the response, when reporting routines grew more explicit, interfaces with host-country authorities became more consistent, and coordination with international humanitarian actors began to follow established systems (Organisation Z, 2016f, 2016g). Responsibilities across actors were gradually delineated, with Chinese social organizations shifting from ad hoc participation toward more defined operational functions. These developments reflect a movement toward procedural regularity and organizational clarity, consistent with broader patterns observed in Ecuador's evolving disaster governance arrangements (Córdova et al., 2023).

At the same time, these structured elements did not coalesce into a fully institutionalized response model. Institutionalization would imply stable mandates, formalized authority, standardized operating procedures transferable across contexts, and reduced dependence on personal or community-based mediation. None of these conditions were fully present in Ecuador. Engagement continued to rely heavily on diaspora-based intermediary infrastructures, trust-based relationships, and negotiated access rather than codified rules or permanent organizational platforms (Organisation Z, 2016h). Coordination remained relational rather than bureaucratic, and learning occurred primarily through case-specific practice rather than through institutionalized training or modular replication (Norris, 2017).

This reliance on relational mechanisms constituted both a strength and a limitation. On the one hand, it enabled flexibility, responsiveness, and contextual adaptation within a complex governance environment. On the other hand, it constrained scalability and transferability. Structured engagement in Ecuador did not generate standardized templates that could be readily applied elsewhere without renewed negotiation and relational investment. The absence of modular replication mechanisms meant that coordination gains remained context-bound rather than systematized (Organisation Z, 2016f, 2016g, 2016h).

17.4.4 Implications for Subsequent Pathways of International Disaster Engagement

The Ecuador Earthquake does not offer a ready-made model for international disaster response, nor does it prescribe a standardized pathway that can be directly replicated across contexts. Its significance instead lies in clarifying a set of enabling conditions through which overseas engagement can transition from episodic mobilization toward more structured participation. Understanding this distinction is critical for situating Ecuador within a broader comparative trajectory of international disaster engagement.

One central implication concerns the role of intermediary infrastructures in enabling structured collaboration without formal institutionalization. The Ecuador case demonstrates that structured overseas engagement can emerge in the absence of permanent mandates or centralized platforms, provided that socially embedded networks, host-country governance stability, and accessible international coordination mechanisms are present (Organisation Z, 2016h; Pacheco-Miranda et al., 2025). Rather than producing a transferable operational template, Ecuador illustrates how certain preconditions—such as reliable diaspora mediation, functioning state authority, and openness to multi-actor coordination—create a space in which structured engagement becomes possible.

A second implication relates to scale and scope. Collaboration in Ecuador remained nationally bounded and case-specific, shaped by the spatial, institutional, and political characteristics of a single disaster context. This boundedness distinguishes Ecuador from later responses characterized by regional coordination dynamics or cross-border operational platforms. In this sense, Ecuador serves as a conceptual bridge rather than an endpoint: it demonstrates how embedded collaboration can be stabilized and routinized, while stopping short of regionalization or institutional consolidation. This positioning helps explain why subsequent cases unfolded along divergent pathways rather than following a linear progression (Baize et al., 2020).

Third, the Ecuador experience underscores the importance of sequencing in overseas engagement. Rapid mobilization alone did not generate structured participation; nor did embedded collaboration automatically evolve into institutionalized response. Instead, structuring emerged gradually through repeated coordination, role clarification, and interaction with host-country and international systems. This temporal dimension suggests that pathway development depends less on singular interventions than on cumulative alignment across multiple response phases (Organisation Z, 2016h).

These implications provide a necessary foundation for understanding subsequent cases examined in this volume. The Sulawesi tsunami response illustrates how collaboration expands across regional scales under different geopolitical and coordination conditions. The Türkiye–Syria Earthquake represents a further shift toward institutionalized international engagement, marked by stronger mandates and standardized mechanisms. The Myanmar Earthquake, by contrast, highlights the constraints under which rapid response unfolds when institutional and political conditions are less permissive. Against this backdrop, Ecuador occupies a distinct position: it marks the transition toward structured overseas participation while leaving institutionalization incomplete.

17.5 Conclusion

This chapter has demonstrated that the Ecuador Earthquake constitutes more than an isolated case of overseas humanitarian response. Collaboration emerged as an embedded practice shaped by the interaction of structured governance, intermediary infrastructures, and evolving modes of participation. Rather than replacing state authority or international coordination, collaborative arrangements addressed persistent gaps generated by spatial unevenness and operational constraints. Local Chinese communities played a pivotal intermediary role, enabling Chinese social organizations to participate as embedded actors across health, shelter, and relief practices. Viewed in temporal perspective, Ecuador represents a transitional moment in overseas engagement, where emergency mobilization was gradually reconfigured into more structured participation. At the same time, the absence of permanent mandates and modular replication underscores the limits of institutionalization at this stage.

References

Baize, S., Nurminen, F., Sarmiento, A., Dawson, T., Takao, M., Scotti, O., Azuma, T., Boncio, P., Champenois, J., Cinti, F. R., Civico, R., Costa, C., Guerrieri, L., Marti, E., McCalpin, J., Okumura, K., & Villamor, P. (2020). A worldwide and unified database of surface ruptures (SURE) for fault displacement hazard analyses. *Seismological Research Letters, 91*(1), 499–520. https://doi.org/10.1785/0220190144

Cordero-Reyes, A. M., Palacios, I., Ramia, D., West, R., Valencia, M., Ramia, N., Egas, D., Rodas, P., Bahamonde, M., & Grunauer, M. (2017). Natural disaster management: Experience of an academic institution after a 7.8 magnitude earthquake in Ecuador. *Public Health, 144*, 134–141. https://doi.org/10.1016/j.puhe.2016.12.003

Córdova, M., Menoscal, J., & Flores, E. M. (2023). Governance and the design of post-disaster policies: A comparative analysis from Latin America and the Caribbean. *Disasters, 47*(3), 766–787. https://doi.org/10.1111/disa.12561

Furtado, A., Rodrigues, H., Arêde, A., & Varum, H. (2021). A review of the performance of infilled RC structures in recent earthquakes. *Applied Sciences, 11*(13), 5889. https://doi.org/10.3390/app11135889

Guerrero-Miranda, P., & Luque González, A. (2021). Social responsibility, sustainability, and public policy: The lessons of Debris management after the Manabí earthquake in Ecuador. *International Journal of Environmental Research and Public Health, 18*(7), 3494. https://doi.org/10.3390/ijerph18073494

Lai, C., & Hsu, Y. (2019). Understanding activated network resilience: A comparative analysis of co-located and co-cluster disaster response networks. *Journal of Contingencies and Crisis Management, 27*(1), 14–27. https://doi.org/10.1111/1468-5973.12224

Lozano, E., & Vaca, C. (2017). Crisis management on Twitter: Detecting emerging leaders. In *2017 Fourth International Conference on eDemocracy & eGovernment (ICEDEG)* (pp. 140–147). https://doi.org/10.1109/ICEDEG.2017.7962524

Norris, W. (2017). Digital Humanitarians: Citizen journalists on the virtual front line of natural and human-caused disasters. *Journalism Practice, 11*(2–3), 213–228. https://doi.org/10.1080/17512786.2016.1228471

Organisation Z. (2016a). *Situation report no. 1: Ecuador earthquake relief (April 16, 2016).*

Organisation Z. (2016b). *Situation report no. 2: Ecuador earthquake relief (April 16, 2016).*

Organisation Z. (2016c). *Situation report no. 3: Ecuador earthquake relief (April 16, 2016).*

Organisation Z. (2016d). *Situation report no. 4: Ecuador earthquake relief (April 16, 2016).*

Organisation Z. (2016e). *Situation report no. 5: Ecuador earthquake relief (April 16, 2016).*

Organisation Z. (2016f). *Situation report no. 6: Ecuador earthquake relief (April 16, 2016).*

Organisation Z. (2016g). *Situation report no. 7: Ecuador earthquake relief (April 16, 2016).*

Organisation Z. (2016h). *Situation report no. 8: Ecuador earthquake relief (April 16, 2016).*

Pacheco-Miranda, J., Wagner, N., & Ponce, J. (2025). Responses in child health to concerted reconstruction efforts in the aftermath of the 2016 earthquake in Ecuador. *Economics and Human Biology, 59*(101), 522. https://doi.org/10.1016/j.ehb.2025.101522

Rodríguez-Ríos, E. (2024). Dataset on recovered waste post-earthquake 2016 in Manabi Province, Ecuador for recycling and reuse. *Data in Brief, 57*(110), 953. https://doi.org/10.1016/j.dib.2024.110953

Testori, G., Janoschka, M., Mena, A. B., & Iuorio, O. (2021). Ecuadorian housing resettlements five years after the 2016 earthquake: A critical analysis. *Habitat International, 117*(102), 433. https://doi.org/10.1016/j.habitatint.2021.102433

Torres, J., & Vaca, C. (2019). Cross-lingual perspectives about crisis-related conversations on Twitter. In *Companion Proceedings of the 2019 World Wide Web Conference* (pp. 255–261). https://doi.org/10.1145/3308560.3316799

Valderrama-Martínez, C., Bates, B. R., & Grijalva, M. J. (2022). Learning communication channel selection lessons from the earthquake in Bahía De Caráquez, Ecuador, 2016. *Southern Communication Journal, 87*(2), 166–180. https://doi.org/10.1080/1041794X.2022.2037062

Vega, C., Torres Santana, A., & Paredes, M. (2021). Crisis reproductiva, cuidados y sostenimiento en contextos de desastre. Experiencias comunitarias tras el terremoto en Ecuador. *Revista Española de Sociología, 30*(2), a34. https://doi.org/10.22325/fes/res.2021.34

Chapter 18
Sulawesi Earthquake-Tsunami: Expanding Roles in Regional Disaster Response

Abstract Set against the backdrop of the 2018 Sulawesi Earthquake and tsunami, this chapter traces the expanding roles of Chinese civil society organizations in regional disaster response. It begins by unpacking the structural barriers within Indonesia's sovereignty-centered humanitarian system and highlights the operational gaps that enabled informal and relational entry by non-traditional actors. Drawing on field-based experiences, it explores how Chinese CSOs navigated access constraints through diaspora mediation, adaptive learning, and decentralized deployment. The analysis then turns to the gradual transition from ad hoc engagement to regionally anchored strategies, before assessing the challenges of proceduralization and institutional integration. Collectively, these trajectories reveal both the limitations and emerging pathways of Chinese CSOs as they navigate the complexities of cross-border humanitarian governance in Southeast Asia.

Keywords Humanitarian access · Diaspora mediation · Adaptive learning · Cross-border response · Institutional integration

18.1 Disaster Governance in Indonesia: Constraints and Openings for Collaboration

The 2018 Sulawesi Earthquake and tsunami exposed both the institutional rigidity and operational cracks within Indonesia's disaster governance system. This section examines how the country's sovereignty-first humanitarian architecture—centered on BNPB and reinforced by regional and cluster mechanisms—structured the formal entry points for international aid. It then traces the operational breakdowns that created field-level vacuums, highlights the fragmented coordination among humanitarian actors, and analyzes how these dynamics opened space for local civil society organizations to emerge as critical intermediaries. These factors created limited but meaningful policy windows through which non-traditional actors, including Chinese social organizations, could engage.

Q. Du and R. Shaw, *Social Organizations in Disaster Response and Recovery*, Disaster Risk Reduction, https://doi.org/10.1007/978-981-92-2846-1_18

18.1.1 The Formal Humanitarian Governance Architecture in Indonesia

In the immediate aftermath of the 2018 Sulawesi Earthquake and tsunami, Indonesia's humanitarian governance structure was activated through a state-led, sovereignty-anchored coordination model. At its core, this system is built around the National Disaster Management Authority (BNPB), which retains exclusive control over both domestic and foreign disaster response operations. BNPB operates under a sovereignty-first policy framework that requires international humanitarian actors to receive explicit authorization before engaging in field activities. Foreign organizations are prohibited from self-deploying and must partner with nationally recognized local organizations for access to affected zones (Trias & Cook, 2021; Triyanti et al., 2023).

This centralized architecture is further reinforced through regional mechanisms. In the Sulawesi response, the AHA Centre—the ASEAN Coordinating Centre for Humanitarian Assistance—played a bridging role between the Indonesian government and international responders. In parallel, the Joint Operations and Coordination Centre for International Assistance (JOCCIA) was established as a temporary multi-agency platform. It included representatives from BNPB, the Indonesian military, UN OCHA, and various cluster leads. However, JOCCIA's operational scope was tightly controlled by national authorities and limited in terms of transparency and inclusivity, especially during the critical early response phase (Organisation Z, 2018b).

International non-governmental organizations (INGOs) were required to register with BNPB and coordinate their activities through Indonesian partner organizations, as mandated under Peraturan Kepala BNPB No. 22/2010. This regulation imposed significant bureaucratic friction, including visa and equipment clearance requirements, thereby slowing down the arrival of international relief teams. While INGOs such as the Red Cross and Médecins Sans Frontières managed to operate under formal partnerships, less institutionalized foreign actors—including many Asian and diaspora-based CSOs—faced severe delays or exclusion due to limited regulatory clarity and lack of pre-existing partnerships (Organisation Z, 2018b; Triyanti et al., 2023).

Moreover, this architecture institutionalized a cluster-based coordination framework, adapted from the UN humanitarian system, with sectoral clusters (e.g., Health, Shelter, WASH) co-led by UN agencies and Indonesian ministries. However, access to these clusters was contingent upon formal accreditation and participation in coordination meetings, where non-traditional actors were largely underrepresented (Organisation Z, 2018b). Consequently, even as the humanitarian needs escalated, many actors—particularly those from China or other Asian countries—remained outside the formal humanitarian pipeline, operating in parallel or peripheral spaces.

Despite its emphasis on state control, the Indonesian model left critical operational gaps, particularly in peripheral regions like Donggala and Sigi, where local governments lacked capacity and communications infrastructure collapsed. These institutional constraints inadvertently opened space for non-traditional humanitarian actors, including Chinese social organizations, to engage through informal channels, often via diaspora networks or faith-based intermediaries (Organisation Z, 2018b; Trias & Cook, 2021).

18.1.2 Operational Challenges in the Sulawesi Context

The operational context of the 2018 Sulawesi Earthquake and tsunami response was marked by severe infrastructural collapse and systemic bottlenecks, which created profound constraints for both domestic and international humanitarian actors. The disaster—measuring 7.5 in magnitude—triggered not only a tsunami but also massive soil liquefaction and landslides in Palu, Donggala, and Sigi, severely damaging the region's logistical backbone. Key transport arteries, including Palu's Mutiara SIS Al-Jufrie Airport, were rendered non-functional in the initial days, while roads connecting affected districts were blocked or destroyed (Organisation Z, 2018a; Wardhani et al., 2024).

This physical disruption was compounded by the total breakdown of communications infrastructure. Telecommunications networks collapsed entirely, leaving affected communities without access to external communication and emergency responders without means to conduct needs assessments or coordinate response efforts (Titov, 2021). Humanitarian agencies, including the UN and the Indonesian Red Cross, reported that the blackout significantly delayed both situational awareness and resource mobilization during the critical 72-h response window (Organisation Z, 2018a).

At the local governance level, disaster management capacities were limited. District-level BPBD (regional disaster agencies) lacked the autonomous authority, pre-positioned resources, and trained personnel to mount an immediate and structured response. Local responders relied heavily on instructions and support from BNPB and the Indonesian military, creating delays in decision-making and implementation (Mas et al., 2020). As a result, humanitarian delivery in the first week of the disaster was uneven, with some communities receiving overlapping assistance and others being left unattended (Wardhani et al., 2024).

The cascading effect of these challenges was an operational vacuum—a gap between urgent humanitarian needs and the functional capacity of the formal response system. This vacuum opened critical spaces for non-traditional actors, including local civil society organizations, faith-based groups, and foreign CSOs operating through informal networks. Chinese social organizations, although not embedded in the official coordination structures, found entry points through diaspora networks and industrial zones that had retained limited operational capacities (Organisation Z, 2018b; Wardhani et al., 2024).

18.1.3 Fragmented Coordination and the Role of Local CSOs

The aftermath of the Sulawesi Earthquake-tsunami revealed a fragmented humanitarian coordination landscape, where overlapping mandates, institutional silos, and weak inter-organizational linkages constrained the overall effectiveness of response efforts. Despite the formal activation of the Joint Operations and Coordination Centre for International Assistance (JOCCIA) under the leadership of BNPB and supported by AHA Centre and OCHA, field-level coordination among actors remained limited, especially during the early response phase (Organisation Z, 2018b; Triyanti et al., 2023).

Multiple platforms operated in parallel—UN-led cluster mechanisms, bilateral military relief missions, the Indonesian Red Cross (PMI), and religious-based charities—all attempting to deliver emergency services, but often lacking structured alignment. In many cases, information sharing was ad hoc, while sector-specific cluster meetings were inconsistently convened or poorly attended in the critical early days (Harnantyari et al., 2020). For instance, although the Health Cluster was technically active, coordination between medical teams—both local and international—was minimal, resulting in duplication in certain IDP camps and service gaps in others (Organisation Z, 2018c).

In this disjointed environment, local civil society organizations emerged as key intermediaries—not only implementing direct relief, but also acting as interfaces between communities, local authorities, and incoming foreign actors. Nationally recognized CSOs such as Aksi Cepat Tanggap (ACT), Pos Keadilan Peduli Umat (PKPU), and Muhammadiyah's disaster units provided rapid on-the-ground response and critical logistical knowledge. Their embeddedness within local networks enabled quicker access to affected populations, often ahead of official mechanisms (Wardhani et al., 2024). These groups also served as "translators" of institutional processes for foreign NGOs unfamiliar with Indonesia's regulatory and social landscape.

This intermediary function created strategic entry points for external actors—including Chinese civil society organizations—which lacked formal registration or prior operational presence in Indonesia. In several instances, Chinese CSOs established linkages with ethnic Chinese business associations, industrial zones, and charity groups already integrated with local CSOs or faith-based organizations (Organisation Z, 2018c). These collaborations often operated outside the UN cluster system but addressed urgent humanitarian gaps in food distribution, shelter support, and medical services.

18.1.4 Policy Windows and Openings for Non-traditional Actors

Despite Indonesia's formal restrictions on unsolicited foreign humanitarian aid, the crisis conditions following the Sulawesi Earthquake-tsunami generated practical governance vacuums that allowed non-traditional actors to enter through emergent policy windows. The Indonesian government's initial stance emphasized national sovereignty and operational control, requiring all international assistance to be channeled through BNPB and its designated coordination platforms, such as the AHA Centre and JOCCIA (Organisation Z, 2018b; Trias & Cook, 2021). However, the scale of devastation, including over 2000 deaths, massive infrastructure collapse, and severely disrupted local capacity, created a mismatch between formal control mechanisms and on-the-ground needs (Organisation Z, 2018a).

In response to mounting logistical bottlenecks and mounting public pressure, the Indonesian President announced on October 1 a temporary relaxation of restrictions on foreign humanitarian aid. This included conditional acceptance of select international assistance, particularly in the areas of air transport, medical support, water purification, and sanitation (Organisation Z, 2018b). Though the policy remained restrictive—requiring foreign actors to partner with local counterparts and prohibiting autonomous foreign deployment—it nonetheless opened critical "gray zones" for non-traditional engagement, especially for actors capable of operating flexibly and informally within local networks (Wardhani et al., 2024).

These conditions aligned with the operational logics of Chinese civil society organizations, which often function outside the formal international humanitarian system and rely on relational networks for access. Unlike INGOs with institutional presence or cluster registration, Chinese CSOs entering Sulawesi typically did so via hybrid configurations—partnering with local Chinese chambers of commerce, industrial parks, and religious or ethnic community associations (Organisation Z, 2018c). These actors provided informal pathways for needs assessment, trust-building, and logistical support, compensating for regulatory and informational asymmetries between the Indonesian government and incoming aid providers.

Crucially, this hybrid model—linking enterprise resources, diaspora networks, and CSO mobilization—proved more adaptive to the constrained humanitarian space than traditional international mechanisms. The ability to leverage community trust, bypass bureaucratic inertia, and embed into existing local supply chains enabled certain Chinese CSOs to deliver rapid relief even while remaining outside formal coordination systems (Organisation Z, 2018d; Wardhani et al., 2024).

18.2 Intermediary Roles of Overseas and Local Chinese Communities

In the Sulawesi response, the operational presence of Chinese civil society organizations was made possible largely through the connective capacities of overseas and local Chinese communities. Rather than operating within formal humanitarian architectures, these communities enabled access, coordination, and adaptation through embedded social, cultural, and commercial networks. Their intermediary roles unfolded along several interconnected dimensions: facilitating entry into local governance and logistical systems; translating language, norms, and institutional expectations to build trust; mobilizing resources and organizing last-mile delivery through informal brokerage; and revealing the structural limits of informality in terms of coordination, legitimacy, and long-term integration. These dynamics illuminate how relational infrastructures shaped the possibilities and constraints of Chinese humanitarian engagement in Southeast Asia.

18.2.1 Diaspora as Gatekeepers: Local Chinese Communities in Indonesia

In the immediate aftermath of the 2018 Sulawesi Earthquake and tsunami, Chinese civil society organizations faced institutional and logistical barriers to direct humanitarian entry. Against this backdrop, local Chinese communities in Indonesia—though numerically limited in Sulawesi—emerged as crucial intermediaries, facilitating access, trust, and coordination for Chinese actors unfamiliar with the local environment. Rather than functioning as formal NGOs, these communities operated through informal networks, including business chambers, family associations, and Buddhist or Christian religious groups, which enabled rapid response beyond official channels (Organisation Z, 2018c).

While the Chinese diaspora in Sulawesi is relatively small, more established Chinese-Indonesian communities in urban centers such as Makassar, Jakarta, and Medan played connective roles. Their long-standing commercial presence and embedded ties with local governance and logistics systems enabled them to serve as "gateways" to disaster-affected areas. During the early response phase, relief materials from Chinese foundations were routed through Makassar and transported into Palu and Donggala by diaspora-affiliated logistics firms—at a time when official aid convoys faced delays due to access restrictions and traffic congestion (Organisation Z, 2018c).

These diaspora actors also facilitated rapid needs assessment and response coordination. With limited access to the formal UN-led coordination mechanisms and BNPB-controlled logistics hubs, Chinese CSOs depended on informal data and relational brokerage to make deployment decisions. Local Chinese communities in South Sulawesi shared situational updates through WeChat groups and direct phone calls, functioning as real-time information relays and trust nodes. In interviews compiled in post-disaster studies, Indonesian civil society leaders noted that while international aid was often slow or rigid, the diaspora-supported Chinese teams were agile and able to reach overlooked zones within 5–7 days of the event (Harnantyari et al., 2020).

Importantly, these intermediary roles were not only functional but also relational. In contrast to Chap. 16's Nepal case, where cultural integration was emphasized, or Chap. 17's Ecuador case, where institutionalized diaspora-CSO partnerships existed, the Sulawesi case demonstrates a more situational and tactical form of diaspora engagement. The Indonesian Chinese communities did not act as formal "partners" but rather as responsive "facilitators," adapting quickly to unfolding needs and leveraging pre-existing social capital to unlock operational access.

18.2.2 Trust, Translation, and Informal Mediation

In the aftermath of the 2018 Sulawesi Earthquake and tsunami, Chinese civil society organizations encountered major barriers in accessing affected areas—not only due to regulatory constraints but also because of language, institutional unfamiliarity, and limited cultural embeddedness. Unlike Western humanitarian actors with long-standing engagement in the UN-led cluster system, Chinese CSOs often lacked prior experience in Indonesian humanitarian governance and had no pre-established liaison mechanisms at the provincial or district level. This created an immediate need for informal mediation and trust-building mechanisms to navigate both state authority and local communities (Organisation Z, 2018b; Trias & Cook, 2021).

Within this context, local Chinese diaspora communities played a pivotal intermediary role. While limited in numbers in the directly affected areas such as Palu or Donggala, Chinese-Indonesian individuals with linguistic, cultural, and social proximity to both Chinese CSOs and Indonesian institutions served as key facilitators. Their roles included ad hoc interpretation during logistics negotiations, acting as liaison officers with local governments, and advising Chinese teams on appropriate customs, norms, and bureaucratic protocols.

Trust-building was particularly essential in remote zones where foreign actors lacked direct access. In some cases, Chinese CSO staff were unable to communicate with affected residents or even identify local leadership without assistance from bilingual intermediaries. The concept of "trust translation" became not merely linguistic but cultural: mediators helped Chinese organizations interpret not just spoken language, but also institutional logics, expectations of reciprocity, and community protocols of giving and receiving aid. This function mitigated the risks of missteps in culturally sensitive environments, thereby increasing the social acceptability of Chinese CSOs among local actors (Harnantyari et al., 2020; Trias & Cook, 2021).

Moreover, informal mediators provided real-time assessments of on-the-ground needs, bypassing the lag of formal reporting systems. Diaspora volunteers in Mamuju and Palu, for example, used mobile communication apps to relay urgent medical supply requests to Chinese teams still in Makassar or Jakarta, accelerating delivery timelines. These informal, relational channels were vital in the first ten days of response, when formal coordination mechanisms were either non-functional or inaccessible to new entrants.

18.2.3 Logistics, Brokerage, and Resource Mobilization

In the early days following the 2018 Sulawesi Earthquake and tsunami, severe disruptions in infrastructure and formal humanitarian coordination created logistical bottlenecks that limited the reach of international relief efforts. Amid these constraints, Chinese civil society organizations, operating largely outside the formal humanitarian architecture, were able to mobilize resources and deliver aid by embedding themselves into local Chinese business networks and leveraging diaspora-linked logistical pathways.

Multiple Chinese CSOs utilized storage and transfer points operated by Chinese-Indonesian business chambers in cities such as Makassar and Manado to reroute incoming supplies, bypassing congestion at Palu Airport and overburdened government distribution depots (Organisation Z, 2018a, 2018b). This decentralized approach allowed Chinese organizations to avoid delays associated with the official Joint Operations and Coordination Centre for International Assistance (JOCCIA), which required formal registration and adherence to centralized protocols (Organisation Z, 2018b).

The flexibility of these informal networks was further demonstrated through their use of pre-existing transport networks. Local Chinese associations coordinated fleets of privately owned vehicles—often commercial trucks repurposed for humanitarian delivery—to reach rural areas in Donggala and Sigi, where international organizations had limited presence (Ahmed et al., 2023). Drivers and volunteers were often recruited from local business communities, enabling rapid deployment of water, food, medical supplies, and temporary shelter kits to underserved communities without needing clearance through state-controlled cluster mechanisms (Steenbergen & Mes, 2020).

In many instances, these decentralized logistics chains operated faster than formal channels. Organization Z noted that one Chinese CSO completed three distribution cycles between October 4–10, 2018, reaching over 2800 households in peripheral districts—while UN-coordinated convoys were still undergoing clearance checks and access assessments (Organisation Z, 2018c). The adaptability of Chinese actors, drawing on diasporic social capital and logistical acumen, filled gaps left by the highly regulated international system.

18.2.4 Limits of Informality and Lessons from Embedded Adaptation

While informal networks of overseas and local Chinese communities played a crucial role in enabling rapid humanitarian engagement after the 2018 Sulawesi Earthquake and tsunami, the sustainability and institutional integration of these efforts faced significant limitations. These challenges were rooted in structural informality, fragmented coordination, and weak legal recognition—factors that constrained the long-term impact of Chinese civil society organizations in Indonesia's humanitarian governance framework.

Most diaspora-led initiatives operated outside the official NGO registration system in Indonesia. Without legal personality or formal memoranda of understanding (MoUs) with local governments, many Chinese actors lacked recognized channels for engagement with provincial and national disaster authorities (Wardhani et al., 2024). This lack of formal recognition limited access to state-managed logistics corridors and restricted their ability to participate in coordinated assessments or distribution planning, particularly during the transition from emergency to recovery phases (Triyanti et al., 2023). Consequently, although material aid reached affected populations quickly in the immediate aftermath, it remained largely disconnected from national recovery planning structures and accountability mechanisms.

The absence of a unified coordination mechanism among Chinese CSOs contributed to operational redundancies. In Central Sulawesi's Sigi and Donggala districts, different Chinese actors—often unaware of each other's activities—delivered similar supplies to neighboring communities within short timeframes, resulting in overlapping distributions and inefficiencies (Organisation Z, 2018c, 2018d). These challenges reflected a broader lack of horizontal coordination among Chinese humanitarian responders abroad, in contrast to the more structured cluster-based systems used by international organizations.

Compared to the Ecuador case discussed in Chap. 17, where long-established diaspora institutions were able to integrate Chinese CSOs into state-civil society partnerships, the Sulawesi experience revealed a more ad hoc and reactive model. The Indonesian Chinese community, though resourceful, mobilized through business, religious, or familial channels without the backing of institutionalized disaster response frameworks. This "event-driven mobilization" lacked continuity and institutional memory, thereby limiting opportunities for follow-up programs or post-disaster development engagement.

18.3 Sectoral Engagement and Adaptive Relief Practices

In the aftermath of the 2018 Sulawesi Earthquake and tsunami, Chinese civil society organizations became active responders across multiple humanitarian sectors. Operating largely outside formal coordination frameworks, their interventions spanned emergency shelter, mobile healthcare, food distribution, and WASH (water, sanitation, and hygiene) services. These efforts were marked by decentralized deployment, community-level collaboration, and reliance on local Chinese-Indonesian networks. By targeting underserved areas and populations, Chinese CSOs helped fill urgent service gaps. At the same time, their informal and fragmented approach revealed limitations in coordination, continuity, and the systematic inclusion of vulnerable groups, highlighting both the adaptability and constraints of their relief model.

18.3.1 Shelter and Emergency Housing: Filling the Gaps in IDP Response

In the wake of the 2018 Sulawesi Earthquake and tsunami, large-scale internal displacement quickly overwhelmed the existing humanitarian shelter infrastructure. The Indonesian government, in coordination with the UN-led cluster system, established formal Internally Displaced Persons (IDP) camps primarily in urban centers such as Palu. However, these camps soon faced capacity limits, logistical bottlenecks, and inconsistent standards of care, leaving significant gaps in shelter provision for communities on the periphery (Triyanti et al., 2023).

Chinese civil society organizations, operating outside the centralized coordination frameworks, intervened to fill these gaps through adaptive and decentralized shelter initiatives. Lacking direct access to government-managed shelter sites, Chinese organizations instead leveraged their connections with local Chinese-Indonesian networks—particularly religious institutions and business associations—to deliver emergency housing support in underserved areas. This included the rapid deployment of tents, waterproof tarpaulins, sleeping mats, and basic non-food items, which were often distributed directly to communities in Palu, Donggala, and surrounding districts (Organisation Z, 2018a, 2018b).

Field records from Organization Z indicate that between October 2–8, 2018, Chinese-supported teams distributed over 2000 temporary shelter kits through localized operations coordinated by Buddhist temples and diaspora-run logistics hubs in Makassar and Manado. This "bypass model" enabled the Chinese response to avoid delays commonly reported in the formal UN-led Joint Operations and Coordination Centre for International Assistance (Organisation Z, 2018b), and reach areas not yet covered by major INGOs.

The Chinese CSOs' engagement was marked by flexibility and responsiveness. Rather than relying on formal camp infrastructure, their shelter strategy focused on community-based distribution and rapid needs assessment. This allowed for quicker response times in semi-urban and rural areas where populations were less visible to centralized actors. However, this same informality also introduced challenges—such as inconsistent shelter quality, duplication of efforts in some districts, and limited integration into national recovery planning (Organisation Z, 2018b; Triyanti et al., 2023).

18.3.2 Medical and Health Services: Mobile Clinics and Rapid Response

In the immediate aftermath of the Sulawesi Earthquake and tsunami, Indonesia's public health infrastructure experienced severe disruptions. With major hospitals in Palu either damaged or overburdened, and rural clinics lacking capacity, affected populations—especially in peri-urban and remote areas—faced a sharp decline in access to basic healthcare. This vacuum created space for Chinese civil society organizations to deploy mobile, lightweight medical interventions, aimed at stabilizing conditions in underserved zones while bypassing the procedural delays common to state or UN-affiliated actors (Organisation Z, 2018c; Yew et al., 2020).

Rather than establishing formal field hospitals, most Chinese CSOs adopted a "mobile clinic" approach, characterized by small, mobile units operated in partnership with local health workers or informal networks. In several instances, Chinese medical volunteers worked alongside staff from local clinics in Donggala and Sigi, offering wound treatment, distributing basic pharmaceuticals, and conducting health screenings for internally displaced persons (IDPs) (Ahmed et al., 2023). Between October 5–10, 2018, five such temporary medical stations were deployed, reaching over 1400 individuals in community-based settings outside the formal camp structure (Organisation Z, 2018c).

Particular attention was given to trauma care, women's health, and psychosocial support—areas underserved by official relief mechanisms. Medical kits included dressings, antiseptics, analgesics, and sanitary pads, responding to immediate needs in post-disaster conditions. In some cases, basic mental health consultations were facilitated through volunteer interpreters embedded within Chinese-Indonesian networks, especially for children and elderly survivors (Ahmed et al., 2023; Organisation Z, 2018c).

The adaptive nature of these health interventions also extended to immunization and public health outreach. Working with the Indonesian Red Cross and local mosques, several Chinese-supported teams contributed to joint vaccine campaigns and hygiene education efforts, particularly in temporary settlements with poor sanitation. While long-term epidemiological data remains unavailable, select areas of intervention reported reduced patient referrals to overwhelmed urban hospitals, suggesting a localized stabilizing effect (Organisation Z, 2018c).

18.3.3 Food and Nutrition: Local Procurement and Community Distribution

In the post-disaster environment of Central Sulawesi, food security became a critical humanitarian concern. The earthquake and tsunami disrupted traditional supply chains, damaged infrastructure, and displaced over 200,000 people, many of whom were relocated to informal settlements or temporary shelters with limited access to food (Organisation Z, 2018c). In response, Chinese civil society organizations adopted a decentralized and locally embedded model of food aid delivery, prioritizing rapid access, community distribution, and the nutritional needs of vulnerable groups.

Rather than relying on international procurement and long-haul shipping, most Chinese CSOs sourced relief supplies—including rice, noodles, canned goods, cooking oil, and bottled water—within Indonesia. This strategy significantly shortened supply timelines and circumvented customs-related delays at entry points like the Makassar and Balikpapan ports (Organisation Z, 2018d). By procuring food locally, these organizations also aligned with Sphere standards' recommendation of supporting local markets during early recovery phases (Steenbergen & Mes, 2020).

Distribution efforts were largely facilitated through existing community-based networks, particularly Chinese-Indonesian religious institutions such as Buddhist associations and charitable kitchens. In Palu, for example, a Buddhist hall affiliated with Organization Z served as a temporary distribution hub, coordinating daily meal services to over 500 IDPs between October 6–12, 2018 (Organisation Z, 2018c). Volunteers from Chinese diaspora groups also collaborated with local mosques and neighborhood committees to ensure equitable food allocation, particularly in semi-formal displacement zones not directly covered by UN-led operations (Opabola et al., 2023).

Attention to vulnerable populations was another defining characteristic of Chinese CSOs' food assistance strategy. Nutrition packs were specifically distributed to pregnant women, young children, and the elderly—often through coordination with local midwives and health posts (Organisation Z, 2018d). In addition to dry rations, some teams prepared culturally familiar hot meals in collaboration with community kitchens, improving both the acceptability and cultural relevance of the aid delivered.

18.3.4 WASH Services and Vulnerable Group Targeting: Adaptive But Uneven

The post-disaster environment in Central Sulawesi presented urgent challenges in water, sanitation, and hygiene (WASH), particularly in informal shelters and overcrowded displacement sites. As public infrastructure was severely damaged, and formal service coverage struggled to meet needs, Chinese civil society organizations mobilized rapidly to fill WASH-related gaps through emergency distributions and localized interventions. However, their efforts, while adaptive, revealed uneven patterns of coverage—especially in addressing the needs of vulnerable groups.

During the early phase of the 2018 Sulawesi Earthquake–tsunami response, Chinese CSOs played a visible role in addressing urgent sanitation gaps in displacement sites around Palu and Sigi. Between October 3 and 12, hygiene kits—including soap, sanitary napkins, toothbrushes, and detergent—were distributed in multiple camps, while volunteers worked with local youth groups in areas such as Lolu and Biromaru to construct temporary bathing facilities and communal latrines using tarpaulin and bamboo (Organisation Z, 2018d). Implemented under conditions of damaged water infrastructure and limited institutional support, these low-cost and rapidly deployable interventions helped mitigate open defecation risks and stabilize environmental hygiene during the critical first two weeks of emergency response, illustrating the functional contribution of civil society actors to public health protection in resource-constrained disaster settings (Ahmed et al., 2023).

In some instances, coordination with local schools enabled CSOs to restore partial WASH functionality in education settings. For example, Organization Z's teams supported the cleaning and repair of toilet facilities in three primary schools in eastern Palu, enabling displaced students to return to classes in minimally safe conditions (Organisation Z, 2018d). However, these initiatives were scattered, project-based, and lacked systematic integration into education or public health recovery plans.

Despite these efforts, significant coverage gaps remained—especially in addressing the needs of women, older adults, and persons with disabilities. Many of the hygiene kits distributed at the time were standardized and failed to accommodate gender- or age-specific requirements, such as incontinence supplies or culturally appropriate menstrual hygiene support (Organisation Z, 2018d). Site-level observations further revealed that WASH facilities in informal settlements often lacked privacy and were inaccessible to those with limited mobility, underscoring broader deficiencies in needs assessment and inclusive design practices (Wardhani et al., 2024).

18.4 Limitations, Learning, and Emerging Pathways for Regional Engagement

While Chinese civil society organizations demonstrated agility and commitment in responding to the Sulawesi disaster, their involvement also revealed deeper structural, institutional, and procedural challenges. This section explores four interrelated dimensions: the constraints imposed by Indonesia's access and coordination systems; the adaptive, experience-based learning practices of Chinese CSOs; the transitional shift from reactive mobilization to regionally anchored strategies; and the emerging need for proceduralization and institutional alignment. These sub-sections map both the limitations of ad hoc engagement and the potential trajectories toward more durable, integrated roles in regional humanitarian governance.

18.4.1 Structural Limitations: Between Access Constraints and Institutional Exclusion

Although Chinese civil society organizations established a presence after the 2018 Sulawesi Earthquake and tsunami, their participation was constrained by structural barriers embedded in Indonesia's humanitarian governance system. Access to affected areas was tightly regulated through visa requirements, registration protocols, and mandatory affiliation with recognized local partners. These conditions led to delays in personnel deployment and equipment clearance, forcing Chinese teams to depend on informal networks—such as religious affiliations and personal connections—for early-stage operational access (Organisation Z, 2018a, 2018b, 2018c, 2018d). While such workarounds enabled limited engagement in the field, they also reduced institutional visibility and constrained both the scale and perceived legitimacy of interventions.

Beyond physical access, institutional exclusion further restricted meaningful participation. Indonesia's response relied heavily on the UN cluster system and the Joint Operations and Coordination Centre for International Assistance, where information exchange, resource allocation, and task division were centralized. Chinese CSOs were largely absent from these platforms due to lack of formal accreditation, limited familiarity with humanitarian coordination procedures, and language barriers, resulting in minimal participation in sectoral meetings and planning processes (Trias & Cook, 2021). As a consequence, Chinese actors operated primarily in parallel spaces, receiving fragmented situational information and lacking access to coordinated logistics pipelines or needs assessment data. This institutional distance increased the risk of duplication and weakened the capacity for strategic alignment with broader response objectives.

Social legitimacy and trust formation constituted a third layer of constraint. In the initial weeks following the disaster, cultural misunderstandings and limited local recognition constrained community engagement and slowed relationship-building with village leaders, health workers, and district authorities. Studies of post-disaster governance in Sulawesi emphasize that trust is a critical determinant of operational effectiveness in fragmented response environments, particularly when external actors lack embedded local presence (Harnantyari et al., 2020; Yew et al., 2020). In several peripheral settlements, Chinese teams encountered cautious or hesitant reception, often requiring mediation by local Chinese volunteers or faith-based intermediaries before gaining basic community acceptance (Organisation Z, 2018b, 2018c).

Taken together, these constraints reveal a structural paradox: while Chinese CSOs demonstrated the capacity to mobilize rapidly and operate flexibly at the margins of the humanitarian system, their exclusion from formal coordination architectures limited their influence on collective decision-making, accountability mechanisms, and recovery planning trajectories. Access was therefore achieved primarily through relational improvisation rather than institutional integration, reinforcing a mode of participation characterized by operational visibility but governance marginality. This configuration shaped both the scope and the limits of Chinese engagement in Sulawesi and underscores the tension between pragmatic entry strategies and the absence of durable institutional embedding in regional humanitarian governance.

18.4.2 Experiential Learning: Adaptive, Relational, and Field-Based Growth

The humanitarian response to the 2018 Sulawesi Earthquake and tsunami highlighted how Chinese civil society organizations, despite their marginal status within formal coordination frameworks, developed context-sensitive learning strategies grounded in field realities. Lacking prior operational presence in Indonesia, these CSOs adapted rapidly by relying on experiential, relational, and situated modes of learning—markedly different from procedural or institutionalized models observed in more established humanitarian actors.

A core pattern emerging from field operations was the predominance of in-situ learning. Operating in highly dynamic conditions without standard operating procedures or pre-established assessment templates, CSOs refined their interventions by learning through action. Volunteers adjusted logistical strategies, distribution methods, and beneficiary targeting on-site, based on real-time feedback from affected communities and local intermediaries. In the first week of deployment, distribution approaches shifted—for example, moving from centralized drop-off points to mobile team-based delivery in response to congestion and unmet needs in Palu's urban peripheries (Organisation Z, 2018b).

Another key dimension of learning involved relational navigation. Chinese CSOs, unfamiliar with Indonesian bureaucratic procedures and linguistic norms, leaned heavily on Chinese-Indonesian intermediaries to gain local access and build trust with affected populations. Buddhist institutions and diaspora-led chambers of commerce provided critical platforms for social embedding, mediating both access to local information and community acceptance. For example, in Biromaru and Banawa, Organization Z teams coordinated food and hygiene distributions with ethnic Chinese-led temples, which acted as both logistical bases and channels for communication with multi-ethnic residents (Organisation Z, 2018c, 2018d).

Over time, elements of cross-cultural adaptation became visible. Volunteers engaged local interpreters and religious leaders to tailor services in line with cultural sensitivities, such as ensuring gender-appropriate hygiene kit contents or coordinating meal services during prayer hours. These adjustments, while informal, reflect early stages of embedded learning and localization (Han et al., 2021).

Despite these gains, the learning remained fragmented and individual-dependent. There were no systemic mechanisms for codifying knowledge across teams or scaling successful practices beyond single operations. While learning was intensive and context-specific, it lacked institutional memory or formalized feedback loops, reinforcing the "adaptive but non-procedural" character of Chinese CSO engagement in Sulawesi (Organisation Z, 2018b, 2018c; Yen et al., 2020).

18.4.3 Transitional Positioning: From Ad Hoc Entry to Strategic Regionalism

The 2018 Sulawesi Earthquake and tsunami marked a critical inflection point in the evolving trajectory of Chinese civil society organizations in regional humanitarian engagement. While their involvement was initially characterized by ad hoc entry and reactive mobilization, the aftermath of the Sulawesi response generated conditions that began to shift their operational logic toward more regionally strategic configurations. Rather than remaining peripheral or episodic actors, certain Chinese CSOs began to explore institutional footholds and transnational linkages within Southeast Asia.

One notable transformation during the post-emergency phase was the emergence of tentative regional anchoring. In the months following the Sulawesi disaster, certain Chinese organizations began exploring semi-permanent coordination mechanisms in Indonesia, often through partnerships with Chinese-Indonesian business associations and local branches of the Indonesian Red Cross. For example, a logistics hub in Makassar—initially used for relief transshipment—was proposed as a long-term platform for cross-border disaster preparedness. Though modest in scale, these initiatives marked a shift from one-off delivery toward sustained presence, with potential to build institutional memory and continuity capacity (Organisation Z, 2018d).

The Sulawesi case also revealed the embryonic formation of a cross-border coordination mechanism among Chinese CSOs. Compared to the Nepal response of 2015, which largely relied on spontaneous dispatch, or the Ecuador case of 2016, which hinged on singular bilateral collaboration, the Indonesian deployment involved multiple actors coordinating across Jakarta, Makassar, and Palu. Despite the lack of a unified command structure, there were early signs of lateral information sharing through digital platforms and informal diaspora networks, indicating a nascent form of horizontal integration (Triyanti et al., 2023).

Perhaps most significantly, the Sulawesi experience contributed to a redefinition of Southeast Asia as a strategic focus area for Chinese civil actors. Interviews conducted during post-disaster reviews noted a perceptible shift in organizational priorities—from global symbolic engagement toward regionally embedded, culturally adjacent operations (Wardhani et al., 2024). This recalibration was driven not only by geographic proximity and logistical feasibility, but also by the presence of trusted intermediaries within the Chinese diaspora, which enabled more sustained and culturally attuned engagement.

18.4.4 Toward Proceduralization: Challenges and Conditions for Institutionalization

While the Sulawesi Earthquake response demonstrated the growing presence of Chinese civil society organizations in regional humanitarian operations, it also revealed the absence of standardized operational procedures that could support long-term institutionalization. Most participating CSOs functioned through improvised decision-making, ad hoc mobilization, and informal networks, lacking formal mechanisms for project management, quality assurance, or documentation.

These internal limitations were compounded by structural challenges in engaging external coordination systems. Chinese CSOs operated largely outside the United Nations cluster framework and had limited access to inter-agency coordination meetings due to language barriers, unfamiliarity with coordination platforms such as OCHA's 3W matrix, and absence of liaison protocols (Mas et al., 2020). Without institutionalized linkages to multilateral mechanisms, these organizations often relied on bilateral improvisation or personal ties to local partners, limiting their visibility and influence in broader response architecture (Shah et al., 2021).

Nevertheless, the Sulawesi experience also surfaced potential entry points for procedural integration. The involvement of Chinese CSOs in localized WASH interventions and school-based hygiene reconstruction suggests possible alignment with cluster-sector standards in early recovery (Ahmed et al., 2023). Efforts to collaborate with local mosques and health centers in hygiene outreach and vaccine distribution further indicate increasing familiarity with grassroots institutional interfaces (Organisation Z, 2018c).

Future institutionalization will require not only internal capacity-building—such as training in humanitarian principles, monitoring and evaluation (M&E), and Sphere standards—but also external system alignment. This includes multilingual coordination officers, participation in pre-disaster planning platforms, and data-sharing compatibility with host governments and multilateral partners. Emerging areas such as early warning systems, post-disaster education, and decentralized WASH infrastructure offer strategic points of entry (Opabola et al., 2023). Some scholars further advocate for the establishment of regional certification schemes or shared response protocols to facilitate cross-border integration (Shah et al., 2021).

18.5 Conclusion

The Sulawesi Earthquake and tsunami response marked a critical transition in the international engagement of Chinese civil society organizations. Operating in a constrained and sovereignty-centered humanitarian space, these actors navigated access barriers through diaspora networks, relational brokerage, and adaptive field-based learning. Their efforts addressed urgent gaps in shelter, health, food, and WASH services, particularly in underserved regions. However, their exclusion from formal coordination mechanisms and lack of standardized procedures exposed structural limitations and coordination deficits. While the response remained largely informal and fragmented, it catalyzed a strategic shift toward regional anchoring and revealed emerging pathways for institutionalization. The Sulawesi case thus represents both a frontier of operational experimentation and a turning point in the procedural maturation of Chinese CSOs within Southeast Asia's humanitarian landscape.

References

Ahmed, I., Garfias Royo, M., Opabola, E., Nurdin, S., Meilianda, E., Idris, Y., Rusydy, I., Joffe, H., & Parikh, P. (2023). Assessment of WASH infrastructure in schools in Central Sulawesi, Indonesia using structured observations and principal interviews. *Journal of Water, Sanitation and Hygiene for Development, 13*(6), 375–390. https://doi.org/10.2166/washdev.2023.147

Han, P., Yu, H., & Yu, X. (2021). A sloshing induced tsunami: 2018 Palu Bay event. *Applied Ocean Research, 117*, 102915. https://doi.org/10.1016/j.apor.2021.102915

Harnantyari, A. S., Takabatake, T., Esteban, M., Valenzuela, P., Nishida, Y., Shibayama, T., Achiari, H., A.G., R. M., Marzuki, M. F. H., Aránguiz, R., & Kyaw, T. O. (2020). Tsunami awareness and evacuation behaviour during the 2018 Sulawesi Earthquake tsunami. *International Journal of Disaster Risk Reduction, 43*, 101389. https://doi.org/10.1016/j.ijdrr.2019.101389

Mas, E., Paulik, R., Pakoksung, K., Adriano, B., Moya, L., Suppasri, A., Muhari, A., Khomarudin, R., Yokoya, N., Matsuoka, M., & Koshimura, S. (2020). Characteristics of tsunami fragility functions developed using different sources of damage data from the 2018 Sulawesi Earthquake and tsunami. *Pure and Applied Geophysics, 177*(6), 2437–2455. https://doi.org/10.1007/s00024-020-02501-4

Opabola, E. A., Galasso, C., Rossetto, T., Meilianda, E., Idris, Y., & Nurdin, S. (2023). Investing in disaster preparedness and effective recovery of school physical infrastructure. *International Journal of Disaster Risk Reduction, 90*, 103623. https://doi.org/10.1016/j.ijdrr.2023.103623

Organisation Z. (2018a). *Situation report No. 3: Sulawesi earthquake and tsunami, Indonesia (September 28, 2018a)*. https://shimo.im/docs/BrcQT1oF7uUopxkz

Organisation Z. (2018b). *Situation report No. 4: Sulawesi earthquake and tsunami, Indonesia (September 28, 2018b)*. https://shimo.im/docs/LqMuMtRngEUezqLu

Organisation Z. (2018c). *Situation report No. 5: Sulawesi earthquake and tsunami, Indonesia (September 28, 2018c)*. https://shimo.im/docs/HizEiPfj8vkHPkDD

Organisation Z. (2018d). *Situation report No. 6: Sulawesi earthquake and tsunami, Indonesia (September 28, 2018d)*. https://shimo.im/docs/XH1JBXo49Ok5rl6G

Shah, S. A., Yahia, S. B., McBride, K., Jamil, A., & Draheim, D. (2021). Twitter Streaming Data Analytics for Disaster Alerts. In *2021 Second International Informatics and Software Engineering Conference (IISEC)* (pp. 1–6) https://doi.org/10.1109/IISEC54230.2021.9672370

Steenbergen, R. V., & Mes, M. (2020). A simulation framework for UAV-aided humanitarian logistics. In *2020 Winter Simulation Conference (WSC)* (pp. 1372–1383). https://doi.org/10.1109/WSC48552.2020.9383977

Titov, V. V. (2021). Hard lessons of the 2018 Indonesian tsunamis. *Pure and Applied Geophysics, 178*(4), 1121–1133. https://doi.org/10.1007/s00024-021-02731-0

Trias, A. P. L., & Cook, A. D. B. (2021). Future directions in disaster governance: Insights from the 2018 Central Sulawesi Earthquake and Tsunami response. *International Journal of Disaster Risk Reduction, 58*, 102180. https://doi.org/10.1016/j.ijdrr.2021.102180

Triyanti, A., Surtiari, G. A. K., Lassa, J., Rafliana, I., Hanifa, N. R., Muhidin, M. I., & Djalante, R. (2023). Governing systemic and cascading disaster risk in Indonesia: Where do we stand and future outlook. *Disaster Prevention and Management: An International Journal, 32*(1), 27–48. https://doi.org/10.1108/DPM-07-2022-0156

Wardhani, T. P. M., Samaddar, S., Tatano, H., Tahir, Z., & Roy, S. (2024). Assessment of post-disaster concerns using social network analysis and machine learning: A study of 2018 Palu disaster. In *2024 International Conference on Information and Communication Technologies for Disaster Management (ICT-DM)* (pp. 1–7) https://doi.org/10.1109/ICT-DM62768.2024.10798935

Yen, E., Lin, S. C., Wu, T.-R., Tsai, Y.-L., & Chung, M.-J. (2020). Knowledge-building approach for tsunami impact analysis aided by citizen science. *Frontiers in Earth Science, 8*, 315. https://doi.org/10.3389/feart.2020.00315

Yew, Y., Arcos González, P., & Castro Delgado, R. (2020). Real-time impact analysis and response using a new disaster metrics: 2018 Sulawesi (Indonesia) Earthquake and tsunami. *Prehospital and Disaster Medicine, 35*(1), 76–82. https://doi.org/10.1017/S1049023X19005247

Chapter 19
Türkiye–Syria Earthquake: Pathways of Institutionalized International Response

Abstract The 2023 Türkiye–Syria Earthquake marked a critical juncture for Chinese social organizations' participation in international humanitarian response. Moving beyond earlier patterns of isolated, short-term engagement, this chapter examines their evolving integration into transnational humanitarian architectures. It analyzes four key dimensions: the shift from ad hoc mobilization to structured institutional interfaces; the emergence of modular collaboration through platform-based governance; the negotiation and localization of global humanitarian norms; and the capacity-building processes that followed crisis engagement. By situating these developments in comparative and historical perspective, the chapter underscores how the Türkiye–Syria response catalyzed a deeper trajectory of professionalization, reflexive reform, and strategic repositioning—revealing both the possibilities and limitations of Chinese social organizations as emerging actors in global disaster governance.

Keywords Institutional integration · Modular collaboration · Normative alignment · Crisis-driven learning · Strategic positioning

19.1 Evolving Modes of Engagement: From Ad Hoc Participation to Structured Humanitarian Interface

The Türkiye–Syria Earthquake in 2023 marked a pivotal moment in the evolution of Chinese social organizations' overseas humanitarian engagement, signaling a departure from earlier patterns of autonomous, short-term responses. This section first outlines the historical trajectory of these organizations' international involvement, characterized by limited coordination and minimal institutional integration. It then examines the Türkiye–Syria case as a watershed, where Chinese actors interfaced with national and multilateral platforms such as AFAD, the UN situation report system, and Organization Z. Through analyzing new mechanisms of coordination,

Q. Du and R. Shaw, *Social Organizations in Disaster Response and Recovery*, Disaster Risk Reduction, https://doi.org/10.1007/978-981-92-2846-1_19

information sharing, and task alignment, the section further explores how these developments reflect a broader shift toward institutional embeddedness in transnational humanitarian governance.

19.1.1 Historical Trajectories of Overseas Engagement

Prior to the Türkiye–Syria Earthquake in 2023, Chinese social organizations' participation in overseas disaster response was largely characterized by ad hoc mobilization, fragmented coordination, and limited interface with international humanitarian systems. Case studies from the Nepal Earthquake (2015), Ecuador Earthquake (2016), and Sulawesi Tsunami (2018) reveal a pattern of short-term, reactive missions driven primarily by organizational initiative rather than strategic integration into global relief architectures.

In the 2015 Nepal Earthquake, Chinese NGOs such as China Foundation for Poverty Alleviation (CFPA) and One Foundation mobilized within days. These responses demonstrated speed and determination but operated largely outside of formal multilateral coordination systems. While some Chinese organizations engaged with local governments and the Chinese embassy, there was limited participation in UN-led cluster meetings, and few organizations reported through the humanitarian situation report system.

Similarly, the 2016 Ecuador Earthquake prompted Chinese diaspora networks and social organizations to deliver relief materials and short-term medical services. Yet, due to the absence of pre-established institutional pathways, coordination with host-country authorities and established INGOs remained largely improvised. Many early operations were routed through personal relationships and diplomatic ties rather than formal humanitarian platforms or shared logistics systems (Organisation Z, 2023a).

The Sulawesi disaster response in 2018 marked a slight evolution, with several Chinese NGOs attempting joint actions and even sharing warehousing space. However, such cooperation remained operationally fragmented, and most Chinese actors were not yet integrated into the Sphere standard frameworks or regional humanitarian coordination bodies (Schwarz et al., 2025). Across these earlier cases, Chinese social organizations generally focused on direct material assistance—delivering tents, food, and medicine—but lacked involvement in multi-agency needs assessments, accountability reporting, or protection mainstreaming, all of which are standard in coordinated international humanitarian response. These missions rarely included mechanisms for after-action review or institutional learning, further limiting their capacity to evolve from one event to the next.

This trajectory of self-reliant but structurally isolated responses set the baseline for understanding the significance of their later evolution. The Türkiye–Syria Earthquake marked a turning point where Chinese social organizations, for the first time, participated in structured coordination platforms such as AFAD, Organization Z, and UN situation report mechanisms, indicating a substantial departure from the earlier, more autonomous modes of operation.

19.1.2 Institutional Interfaces in the Türkiye–Syria Earthquake

The 2023 Türkiye–Syria Earthquake marked a critical juncture in the evolution of Chinese social organizations' cross-border humanitarian participation. Unlike earlier engagements where Chinese actors primarily operated in parallel to established global coordination frameworks, this response was distinguished by their increasing institutional interfacing with formal humanitarian architectures, both national and multilateral.

One of the most significant developments was the engagement with Türkiye's Disaster and Emergency Management Authority (AFAD), which served as the central national coordination body. Chinese NGOs participating in the emergency response were officially registered with AFAD, enabling them to receive operational permissions, task assignments, and access to on-site coordination mechanisms. This institutional pathway not only legitimized their presence but also embedded them in the government-led humanitarian workflow—something not observed in previous cases such as Nepal or Sulawesi.

In parallel, Chinese organizations increasingly aligned with United Nations–led coordination mechanisms. At least three NGOs submitted structured needs assessments and activity updates between February and March 2023, contributing to greater information transparency and operational visibility within the broader humanitarian system (Organisation Z, 2023i).

Moreover, the Organization Z coordination platform played a pivotal intermediary role. Functioning as a Chinese-led humanitarian coordination hub, it facilitated regular task group meetings, maintained a central database for resource allocation, and ensured alignment between participating NGOs and host-country standards. Chinese organizations involved in Organization Z's platform reported improved efficiency in warehouse logistics, camp deployment, and medical dispatching, largely due to clear role differentiation and centralized scheduling systems.

These developments point to a broader shift: Chinese NGOs are no longer peripheral humanitarian actors but are increasingly integrated into both national and multilateral response systems. This contrasts with many Global South NGOs that often remain outside formal UN coordination structures due to capacity or recognition constraints (Schwarz et al., 2025; Tayfur et al., 2024). The Türkiye–Syria case thus signals a deepening institutional maturity and a growing capacity among Chinese social organizations to navigate transnational humanitarian governance.

19.1.3 Modes of Interaction: Coordination, Information Sharing, and Operational Alignment

During the Türkiye–Syria Earthquake response, Chinese social organizations began engaging in more structured and multilateral forms of humanitarian coordination. A notable shift occurred in how these organizations interacted with international platforms, leveraging coordination mechanisms such as situational reporting systems, inter-agency task forces, and digital information-sharing platforms to align with broader operational architectures.

One of the most significant modalities was participation in the situation report system, which facilitated real-time updates on operational activities and needs assessment. Multiple Chinese organizations contributed to the information flow by regularly updating their service status, needs evaluations, and logistic capacity through the situation report platform hosted by Organization Z. These reports, collected from field teams, covered distributions of winterization kits, food parcels, and health-related services, particularly in hard-hit provinces such as Hatay and Gaziantep.

In addition to information systems, Chinese NGOs engaged in multilateral coordination meetings, particularly within Organization Z's modular coordination framework. Field task forces were organized by thematic clusters—such as logistics, medical aid, shelter, and psychosocial support—with Chinese groups embedded in different segments depending on their expertise. For example, one medical NGO actively contributed to the health and trauma response cluster, while another focused on warehouse management and distribution logistics. These meetings allowed organizations to map activities, identify overlaps, and fill operational gaps collaboratively.

A critical feature of this engagement was the use of cross-organizational communication platforms, such as WeChat groups and shared online documents, where field teams uploaded real-time updates and received centralized coordination from Organization Z. This allowed for quick redeployment of resources in response to shifting needs, as evidenced by the rapid redirection of hygiene kits from one warehouse in Kahramanmaraş to another in Şanlıurfa based on emerging shortages.

Despite these improvements in interactional mechanisms, several operational challenges persisted. Language barriers limited participation in international coordination meetings, often requiring translation support. Time-zone differences between remote coordination staff in China and field teams in Türkiye sometimes led to misalignments in reporting cycles. Furthermore, unfamiliarity with UN-led humanitarian protocols created hesitancy in joining certain thematic clusters, especially those with highly technical requirements such as shelter engineering or gender-based violence response.

19.1.4 From Participation to Integration: Shifting Institutional Logics

The 2023 Türkiye–Syria Earthquake marked a significant turning point in the evolving engagement logic of Chinese social organizations in overseas humanitarian response. While previous interventions—such as during the 2015 Nepal Earthquake, 2016 Ecuador Earthquake, and 2018 Sulawesi Tsunami—were often characterized by ad hoc mobilization, minimal cross-organizational coordination, and limited interface with international humanitarian systems, the Türkiye–Syria case witnessed a deeper shift toward institutional embedding. Chinese NGOs were no longer simply participating as external responders but began integrating into the architecture of global humanitarian governance.

This transition was particularly evident in the way Chinese organizations interfaced with host country and multilateral coordination platforms. For example, several organizations formally registered under the Türkiye Disaster and Emergency Management Authority (AFAD), thus aligning themselves with the official disaster management framework of the host country. This registration not only enabled legal recognition and access to affected areas, but also facilitated structured task distribution in line with AFAD's national contingency planning.

Moreover, participation in Organization Z's coordinated mechanism marked a further institutional step. Chinese NGOs joined task-based working groups—such as shelter, medical support, and warehousing—coordinated through digital platforms and weekly planning meetings. Organization Z's situation report system provided a standardized method of information input, reinforcing procedural discipline among participating organizations (Organisation Z, 2023f, 2023g). Some Chinese organizations, for the first time, were responsible not only for data submission, but also for jointly drafting situation reports that were circulated among UN agencies, national governments, and INGOs.

This transition was not only logistical but also conceptual. Moving beyond project-based thinking, organizations began adopting the institutional logic of "embedded humanitarianism," which prioritizes predictability, accountability, and multilateral coordination (Durmus et al., 2025; Tayfur et al., 2024). Rather than viewing themselves as temporary service providers, several organizations demonstrated aspirations to become stable actors within the global humanitarian system. This shift was supported by strategic investments in organizational infrastructure, including standardized training, translation of humanitarian codes of conduct, and the establishment of internal reporting mechanisms (Organisation Z, 2023h).

Nonetheless, the integration was partial and uneven. Many organizations still lacked personnel with humanitarian coordination experience, and language barriers frequently hampered real-time decision-making in multilateral settings. Furthermore, resource asymmetry and limited access to core funding constrained their ability to participate in long-term planning.

19.2 Modular Collaboration and Platform-Based Governance in Transnational Disaster Relief

The response of Chinese social organizations to the 2023 Türkiye–Syria Earthquake marked a significant departure from previous fragmented interventions, giving rise to a more modular and institutionally embedded mode of transnational disaster engagement. This section examines how Chinese actors transitioned from isolated efforts to coordinated participation through a platform-based governance model anchored by Organization Z. It explores the modular organization of humanitarian functions, the operational role of Organization Z in clustering and task allocation, the mechanisms enabling multi-organizational coordination in the field, and the trade-offs between improved efficiency and emerging bureaucratic and cultural frictions. These dynamics illuminate evolving pathways for professionalized international collaboration.

19.2.1 From Fragmentation to Modularity: Organizing Chinese Humanitarian Actors

In the wake of the 2023 Türkiye–Syria Earthquake, a notable structural shift emerged in the organization of Chinese social organizations operating overseas: a transition from fragmented individual responses toward a modular, inter-organizational form of collaboration. Unlike earlier humanitarian deployments in Nepal (2015), Ecuador (2016), or Sulawesi (2018)—where most Chinese NGOs operated in isolation and without a unified framework—this response featured a clear attempt to embed Chinese actors within a decentralized yet highly structured system of coordination.

This transformation was neither accidental nor solely reactive. Rather, it reflected a growing institutional consciousness among Chinese humanitarian actors about the limitations of past fragmentation. Drawing on prior operational setbacks, several organizations proactively linked their capacities into a larger network structure, coordinated primarily through the platform Organization Z. This platform played a crucial governance role by facilitating modular clustering of functions—such as shelter, medical services, warehousing, and needs assessment—and assigning organizations into thematic working groups with predefined responsibilities (Organisation Z, 2023b, 2023d). Within each cluster, Chinese NGOs were encouraged to nominate focal points, follow standardized reporting templates, and synchronize action plans across actors.

This model of modular engagement was not merely logistical—it reflected a deeper evolution in how Chinese humanitarianism conceptualized its role in transnational crisis settings. Thematically clustered teams replaced isolated interventions. Coordination logs show that at least five Chinese organizations participated in the "Medical Relief" and "Warehousing & Logistics" modules (Organisation Z, 2023d), collaborating with both international and Turkish partners. These groups worked semi-autonomously but within a larger operational grid, enabling more efficient resource allocation and clearer accountability pathways.

This modular turn echoes broader shifts observed in the humanitarian landscape, where decentralization does not imply disorganization but instead leverages distributed expertise for collective response (Acharya et al., 2024; Tayfur et al., 2024). As Organization Z's coordination records demonstrate, this approach enabled Chinese organizations to quickly scale task-specific deployments—e.g., transferring hygiene kits between cities or establishing field medical units—without duplicating efforts or overstretching logistical capacities (Organisation Z, 2023b).

19.2.2 The Role of Organization Z: Platform Governance and Operational Clusters

In the wake of the Türkiye–Syria Earthquake, Organization Z emerged as a central coordinating platform that structured the participation of Chinese social organizations through a modular and platform-based governance model. Unlike earlier responses marked by fragmented deployments, this framework enabled a coordinated entry of diverse actors into a shared operational architecture, improving task differentiation, information flow, and overall humanitarian coherence.

Organization Z assumed a bridging role between international coordination systems and domestic Chinese humanitarian actors. It facilitated key administrative functions including NGO registration, task assignment, and interface with data reporting systems. By February 12, over 14 Chinese organizations had been registered and embedded within Organization Z's coordination network, operating across multiple provinces such as Hatay, Kahramanmaraş, and Malatya (Organisation Z, 2023f). This registration process ensured organizations met minimum readiness standards—logistical capacity, reporting mechanisms, and safety protocols—prior to field deployment.

A critical feature of Organization Z's governance structure was its division into operational clusters. These included Medical Relief, Warehousing and Logistics, Rapid Assessment, and Shelter Management. Each cluster operated semi-independently but was tied together by a shared situational reporting protocol and centralized scheduling mechanism. Within these clusters, focal points were appointed to coordinate daily operations, standardize data submission formats, and maintain cross-organizational communication channels (Organisation Z, 2023c, 2023d, 2023e, 2023f, 2023g, 2023h). For instance, the Warehousing cluster managed shared storage facilities in Iskenderun and Kirikhan, and facilitated resource redeployment based on situation report feedback regarding local shortages and supply mismatches.

Chinese NGOs were assigned based on their capacities: medical teams participated in trauma response clusters, logistics actors coordinated bulk relief material transportation, while assessment groups conducted early needs evaluations in collaboration with UNDAC and local agencies. Through Organization Z's task mapping, overlapping functions were reduced and spatial distribution of efforts was optimized, especially in urban hotspots like Antakya and Adiyaman (Organisation Z, 2023e, 2023g).

This mode of platform governance also enabled horizontal linkages with multilateral agencies. Several Organization Z clusters regularly interfaced with IFRC, UN-OCHA, and local AFAD coordinators, aligning response strategies with host-country systems. The platform's centralized communication mechanism—consisting of bilingual situation reports, shared WeChat groups, and live coordination dashboards—ensured that Chinese actors could maintain both operational autonomy and institutional visibility (Schwarz et al., 2025).

19.2.3 Coordination in Action: Multi-organizational Interface and Task Integration

The 2023 Türkiye–Syria Earthquake response demonstrated a notable evolution in the operational coordination capacities of Chinese social organizations, transitioning from parallel actions to embedded, modular collaboration. A key mechanism enabling this transformation was the multi-organizational interface structured by Organization Z, which integrated actors through shared systems for information exchange, logistics management, and situational reporting.

One of the most important coordination tools was the structured situation report system, where multiple Chinese NGOs began contributing daily operational updates. Joint reports issued in late February captured not only beneficiary numbers and supply movement but also flagged unmet needs across zones such as Hatay, Gaziantep, and Adiyaman (Organisation Z, 2023f, 2023h). These reports provided standardized data across medical assistance, winterization kits, and hygiene distributions—allowing for real-time prioritization of actions and task reallocation across clusters.

The warehouse and logistics clusters illustrated another dimension of integrated coordination. Warehousing operations in Kahramanmaraş and Şanlıurfa were co-managed by Chinese NGOs and local partners, using shared spatial allocation protocols to prevent congestion and duplication. As detailed in Organization Z's mid-March update, Chinese NGOs coordinated unloading schedules and assigned zones within warehouses, facilitating smooth cross-loading and dispatch of relief materials (Organisation Z, 2023g). This level of interoperability extended to truck fleet sharing and modular packaging of goods based on downstream distribution needs, which were updated daily through digital dashboards.

Beyond logistics, coordination extended into the physical rebuilding phase. In at least two instances, Chinese social organizations jointly engaged in the set-up of temporary shelter compounds and community camps, using spatial data collected by rapid assessment teams to determine layout, sanitation placement, and security needs. Cross-organizational task teams—organized around thematic areas such as health, protection, and shelter—were deployed under Organization Z's weekly coordination directives, ensuring that field operations matched emerging priorities (Organisation Z, 2023i).

These coordination mechanisms also incorporated digital tools. Mobile applications and cloud-based collaborative platforms allowed organizations to upload field reports, photos, and geo-tagged needs in real-time (Oliveira et al., 2024). However, while these tools enhanced information accuracy and speed, they also exposed gaps in digital literacy and standardization, particularly among first-time humanitarian responders.

19.2.4 Assessing Gains and Trade-Offs: Efficiency, Redundancy, and Bureaucratic Costs

The transition toward modular collaboration and platform-based governance in the Türkiye–Syria Earthquake response significantly improved the operational visibility and efficiency of participating Chinese humanitarian organizations. Through standardized reporting, task coordination, and data interoperability, Chinese actors were able to engage more systematically with transnational humanitarian structures. These shifts contributed to faster response cycles, clearer division of labor, and increased recognition within the international humanitarian architecture (Organisation Z, 2023f; Schwarz et al., 2025).

Efficiency gains were particularly visible in logistics and information sharing. The shared situation report system reduced information gaps by centralizing needs assessments and response tracking. Additionally, platform-managed logistics networks avoided duplication in delivery routes and enhanced the equitable distribution of aid across impacted regions, particularly in Hatay and Gaziantep (Binici et al., 2025; Organisation Z, 2023g). However, these gains came with measurable trade-offs, particularly in terms of coordination costs and institutional burdens.

One trade-off was the emergence of bureaucratic layering. The platform governance model introduced multiple levels of reporting, approval, and coordination—such as daily focal point check-ins, cluster-specific updates, and biweekly coordination briefings. These added procedural demands created operational bottlenecks, especially for smaller NGOs with limited personnel (Organisation Z, 2023h). Moreover, the insistence on uniform formats for reporting and task planning occasionally clashed with the more adaptive, informal practices used by some domestic Chinese actors, creating a misalignment in working cultures (Westman et al., 2024).

Redundancy also surfaced as a systemic concern. While coordination was intended to streamline efforts, some situation report data noted repeated deliveries of the same types of supplies—such as hygiene kits or winter clothing—within overlapping zones. This stemmed partly from delays in data updates or the lack of real-time synchronization among partner organizations (Görkemli et al., 2024; Organisation Z, 2023g). In digital coordination systems, latency and user inexperience contributed to the circulation of outdated or conflicting needs assessments.

Beyond logistics and reporting, cultural and linguistic barriers posed an additional layer of friction. Some Chinese NGOs faced difficulty fully engaging with international cluster meetings due to translation gaps or unfamiliarity with sector-specific terminology. These challenges underscored the need for future investments in cross-cultural training and digital literacy to enhance the institutional maturity of participating actors (Aycicek & Özdemir, 2025; Organisation Z, 2023e).

19.3 Negotiating Global Humanitarian Norms: Practices of Standard Alignment and Local Adaptation

The Türkiye–Syria Earthquake response in 2023 marked a turning point for Chinese social organizations' engagement with the global humanitarian system. Rather than operating on the periphery, these actors began to actively navigate, internalize, and reinterpret internationally recognized standards. This section examines their evolving professionalization through four analytical dimensions: initial encounters with the humanitarian normative landscape; concrete practices of institutional alignment; the implementation challenges arising from technical, cultural, and procedural gaps; and emerging strategies of contextual adaptation that localized global norms for greater effectiveness and legitimacy. These dimensions illuminate how normative integration is shaped by both compliance and situated negotiation.

19.3.1 Understanding the Normative Landscape: Global Humanitarian Standards in Practice

The Türkiye–Syria Earthquake response in 2023 marked a critical entry point for Chinese social organizations into the normative infrastructure of global humanitarianism. Unlike previous deployments where engagement with international standards remained minimal, the earthquake response environment was shaped by pre-established frameworks such as the SPHERE Handbook, the Core Humanitarian Standard, the Red Cross Code of Conduct, and Inter-Agency Standing Committee (IASC) guidelines (Kalyoncu Atasoy et al., 2024; Tarnas et al., 2023). These frameworks were not merely rhetorical—they were institutionally embedded in operational mechanisms coordinated by Organization Z and reflected in routine situation report templates and coordination briefs (Organisation Z, 2023a, 2023c).

Organization Z played a vital role in mainstreaming these standards among newly participating Chinese actors. In early February, joint orientation sessions were held in coordination hubs such as Gaziantep and Adana, where actors were introduced to minimum humanitarian standards for shelter density, water and sanitation ratios, and child protection protocols. The situation reports issued between February 10 and 18 explicitly cited SPHERE metrics as benchmarks for evaluating field activities, particularly in shelter and hygiene distribution (Organisation Z, 2023b, 2023c). These documents also included annexes translating key SPHERE indicators into simplified bilingual checklists for field teams, an important step in operationalizing compliance in multilingual environments.

For many Chinese organizations, these encounters represented a first sustained interaction with global norms in practice. The process of alignment began with observational learning—through participating in cluster meetings, shadowing established INGOs, and reviewing standard-aligned documentation templates. Multiple Chinese NGOs were encouraged to follow SPHERE-based registration forms for aid beneficiaries and to integrate CHS indicators into their internal monitoring and reporting routines (Organisation Z, 2023a, 2023c). Notably, task logs show that in Hatay and Şanlıurfa, Chinese medical teams began adapting triage protocols and patient tracking tools in line with WHO and IASC emergency health standards (Organisation Z, 2023c).

While standard adoption remained uneven across actors, the Türkiye–Syria deployment clearly catalyzed a deeper normative exposure for Chinese responders. Rather than engaging as isolated contributors, these organizations began operating within a shared ethical and technical language—an essential prerequisite for institutional legitimacy in the global humanitarian system (Kalyoncu Atasoy et al., 2024). The section that follows examines how these normative frameworks were not only acknowledged but actively implemented—albeit with varying degrees of institutionalization and adaptation across operational domains.

19.3.2 Aligning with Standards: Institutional Practices of Compliance and Adaptation

In the Türkiye–Syria Earthquake response, Chinese social organizations demonstrated a marked shift toward operational alignment with international humanitarian standards. This alignment was neither symbolic nor superficial—it was reflected in a growing body of compliance practices structured around key global frameworks such as the SPHERE Handbook and the Core Humanitarian Standard. Coordination documents and situation reports from Organization Z regularly referenced these benchmarks in areas ranging from shelter setup to child protection protocols (Organisation Z, 2023d, 2023e, 2023f).

One of the clearest examples of standard alignment occurred in the establishment of temporary shelters. Chinese actors collaborated with local and international stakeholders to ensure site selection adhered to SPHERE guidelines on household spacing, water access, and drainage management. Camps in Hatay and Kahramanmaraş were reportedly evaluated using checklists based on SPHERE's minimum standards (Organisation Z, 2023e). The layout of sleeping areas, sanitation blocks, and community kitchens increasingly followed international design templates provided in coordination clusters, reflecting an internalization of these norms among Chinese teams.

In the distribution of relief materials, Chinese organizations also attempted to comply with internationally recommended quotas and tracking procedures. For instance, NFI (non-food item) distributions were often conducted based on household registration systems, using personal ID verification, family size records, and gender-based priority coding. Distribution kits reportedly included winter blankets, hygiene supplies, and cooking sets, with their contents and quantities shaped by Sphere standards (Organisation Z, 2023f). These provisions also reflected broader efforts to prevent aid duplication and strengthen accountability, consistent with lessons drawn from cross-border operational evaluations (Durmus et al., 2025; Mercan Annak et al., 2025).

Efforts to address the needs of vulnerable groups were especially evident in child protection and gender-sensitive service provision. Multiple Chinese NGOs established maternal-infant tents, safe spaces for women, and psychosocial support modules in collaboration with UN partners. These services were guided by internationally recognized protection principles, including the Inter-Agency Standing Committee's (IASC) guidelines on gender-based violence in emergencies (Organisation Z, 2023d). In Hatay, mobile support teams deployed by two Chinese NGOs delivered trauma counseling for displaced children and adolescents, integrating psychosocial first aid protocols derived from global toolkits (Organisation Z, 2023e).

19.3.3 Challenges in Implementation: Gaps, Misalignments, and Cultural Dissonance

Despite meaningful progress toward aligning with global humanitarian standards, Chinese social organizations encountered multiple implementation challenges during the Türkiye–Syria Earthquake response. These challenges did not merely reflect technical or operational deficits but underscored deeper issues of norm misalignment, institutional incompatibility, and cultural dissonance.

A primary difficulty arose from the administrative burden of compliance and reporting. The humanitarian architecture organized through Organization Z required NGOs to submit daily situation reports, attend multi-cluster briefings, and utilize standardized templates derived from international coordination practices (Organisation Z, 2023f). For many participating Chinese actors—particularly grassroots organizations with limited staff—this imposed significant strain. Several coordination notes from February and March flagged delayed or inconsistent reporting from some Chinese NGOs, largely due to unfamiliarity with English-language technical terms and sector-specific humanitarian lexicons (Organisation Z, 2023g; Westman et al., 2024). These limitations not only hindered data accuracy but also affected organizational visibility and integration within the larger response framework.

Cultural sensitivity emerged as another critical fault line. Although efforts were made to meet global protection standards, including the provision of maternal and child-friendly spaces, gaps remained in addressing gender-specific and religiously sensitive needs. Reports from Hatay and Adıyaman noted limited availability of menstrual hygiene materials in some early-stage NFI distributions (Organisation Z, 2023e), despite Sphere guidelines emphasizing their necessity. Furthermore, field assessments revealed that certain Chinese organizations failed to fully recognize local norms regarding gender segregation and religious taboos, resulting in discomfort among some displaced women and girls (Aycicek & Özdemir, 2025).

Accountability mechanisms also presented systemic friction. International guidelines such as the Core Humanitarian Standard emphasize transparency in financial reporting, community engagement, and feedback loops. However, many Chinese NGOs lacked prior experience with these procedures. For example, while hotlines and feedback boxes were deployed in several camps, few were maintained with regular response protocols or language support tailored to Arabic- or Turkish-speaking populations (Organisation Z, 2023h). Some organizations treated these mechanisms as symbolic rather than functional tools, highlighting a broader challenge of normative internalization (Görkemli et al., 2024).

19.3.4 Toward Contextual Adaptation: Localizing Standards for Cultural and Social Fit

As Chinese social organizations deepened their engagement in the Türkiye–Syria Earthquake response, a new phase of adaptation emerged—one that sought not merely to follow international humanitarian standards, but to interpret and implement them in ways that accounted for local cultures, capacities, and institutional logics. This process of contextualization moved beyond compliance to the realm of situated practice, reflecting a growing awareness that effective humanitarianism requires both norm alignment and cultural responsiveness.

One major step in this localization process involved the simplification and translation of key normative frameworks, especially the SPHERE Handbook. Several Chinese organizations, coordinated under Organization Z, produced abridged versions of Sphere content with added visual elements for field volunteers (Organisation Z, 2023g). These materials were used during volunteer briefings in Adıyaman and Şanlıurfa, and included locally relevant examples to explain standards on water access, shelter density, and hygiene kits. Importantly, this translation was not purely linguistic—it involved a re-encoding of norms into accessible, operational formats aligned with on-the-ground constraints.

Localization also extended into the actual design and delivery of services. In child protection programming, initial adherence to global guidance gradually gave way to adaptive strategies. For example, while initial psychosocial service modules relied on pre-scripted curricula, several teams revised content after interacting with

Syrian and Turkish educators, incorporating local storytelling, art-based play, and faith-sensitive approaches to trauma recovery (Organisation Z, 2023h). Similarly, women-focused services evolved in response to local feedback: mother-and-infant tents were repositioned based on cultural preferences for privacy, and hygiene kits were revised to include culturally appropriate items such as headscarves and ablution materials (Aycicek & Özdemir, 2025; Görkemli et al., 2024).

The localization of standards was not only about cultural adaptation but also about institutional negotiation. Several Chinese NGOs engaged local municipal actors and religious leaders to coordinate shelter planning and community engagement strategies. This not only improved trust and legitimacy but allowed for the hybridization of international norms with existing governance systems (Organisation Z, 2023i). In Kahramanmaraş, one project adjusted its site planning approach to integrate local security protocols and spatial customs, balancing SPHERE shelter guidelines with municipal zoning regulations (Durmus et al., 2025).

Digital tools also played a facilitative role in enabling flexible, context-responsive coordination. Mobile-based reporting platforms developed by Organization Z were updated to include Turkish and Arabic interfaces, allowing local staff and community members to contribute real-time feedback on service adequacy (Oliveira et al., 2024). This feedback loop helped organizations to iteratively refine their practices in line with both normative expectations and socio-cultural sensitivities.

19.4 Institutional Learning and Strategic Capacity-Building: Toward Long-Term Engagement

The Türkiye–Syria Earthquake in 2023 marked a critical juncture for Chinese social organizations, revealing both the limits of their international operational readiness and the transformative potential of crisis-driven engagement. This section examines how these actors transitioned from reactive participation to reflective learning, organizational reform, and strategic repositioning. It begins by analyzing the institutional stressors that catalyzed on-the-ground innovation, followed by an exploration of post-crisis internal capacity-building efforts. It then assesses persistent structural constraints in the global humanitarian field before outlining pathways toward long-term, rules-shaping participation in international humanitarian governance.

19.4.1 Learning Through Crisis: Operational Pressure as a Driver of Institutional Growth

The Türkiye–Syria Earthquake of February 2023 imposed significant institutional stress on Chinese social organizations, many of which lacked prior experience in high-intensity, transnational disaster settings. Under this pressure, these organizations were compelled to engage in rapid, adaptive learning processes, transforming operational constraints into opportunities for structural development and reflexive governance.

Coordination during the initial phase encountered critical bottlenecks in multilingual communication, real-time resource allocation, and cluster-based decision-making (Organisation Z, 2023d, 2023f). Several organizations faced difficulties in navigating sector-specific terminologies, submitting timely updates, and participating in cross-agency briefings. These challenges—though initially seen as liabilities—triggered a "learning-by-doing" dynamic, whereby field teams began improvising internal protocols for translation, logistics, and inter-agency communication. This aligns with broader observations in the literature that institutional learning often emerges through the iterative cycles of crisis response and procedural adjustment (Maddah et al., 2025; Zengin & Saglik Tugrul, 2025).

One area of emergent capacity involved the establishment of real-time inventory systems for frontline relief materials. Chinese teams in Hatay and Adıyaman adopted improvised digital tracking tools—often spreadsheet-based—to monitor supply flows, drawing from models observed in cluster coordination meetings (Organisation Z, 2023f). These systems later informed post-crisis standardization efforts and were cited in internal review notes as key learning outputs (Organisation Z, 2023g).

Volunteer management also experienced qualitative transformation. Initially reliant on ad hoc deployment, several organizations began introducing tiered screening and training procedures. Field updates from Şanlıurfa detail how new volunteers were grouped by language ability and assigned shadow roles under experienced staff, enabling both capacity transfer and supervision (Organisation Z, 2023e). The resulting approach, though unplanned, evolved into a semi-formal training ladder that was retained in post-disaster operations.

Notably, the crisis also prompted experimentation in rapid needs assessment. Lacking standardized tools, Chinese actors collaborated with Turkish municipalities to conduct door-to-door household profiling using basic demographic templates. This rudimentary mechanism—though lacking statistical rigor—enabled responsive distribution planning and laid the groundwork for future adaptation of international assessment protocols (Maddah et al., 2025).

19.4.2 Organizational Reform and Internal Capacity-Building Post-disaster

The Türkiye–Syria Earthquake not only tested Chinese social organizations' frontline response capacities, but also catalyzed a wave of institutional reforms and internal capacity-building efforts in the post-disaster phase. As the initial shock of deployment subsided, many organizations shifted attention toward consolidating lessons, formalizing internal procedures, and professionalizing their governance frameworks—a transition that marked a move from reactive mobilization to reflexive institutionalization.

A key area of reform focused on improving multilingual coordination and reporting. In situation reports compiled after March, several Chinese NGOs began adopting standardized bilingual templates for situation updates, task records, and distribution reports, facilitating more effective participation in inter-agency mechanisms (Organisation Z, 2023g, 2023h). In response to communication barriers encountered earlier in Adana and Hatay, some organizations also established internal translation teams and initiated staff training in humanitarian sector terminology, helping bridge linguistic and procedural gaps in international coordination (Organisation Z, 2023f; Westman et al., 2024).

Reforms extended into internal governance structures. In response to role ambiguity and coordination breakdowns during initial deployments, several organizations introduced clearer role descriptions, accountability chains, and rapid decision-making protocols. These adjustments were particularly evident in teams that managed camp logistics, where personnel flowcharts and duty rosters were introduced to reduce duplication and increase efficiency (Organisation Z, 2023g). Internal after-action reviews conducted in April 2023 by at least four Chinese organizations included structured feedback mechanisms, internal evaluation checklists, and staff debriefing sessions—practices previously absent from their operational routines (Acharya et al., 2024).

Volunteer management also became a focal point for institutional upgrading. Organizations that had previously relied on spontaneous volunteers began introducing tiered training systems, distinguishing between generalist volunteers, logistics staff, and technical responders. Training content increasingly included modules on cultural sensitivity, trauma-informed care, and Turkish and Arabic language basics, as well as context-specific protocols such as gender-segregated space management (Aycicek & Özdemir, 2025; Organisation Z, 2023h). These reforms helped reduce friction in cross-cultural interactions and increased overall acceptance by local communities.

Crucially, the reform efforts extended beyond technical fixes toward building organizational memory. Several NGOs institutionalized post-disaster learning through internal knowledge repositories, digital documentation of deployment processes, and recurring drills for overseas coordination (Organisation Z, 2023i). As highlighted in regional humanitarian literature, such internalization of learning is a key indicator of strategic maturity and forward-looking governance in NGO development (Aldamen & Hacimic, 2023; Tayfur et al., 2024).

19.4.3 Gaps and Constraints: Comparative Limitations in Global Field

Despite growing participation in international humanitarian operations, Chinese social organizations still face notable structural disadvantages when positioned within the broader global emergency relief system. These limitations are particularly evident when contrasted with established Western NGOs, which have long histories of international engagement, well-developed communication protocols, and robust accountability mechanisms.

One recurring constraint is linguistic and communicative capacity. Although several Chinese organizations managed to issue situation reports in English during the Türkiye–Syria Earthquake (Organisation Z, 2023d, 2023e), delays in information dissemination and difficulties in articulating needs and actions in real-time often hindered effective coordination with global partners. These challenges resonate with broader findings on Global South NGOs, which frequently struggle to maintain consistent visibility in international forums due to language and media constraints (Westman et al., 2024). The absence of bilingual staff and lack of experience in real-time English situation report drafting slowed responses during multilateral coordination meetings (Organisation Z, 2023f).

Moreover, the lack of pre-established overseas infrastructure significantly constrained the adaptability and reach of Chinese NGOs. Unlike many INGOs that maintain long-term field offices or local partnerships in disaster-prone regions, most Chinese organizations relied on temporary deployment models during the Türkiye–Syria response (Organisation Z, 2023b, 2023e). This reliance on episodic mobilization reduced institutional memory, undermined continuity, and limited the effectiveness of cross-border collaboration. Comparative assessments also highlight that these organizations typically lacked robust rosters of local interpreters, community liaisons, or culturally embedded partners (Taskaldiran et al., 2025).

Transparency and accountability mechanisms form another critical gap. Although Organization Z and its affiliates implemented post-disaster review mechanisms (Organisation Z, 2023h), externally verifiable reporting frameworks—particularly regarding financial flows, procurement decisions, and community feedback—were not systemically developed. This stands in contrast with the audit-backed protocols of many Western NGOs, which are often essential for accessing multilateral funding or forming joint operations with UN agencies (Orak, 2024).

Finally, the limited presence of Chinese NGOs in global humanitarian governance platforms—such as the Inter-Agency Standing Committee (IASC) or Sphere project—further weakens their ability to shape agenda-setting processes and normative frameworks. As observed during the Türkiye–Syria response, organizations without established affiliations found it more difficult to gain timely access to coordination hubs or participate in cluster-based decision-making (Organisation Z, 2023d; Tayfur et al., 2024).

19.4.4 Strategic Outlook: From Episodic Response to Sustained Global Engagement

The Türkiye–Syria Earthquake response marked a turning point for Chinese social organizations in terms of international operational engagement, yet their overall trajectory remains shaped by episodic interventions and reactive deployment. Looking ahead, sustainable integration into the global humanitarian system requires a fundamental shift—from project-driven responses toward long-term strategic positioning, underpinned by institutional anchoring, multi-scalar networks, and proactive agenda-setting.

One key area of transition lies in the formalization of international partnerships. Although ad hoc collaboration with local Turkish NGOs and international coordination clusters was observed during the earthquake (Organisation Z, 2023e, 2023h), these efforts remained transactional rather than embedded. Sustained engagement necessitates institutionalized agreements, such as Memoranda of Understanding (MOUs), joint contingency planning, and shared operational platforms. Existing literature suggests that emerging humanitarian actors must invest in "relational infrastructure" that persists beyond specific crises to secure influence in global governance frameworks (Oliveira et al., 2024).

Active participation in international norm-making bodies is another critical frontier. Chinese NGOs have historically been absent from global standard-setting processes such as those organized by the Inter-Agency Standing Committee (IASC) or the Sphere Project. However, earthquake-related situation reports acknowledged internal reflection on the need to "learn global languages" in terms of accountability, performance measurement, and humanitarian principles (Organisation Z, 2023i). Strategic inclusion in such platforms would not only enhance professionalization but also offer opportunities to shape norms from the perspective of non-Western humanitarian actors (Binici et al., 2025; Schwarz et al., 2025).

Furthermore, leveraging diplomatic channels and diaspora networks—especially under frameworks such as the Belt and Road Initiative—can provide Chinese organizations with semi-formal operational footholds in key regions. This may include embedding liaison officers in Chinese embassies, developing joint social service programs with overseas Chinese associations, or co-hosting humanitarian training with universities in the Global South. These pathways enable transition from one-off interventions to continuous presence, supported by both state-adjacent and grassroots mechanisms.

Crucially, strategic transformation also requires internal paradigmatic shifts. Many Chinese organizations still operate within a project logic, focusing on deliverables and short-term KPIs. To evolve as credible humanitarian actors, there must be a movement toward what Organization Z describes as "strategic co-construction"—a model where organizations not only implement humanitarian programs but contribute to shaping the global field's architecture itself (Organisation Z, 2023i).

19.5 Conclusion

The Türkiye–Syria Earthquake response marked a transformative moment for Chinese social organizations, shifting them from fragmented, reactive actors to increasingly structured participants in global humanitarian governance. Through modular collaboration, normative alignment, and post-crisis institutional reforms, these organizations demonstrated growing professionalization and strategic ambition. Persistent gaps in infrastructure, accountability, and global integration underscore the limitations of episodic engagement. As Chinese actors move toward sustained transnational presence, their trajectory will hinge on deepened partnerships, normative co-construction, and internal capacity-building—positioning them not only as responders, but as evolving contributors to the architecture of international humanitarianism.

References

Acharya, P., Sharma, K., & Liu, F. (2024). Türkiye's road to recovery after the 2023 Kahramanmaras earthquake: Lessons from Chile, Japan, and Nepal. *Natural Hazards Review, 25*(4), 04024031. https://doi.org/10.1061/NHREFO.NHENG-1985

Aldamen, Y., & Hacimic, E. (2023). Positive determinism of twitter usage development in crisis communication: Rescue and relief efforts after the 6 February 2023 earthquake in Türkiye as a case study. *Social Sciences, 12*(8), 436. https://doi.org/10.3390/socsci12080436

Aycicek, H. B., & Özdemir, E. C. (2025). The essential role of early rehabilitation in disasters: A single center experience in Türkiye-Syria earthquake. *Disaster Medicine and Public Health Preparedness, 19*, e188. https://doi.org/10.1017/dmp.2025.10111

Binici, S., Choi, J., Mitchell, S. M., & Pizzi, E. (2025). A Text Analysis of News Media Framing of Government Response to the 2023 Türkiye–Syria Earthquake. *Journal of Contingencies and Crisis Management, 33*(1), e70023. https://doi.org/10.1111/1468-5973.70023

Durmus, G., Mutlu, M. D., & Turkmenoglu, M. A. (2025). Examining corporate social responsibility activities of BIST 100 companies following the 2023 Türkiye–Syria earthquakes. *Social Responsibility Journal, 21*(9), 2053–2071. https://doi.org/10.1108/SRJ-09-2024-0664

Görkemli, H., Yayla, H., & Demirezen, B. (2024). Political communication in disaster: Analysis of two Main political alliances' tweets after 2023 great earthquakes in Türkiye. *Siyasal: Journal of Political Sciences, 33*(1), 111–132. https://doi.org/10.26650/siyasal.2024.33.1387832

Kalyoncu Atasoy, Z. B., Kurt, S. C., Beckerman-Hsu, J., Soyunen, S., Kalyoncu, M. F., & Kaptanoglu, A. (2024). Evaluation of the emergency nutrition response in Malatya after Türkiye–Syria earthquake. *British Journal of Nutrition, 131*(7), 1244–1258. https://doi.org/10.1017/S0007114523002672

Maddah, D., Vadi, R., Aloğlu, N., Boz, M., Wightwick, D., Istanbouly, G., Tannous, R. M., Morelli, A., Tawfeeq, D. A., & Von Schreeb, J. (2025). What do patients seek care for at a health Facility in the Aftermath of an earthquake? Experiences from an emergency medical team in Türkoğlu, Türkiye 2023. *Disaster Medicine and Public Health Preparedness, 19*, e12. https://doi.org/10.1017/dmp.2024.343

Mercan Annak, İ., Erdoğan, B., & Yıldız Emre, N. (2025). Experiences of Turkish nurses volunteering in the disaster zone following the 2023 Türkiye–Syria earthquake. *International Nursing Review, 72*(3), e13056. https://doi.org/10.1111/inr.13056

Oliveira, C. S., Ferreira, M. A., & O'Neill, H. (2024). The role of video cameras and emerging technologies in disaster response to increase sustainability of societies: Insights on the 2023 Türkiye–Syria earthquake. *Sustainability, 16*(17), 7618. https://doi.org/10.3390/su16177618

Orak, Y. (2024). Earthquake in Türkiye: Impact on health services and infection threats. *Journal of Clinical Practice and Research*, 549–557. https://doi.org/10.14744/cpr.2023.59244

Organisation Z. (2023a). *Situation report no. 1: Türkiye–Syria earthquake relief (February 6, 2023).* https://shimo.im/docs/XKq4MZnYerI76BkN

Organisation Z. (2023b). *Situation report no. 2: Türkiye–Syria earthquake relief (February 6, 2023).* https://shimo.im/docs/wV3VVnz407hMVp3y

Organisation Z. (2023c). *Situation report no. 3: Türkiye–Syria earthquake relief (February 6, 2023).* https://shimo.im/docs/erAdPrMbZxf64rAG

Organisation Z. (2023d). *Situation report no. 4: Türkiye–Syria earthquake relief (February 6, 2023).* https://shimo.im/docs/L9kBMGmrD2TV0LqK

Organisation Z. (2023e). *Situation report no. 5: Türkiye–Syria earthquake relief (February 6, 2023).* https://shimo.im/docs/A3n8VuuQjs4MbsGt

Organisation Z. (2023f). *Situation report no. 6: Türkiye–Syria earthquake relief (February 6, 2023).* https://shimo.im/docs/rp3OVbYWPxhJRYAm

Organisation Z. (2023g). *Situation report no. 7: Türkiye–Syria earthquake relief (February 6, 2023).* https://shimo.im/docs/TS5ZICBKpRq3C8G4

Organisation Z. (2023h). *Situation report no. 8: Türkiye–Syria earthquake relief (February 6, 2023).* https://shimo.im/docs/5rk9dx16rjfjyKqx

Organisation Z. (2023i). *Situation report no. 9: Türkiye–Syria earthquake relief (February 6, 2023).* https://shimo.im/docs/8Nk6MXl0LVSD5gqL

Schwarz, A., Sellnow, D. D., Diers-Lawson, A., & Sellnow, T. L. (2025). The crisis response to the Türkiye-Syria earthquakes in 2023: Instructional emergency risk communication by the government, the national disaster management agency, and NGOs via press releases and microblogs. *International Journal of Disaster Risk Reduction, 129*, 105770. https://doi.org/10.1016/j.ijdrr.2025.105770

Tarnas, M. C., Karah, N., Almhawish, N., Aladhan, I., Alobaid, R., & Abbara, A. (2023). Politicization of water, humanitarian response, and health in Syria as a contributor to the ongoing cholera outbreak. *International Journal of Infectious Diseases, 131*, 115–118. https://doi.org/10.1016/j.ijid.2023.03.042

Taskaldiran, E. S., Saracoglu, M. B., Meseli, S. E., & Koyuncuoglu, C. Z. (2025). A cross-sectional assessment of oral health status and hygiene practices of the Turkish population following the 2023 Turkiye-Syria earthquakes. *BMC Oral Health, 25*(1), 1462. https://doi.org/10.1186/s12903-025-06833-2

Tayfur, İ., Bayramoğlu, B., Şimşek, P., & Gunduz, A. (2024). Medical response to the February 6, 2023, earthquakes in Hatay: Challenges faced in the deadliest disaster in the history of Türkiye. *Disaster Medicine and Public Health Preparedness, 18*, e45. https://doi.org/10.1017/dmp.2024.21

Westman, A., Kurland, L., & Hugelius, K. (2024). Valued technical and non-technical skills among disaster responders: A cross sectional study of disaster responders involved in the earthquake in Türkiye and Syria January 2023. *BMC Emergency Medicine, 24*(1), 171. https://doi.org/10.1186/s12873-024-01083-x

Zengin, O., & Saglik Tugrul, G. N. (2025). Between survivor and social worker: The experiences of social workers in the 2023 Türkiye-Syria earthquake. *The British Journal of Social Work*, bcaf210. https://doi.org/10.1093/bjsw/bcaf210

Chapter 20
Myanmar Earthquake: Rapid Response and Institutional Limitations

Abstract The 2025 Myanmar Earthquake presented an extreme test case for international humanitarian action, exposing the limits of conventional coordination, institutional absorption, and field-level operability in fragile governance settings. The crisis unfolded amid profound political fragmentation and infrastructural collapse, rendering many multilateral mechanisms ineffective and leaving large segments of the population beyond the reach of formal aid. In this context, Chinese social organizations mounted an agile response, leveraging digital platforms, modular deployment models, and transnational networks to circumvent institutional bottlenecks and deliver frontline support. However, these efforts encountered deep structural frictions—ranging from coordination failures and legal ambiguity to operational burnout—ultimately revealing the constraints of short-term, project-based interventions. In response, a strategic reorientation has begun to take shape: shifting from rapid action to institutional embeddedness through local partnerships, resilience-oriented programming, and regional governance integration.

Keywords Operational maturity · Digital platforms · Transnational networks · Institutional limits · Governance embedding

20.1 A Stress-Tested Humanitarian Environment: Myanmar as a Limit Case of Fragile Governance

The 2025 Myanmar Earthquake unfolded in a uniquely fragile humanitarian environment shaped by deep governance fragmentation, systemic coordination breakdowns, infrastructural collapse, and institutional incapacity. This section maps Myanmar as a "limit case" for international emergency response, where conventional humanitarian mechanisms—ranging from multilateral coordination to operational logistics—proved insufficient in the face of political contestation and structural dysfunction. Through four interlinked sub-sections, it examines the collapse of centralized

Q. Du and R. Shaw, *Social Organizations in Disaster Response and Recovery*, Disaster Risk Reduction, https://doi.org/10.1007/978-981-92-2846-1_20

authority, the underperformance of UN and ASEAN frameworks, the extreme operational hazards, and the inability of the state to absorb external aid, setting the stage for analyzing non-state actors' adaptive engagement.

20.1.1 Fragile Governance and State Fragmentation in Post-Earthquake Myanmar

The 2025 Myanmar Earthquake struck not a cohesive state, but what can be more accurately described as a "fragmented governance field" in which military authorities, opposition coalitions, and ethnic armed organizations (EAOs) simultaneously asserted power across divergent regions (Heinkel et al., 2025; Samad et al., 2025). This fragmentation created a highly unstable operating context where humanitarian actors had to navigate competing authority systems and overlapping claims to territorial control.

One of the most consequential effects of this governance fragmentation was the politicization of humanitarian access. In regions such as Sagaing, Kachin, and northern Shan, the enforcement of "restricted humanitarian access zones" by military or militia groups resulted in the effective exclusion of entire communities from international assistance efforts (Kim et al., 2025; Organisation, 2025b). Even where access was technically permitted, security concerns—including checkpoints, active hostilities, and presence of landmines—made physical entry highly volatile and unpredictable. In the first two weeks following the quake, more than 40% of critically affected locations remained classified as "inaccessible" due to these constraints (Organisation Z, 2025c).

Compounding the issue was the failure of national data and coordination systems. With official damage assessments delayed or absent, and no consistent reporting channels between central and local authorities, humanitarian actors operated in what several reports described as a "data vacuum" (Nakasu, 2025; Organisation Z, 2025c). This vacuum resulted in significant delays in sectoral needs assessments, and limited the ability to apply targeting frameworks for vulnerable groups such as women, elderly persons, children, and persons with disabilities.

Moreover, the absence of a legitimate and functioning central authority prevented the formation of integrated relief coordination platforms. While ad hoc local arrangements existed, these were often informal, fragmented, or unstable. As a result, humanitarian response actors—including international NGOs, diaspora networks, and Chinese social organizations—found themselves navigating a space that lacked predictable institutional interfaces, making coordination efforts both inefficient and politically sensitive (Organisation Z, 2025b; Samad et al., 2025).

20.1.2 Coordination Failures and Multilateral Limitations

The post-earthquake humanitarian response in Myanmar exposed a series of profound limitations within established multilateral coordination systems. Despite the formal activation of global humanitarian protocols, institutional mechanisms such as the UN cluster system and regional frameworks like ASEAN proved unable to ensure timely, effective, and equitable delivery of aid across the fragmented crisis landscape.

Although the UN launched a Flash Appeal within the first week following the earthquake, actual inter-agency coordination remained severely constrained. Operational overlaps among cluster leads, delays in data-sharing, and inconsistent situational reporting led to duplication of efforts in some zones while leaving others underserved (Zhao et al., 2026). For instance, the health and shelter clusters both deployed separate needs assessments in the Mandalay region, producing conflicting figures on displacement and access needs (Organisation Z, 2025b). This lack of harmonization was further exacerbated by the absence of real-time coordination platforms and limited local access to standard reporting tools.

In theory, ASEAN's Agreement on Disaster Management and Emergency Response (AADMER) provides a regional architecture for response coordination. However, the Myanmar case revealed its practical limitations when facing high-conflict, low-consensus contexts. ASEAN's engagement remained largely symbolic, with no visible operational leadership or mediation capacity to bridge the gaps between the Myanmar military authorities, ethnic armed groups, and humanitarian actors (Phattharapornjaroen et al., 2025). As a result, regional coordination defaulted to diplomatic rhetoric rather than operational convergence.

Beyond these formal structures, traditional international surge mechanisms such as the On-Site Operations Coordination Centre (OSOCC) and International Search and Rescue Advisory Group (INSARAG) were either slow to mobilize or bypassed Myanmar entirely. Only two external Urban Search and Rescue (USAR) teams were formally registered with OSOCC as of Day 10 post-quake, both operating in areas already covered by local responders (Organisation Z, 2025b). The resulting vacuum of structured coordination left frontline NGOs and local civil society networks to operate in parallel, with minimal guidance or resource pooling.

20.1.3 Infrastructure Collapse, Security Risks, and Operational Hazards

The operational environment following the March 2025 Myanmar Earthquake was characterized by widespread infrastructure failure and layered security threats, creating an exceptionally hostile terrain for humanitarian engagement. Unlike previous disaster contexts with functional logistical corridors and basic safety assurances, Myanmar presented a compounded scenario of physical collapse, restricted mobility, and armed insecurity.

Damage to core infrastructure was both extensive and strategic. Over 52,000 residential buildings were reported destroyed across Magway, Sagaing, and Mandalay regions, with an additional 18,000 classified as uninhabitable (Organisation Z, 2025a). Major arterial roads, including sections of the Yangon-Mandalay Expressway, suffered critical cracks and landslides, while several key bridges in the Ayeyarwady basin were rendered unusable (Wang et al., 2025). Civilian airports in Magway and Meiktila sustained moderate structural damage, further constraining aerial relief logistics (Organisation Z, 2025a).

Beyond physical damage, the collapse of public service infrastructure—particularly in health and communications—significantly degraded response capacity. According to post-disaster assessments, 640 medical facilities were either partially or fully incapacitated, including all regional psychiatric units (Sigua et al., 2025). Communications infrastructure experienced partial blackouts, with mobile coverage dropping below 40% in quake-affected rural areas during the first ten days (Organisation Z, 2025d), severely delaying situational reporting and beneficiary tracking.

Compounding these physical barriers were escalating security risks. Several areas remained inaccessible due to active conflict, roadblocks, or control by ethnic armed organizations, particularly in northern Sagaing and Kachin states (Organisation Z, 2025d). In certain districts, humanitarian convoys reported looting, forced inspections, and extortion at informal checkpoints. Moreover, the presence of unexploded landmines, especially in previously contested areas, posed a constant risk to logistics teams and displaced populations (Organisation Z, 2025d).

Finally, ongoing seismic activity and secondary hazards further complicated operations. Over 90 aftershocks were recorded within three weeks, intensifying landslide risks in the Chin hills and eroding temporary shelters built on unstable slopes (Lai et al., 2025). The combination of damaged access routes, volatile weather, and environmental degradation created conditions where even standard relief protocols became difficult to execute.

20.1.4 A Limit Case for Humanitarian Engagement: Beyond Operational Capacity

The 2025 Myanmar Earthquake highlighted a critical inflection point in humanitarian crisis management—not in the availability of response capacities, but in the institutional capacity to absorb and coordinate them. While substantial resources were mobilized within days, the fragmented governance structures, contested legitimacy of actors, and lack of operational coherence rendered much of this capacity ineffective within the national context.

Despite the activation of multiple emergency response protocols, including bilateral deployments and non-governmental initiatives, institutional absorption of aid remained severely restricted. United Nations cluster coordination suffered from both delayed activation and poor inter-agency communication, as access negotiations in areas like Sagaing and northern Shan were blocked by local militias or contested administrative authorities (Organisation Z, 2025c). Even when aid reached staging hubs, onward distribution was frequently bottlenecked by unclear jurisdictional authority and limited logistical handover mechanisms (Samad et al., 2025).

The humanitarian space itself was politically fractured. Humanitarian actors were routinely subject to arbitrary restrictions, informal taxation, or demands for loyalty declarations in exchange for access guarantees. Several INGOs faced withdrawal or suspension from active zones due to political entanglements or perceived affiliations with either the military government or ethnic armed organizations (Chiu, 2025; Organisation Z, 2025c). This undermined not only the operational neutrality of aid delivery, but also the long-term trust necessary for localized recovery planning.

Chinese civil society actors, like other transnational responders, found themselves not as integrated partners within a shared humanitarian framework, but rather as "gap-fillers" operating in institutional vacuums. Their technical capacities—search and rescue, medical support, and logistical assistance—were often welcomed by local communities, yet rarely incorporated into formal coordination channels. This disconnect exposed the limits of assuming that operational maturity can substitute for systemic coherence.

Compared with the relatively structured environments described in Chaps 16–19, Myanmar revealed a fundamentally different constraint: the inability of institutions to function as absorptive platforms for external support. This marked a shift from previous concerns with improving capacity, toward a deeper recognition of the structural fragility of humanitarian ecosystems. The Myanmar case underscores that humanitarian effectiveness depends not only on the mobilization of capable actors, but on the presence of a governance substrate capable of mediating, coordinating, and legitimating their interventions (Chiu, 2025; Organisation Z, 2025c; Samad et al., 2025).

20.2 Rapid Response at Scale: Digital Mobilization, Cross-Border Deployment, and Networked Action

The 2025 Myanmar Earthquake response demonstrated a new level of operational maturity among Chinese social organizations, characterized by rapid mobilization, digital coordination, and modular deployment. This section outlines how actors across national and local levels responded within 72 h, leveraged digital platforms for information management, and activated transnational networks to ensure access and legitimacy. It also highlights the use of functionally specialized teams—from USAR to psychosocial support—signaling a shift toward systematized, cross-border humanitarian operations despite institutional constraints on formal international coordination.

20.2.1 Swift and Scaled Mobilization: Speed, Scale, and Synchronization

The immediate aftermath of the 2025 Myanmar Earthquake witnessed an unprecedentedly rapid and multi-layered mobilization by Chinese social organizations, underscoring a matured capacity for transnational humanitarian response. Within the critical 72-h window following the quake, multiple organizational tiers—from national to grassroots—initiated parallel yet coordinated deployments. Teams such as Blue Sky Rescue and the Civil Aid Corps had completed preliminary needs assessments and established coordination links with border entry points by March 28, less than two days after the event—reflecting not an anomaly, but the operationalization of rapid dispatch mechanisms developed through earlier crises in Nepal, Sulawesi, and Türkiye (Organisation Z, 2025a).

What distinguished this response was not only its pace but its simultaneity across organizational scales. National-level actors like the China International Rescue Team operated alongside provincial platforms from Guangdong and Yunnan, as well as independent Buddhist charity groups and informal volunteer coalitions from Shenzhen and Chengdu. These diverse actors converged operationally without awaiting centralized command, enabled by real-time digital synchronization platforms that aggregated needs data, coordinated deployments, and streamlined resource flows (Kim et al., 2025; Zhao et al., 2026).

Digital infrastructures played a decisive role in enabling this "networked synchrony." Platforms such as WeChat, ByteDance Public Welfare, and Tencent Charity served as coordination backbones, linking field actors with backend logistics teams and donors. Early situational updates were crowdsourced through social media and integrated with geospatial analysis tools, allowing organizations to bypass traditional data bottlenecks and directly target high-need zones in Mandalay, Magway, and northern Shan (Chiu, 2025; Organisation Z, 2025a).

This distributed-yet-synchronized model diverged notably from the more hierarchical, delayed deployments observed during the Türkiye–Syria response. In Myanmar, action was guided not by top-down dispatch but by modular alignment across independent actors—enabled by a shared operational language and interoperable toolkits. As such, the Myanmar case illustrates the increasing institutionalization of Chinese civil society's emergency capabilities: not merely reactive but proactive, not singular but networked, and not isolated but digitally embedded. This level of responsiveness foregrounds the evolving agency of non-state humanitarian actors in fragile contexts, even when operating outside of formal intergovernmental coordination structures.

20.2.2 Digital Command and Information Coordination: From Social Media to Decision Platforms

In the aftermath of the 2025 Myanmar Earthquake, Chinese social organizations leveraged digital ecosystems not merely as communication tools, but as operational infrastructures for real-time coordination. Amid information blackouts in many of the hardest-hit regions, organizations turned to a combination of satellite imagery, social media inputs (including WeChat, Weibo, and TikTok), and GIS-based crisis mapping tools to assess damage and track emerging needs (Laosunthara et al., 2025; Organisation Z, 2025b). These digital feeds served as proxies for on-the-ground data were formal situational reporting was absent.

Collaborative efforts with major domestic tech platforms were instrumental. The Tencent Charity Platform and ByteDance Charity Platform jointly supported the establishment of remote coordination hubs in Guangzhou and Kunming. These hubs centralized incoming data streams and linked them to deployment decisions, enabling task-sharing across regions. Through cloud-based dashboards and shared logistics sheets, organizations could view live status updates of material flows, team locations, and site access conditions (Chiu, 2025).

In locations where communication infrastructure had collapsed entirely, Chinese humanitarian actors employed predictive models to approximate damage zones. Supported by academic partnerships and open-source algorithms, remote sensing data was combined with prior risk exposure layers to simulate impact zones and inform dispatch decisions (Samad et al., 2025). This use of "digital shadow modeling" enabled more accurate targeting of relief materials, especially during the critical 72-h window.

Importantly, this digital command architecture was not purely centralized. Volunteer data collectors in Mandalay, Yangon, and Muse operated in parallel to institutional systems, feeding community-level updates into the national data pool via mobile applications and encrypted chat groups. The integration of crowdsourced and institutional data gave rise to a "hybrid command chain," in which virtual information centers interacted dynamically with frontline responders (Chiu, 2025; Organisation Z, 2025b).

20.2.3 Transnational Humanitarian Networks: Diaspora, Religious and Commercial Linkages

In the absence of stable state-led humanitarian pathways, the 2025 Myanmar Earthquake response saw the activation of transnational networks rooted in diaspora ties, religious affiliations, and commercial partnerships. These networks not only enabled rapid access but also created a de facto cross-border operational architecture that circumvented conventional government-to-government or INGO coordination mechanisms.

A primary vector of entry was through the Yunnan–Myanmar corridor. Leveraging long-standing socio-economic linkages and linguistic proximity, multiple Chinese humanitarian actors established logistical and personnel transit via Ruili and Muse, bypassing congested or militarized checkpoints in central Myanmar (Organisation Z, 2025c). This route was further reinforced by Buddhist institutional networks that had maintained transnational linkages through joint religious exchanges and monastic cooperation. Temples affiliated with Mahayana traditions in both Yunnan and Mandalay functioned as temporary warehouses, medical staging sites, and language mediation centers, serving as critical connectors between Chinese aid providers and local communities (Chiu, 2025).

Diaspora-led business associations, particularly Chinese chambers of commerce in Yangon, also played a strategic role. These entities coordinated bulk procurement of food, medicine, and shelter materials from regional supply chains and facilitated customs clearance through pre-existing informal agreements with border officials (Zhao et al., 2026). In the first ten days post-quake, over 300 tons of relief goods were transported through these channels, many reaching areas deemed "hard-to-access" by international agencies (Organisation Z, 2025c).

Crucially, this networked response was not limited to material transfers. Buddhist and ethnic Chinese volunteers provided critical cultural and linguistic bridges that supported needs assessments and localized triage. Religious actors also enabled trust-building in communities skeptical of military or external aid, while commercial actors provided logistical scalability typically lacking in civil society deployments.

20.2.4 Functional Modularity and Operational Specialization

The 2025 Myanmar Earthquake response marked a significant evolution in the operational architecture of Chinese social organizations, highlighting their capacity to deploy modular, functionally specialized teams across multiple domains. Unlike ad hoc or generalist responses seen in earlier disaster scenarios, this operation showcased a maturing system of task-specific intervention units operating in synchronized clusters.

USAR functions were among the first activated. Professionalized teams such as Blue Sky Rescue and regional disaster task forces deployed to affected zones in Sagaing and Magway within 72 h, bringing with them thermal imaging equipment, canine units, and structural assessment capabilities (Organisation Z, 2025a). Despite infrastructural collapse and security barriers, these units completed operations in at least 31 collapse sites over ten days (Organisation Z, 2025c), evidencing the increasing technical sophistication and field readiness of Chinese grassroots response actors.

Parallel to USAR, medical modules were activated with a similarly high degree of specialization. Teams composed of trauma surgeons, public health experts, and epidemiologists were dispatched to temporary field hospitals in Mandalay and Meiktila, offering both surgical and preventive services. Mental health response teams provided psychological first aid to over 1200 individuals, including frontline responders, using protocols adapted from previous deployments in Türkiye and Nepal (Organisation Z, 2025c; Sigua et al., 2025).

Water, sanitation, and hygiene (WASH) interventions followed a similarly structured model. Modular WASH kits—comprising portable latrines, water purification units, and hygiene packs—were distributed to over 8,600 households across eight high-density displacement sites (Organisation Z, 2025d). These efforts were coordinated with community-based hygiene education campaigns, reducing the risk of cholera and other waterborne diseases in overcrowded shelters (Wang et al., 2025).

Supporting these field modules was a rapidly constructed logistical backbone. Drone-assisted aerial surveys, mobile warehouses at the Yunnan-Myanmar border, and localized distribution nodes enabled efficient supply chain management even in fragmented and insecure regions. Commercial logistics providers, activated through diaspora and business networks, ensured rapid customs clearance and last-mile delivery (Zhao et al., 2026).

20.3 When Capacity Meets Institutional Friction: Limits of Coordination, Legitimacy, and Sustainability

The 2025 Myanmar Earthquake response highlighted a paradox: while Chinese social organizations exhibited enhanced transnational capacity, their efforts were repeatedly constrained by structural frictions within the host environment. This section moves beyond capacity analysis to examine the institutional barriers that undermined effective humanitarian engagement. It identifies three interrelated forms of friction—coordination breakdowns, legitimacy constraints, and sustainability limits—that collectively point to a deeper systemic issue: the host country's limited institutional absorptive capacity. By analyzing these layered challenges, the section offers a theoretical reframing of international humanitarian performance in fragile governance contexts.

20.3.1 Coordination Frictions: Platform Mismatch, Institutional Lag, and Fragmented Command

Despite the swift mobilization and growing technical capacity of Chinese social organizations during the 2025 Myanmar Earthquake response, significant coordination frictions emerged that undermined overall operational coherence. These frictions did not stem from a lack of action, but rather from the failure to integrate distributed capabilities into a shared command structure.

First, international coordination mechanisms proved sluggish and uneven in their activation. The establishment of the On-Site Operations Coordination Centre, Reception/Departure Centre, and Urban Coordination Centre was either significantly delayed or not fully operationalized during the initial days of response (Organisation Z, 2025b). The UN cluster system also struggled to provide real-time tasking, with situational reports lagging behind developments on the ground. As a result, incoming actors lacked centralized guidance and operated in informational silos (Zhao et al., 2026).

Second, platform incompatibility created persistent disconnects between Chinese responders and international systems. Many Chinese social organizations relied on domestic platforms—such as WeChat coordination groups, proprietary dispatch tools, and local charity apps—that were not interoperable with the UN's VOSOCC system or INSARAG standard operating procedures. This lack of data standardization and mutual recognition led to fractured information flows and decision-making blind spots, effectively creating "digital islands" within the broader humanitarian system (Organisation Z, 2025b).

Third, command and site-level coordination were fragmented due to the absence of integrated field governance. Multiple search-and-rescue and medical teams from different provinces entered the same operational zones without centralized site allocation protocols or rotational tasking mechanisms. In certain cases, overlapping deployments led to duplicated efforts, uneven resource distribution, and even safety risks in unstable rubble zones (Phattharapornjaroen et al., 2025). For example, in Magway and northern Shan, three separate rescue teams reported to different domestic dispatch centers, while none were formally integrated into the UCC structure.

20.3.2 Legitimacy Frictions: Regulatory Ambiguity, Political Sensitivity, and Informal Access Channels

While Chinese social organizations demonstrated increased agility and operational maturity in the Myanmar Earthquake response, their efforts were repeatedly constrained by issues of legal ambiguity, political sensitivity, and informal access arrangements. These legitimacy frictions did not merely complicate logistics—they shaped the fundamental conditions under which humanitarian engagement could occur, often determining the sustainability and scale of operations.

At the legal-administrative level, the status of foreign NGOs in Myanmar remained unsettled. The lack of a unified registration framework for cross-border humanitarian actors led to case-by-case approvals, often contingent upon opaque political assessments. Short-term activity permits were granted inconsistently, and multiple Chinese organizations reported delays at customs and immigration due to unclear documentation requirements (Organisation Z, 2025d). This legal vagueness translated into constant uncertainty, limiting the ability to plan medium- or long-term interventions.

Political sensitivities further compounded the regulatory instability. Following the 2021 military takeover, Myanmar's junta treated many international and cross-border actors with suspicion. Humanitarian operations were frequently subjected to security vetting and geographic restrictions. In northern Shan and parts of Sagaing,

Chinese teams were required to liaise with local militia authorities or operate under the sponsorship of religious intermediaries to gain access (Chiu, 2025). This created a fragmented and politically mediated entry process, exposing actors to reputational and operational risk.

In response to these barriers, Chinese organizations increasingly relied on informal access channels. Networks of diaspora communities, Buddhist monasteries, and Chinese commercial associations provided crucial logistical routes, shelter points, and translation services. While effective in bypassing bureaucratic delays, these channels lacked institutional protections and were highly sensitive to shifts in the political climate (Kim et al., 2025). Without official recognition or legal indemnity, actors operating through these pathways faced limits on scalability, insurance coverage, and formal coordination with multilateral systems.

20.3.3 Sustainability Frictions: Short Operational Windows, Resource Fatigue, and Organizational Burnout

The 2025 Myanmar Earthquake response illuminated the increasing capacity of Chinese social organizations to mobilize quickly and operate across borders. This surge in responsiveness came at a cost: the operational tempo, while effective in the immediate aftermath, proved difficult to sustain beyond the initial relief phase. This section examines how the structural absence of long-term support frameworks resulted in rapid exhaustion of human, material, and technological resources, ultimately limiting continuity and institutional learning.

One key issue was the compression of the operational window. Urban Search and Rescue deployments reached peak intensity within the first 72 h, with over 30 coordinated interventions recorded in Magway and Sagaing (Organisation Z, 2025c). However, as focus shifted toward mid-term recovery and community rebuilding, the lack of pre-established project pipelines and sustained funding flows hindered follow-up support. Many Chinese teams were unable to transition into reconstruction roles due to expired permissions and logistical withdrawal orders, curtailing impact durability (Samad et al., 2025).

Resource sustainability also proved challenging. Funding mechanisms for cross-border operations remained ad hoc and heavily reliant on short-cycle donations. Teams lacked financial buffers for prolonged presence or adaptive scaling. Equally problematic was the human toll: volunteer cohorts endured prolonged stress, rapid redeployment cycles, and minimal psychological support. In several cases, returning volunteers showed signs of post-traumatic stress—exacerbated by the cumulative burden of repeated deployments across multiple countries in recent years (Sigua et al., 2025).

Technological infrastructure, while essential in enabling fast coordination, further revealed sustainability limitations. The digital dashboards and data synchronization platforms developed for this mission—particularly those adapted from domestic social media and mapping tools—required continuous IT support, cybersecurity protocols, and interface management. Maintaining this architecture beyond the immediate response proved costly and impractical, especially in areas with low digital penetration or unreliable connectivity (Lai et al., 2025).

20.3.4 From Friction to Theory: Myanmar as a Case of Institutional Absorptive Failure

The 2025 Myanmar Earthquake response offers not only an empirical account of Chinese social organizations' transnational capacity but also a theoretical lens into how such capacity becomes constrained in fragile governance contexts. Despite demonstrable improvements in speed, scale, and modular deployment, the Myanmar case reveals a deeper institutional challenge: the inability of the host environment to absorb, align with, and sustain incoming humanitarian efforts. This mismatch between operational supply and institutional receptivity constitutes what may be termed an institutional absorptive failure.

Across the previous sections, three layers of systemic friction emerged. First, coordination failures resulted not from actor absence but from incompatible operational platforms, lagging international mechanisms, and fragmented local command structures. The disconnect between domestic dispatch systems (e.g., WeChat-based coordination) and multilateral clusters (e.g., INSARAG, VOSOCC) produced siloed responses and resource duplication (Organisation Z, 2025b; Phattharapornjaroen et al., 2025; Zhao et al., 2026).

Second, legitimacy frictions constrained formal access and regulatory clarity. Foreign NGOs operated under temporary or unclear authorization frameworks, and some organizations had to rely on informal intermediaries—such as religious institutions or local militias—for entry and protection. While these informal channels enabled initial access, they lacked legal stability and exposed actors to shifting political risks (Chiu, 2025; Kim et al., 2025; Organisation Z, 2025d).

Third, sustainability limitations emerged as rapid deployments gave way to organizational burnout and logistical withdrawal. The absence of long-term program pipelines, weak financial continuity, and the psychological toll on personnel eroded the sustainability of civil society engagement. Even technological assets, such as real-time coordination dashboards, proved difficult to maintain beyond the acute phase due to high resource demands and lack of institutionalized support (Lai et al., 2025; Organisation Z, 2025c; Samad et al., 2025; Sigua et al., 2025).

These three frictions are not isolated deficiencies but structurally linked indicators of a fragile system with limited absorptive capacity. Compared to the Türkiye–Syria response—where institutional frameworks allowed for greater interoperability, authorization, and continuity—Myanmar illustrates a case where capacity met friction, not opportunity. As argued in broader governance studies, humanitarian effectiveness is as much a function of institutional interfaces as it is of operational skill (Heinkel et al., 2025; Nakasu, 2025).

20.4 From Rapid Action to Institutional Embeddedness: Strategic Reorientation and Future Pathways

The 2025 Myanmar Earthquake marked a turning point in the international engagement of Chinese social organizations, revealing both the strengths of rapid mobilization and the limitations of short-term, project-driven interventions. This section argues for a strategic shift from operational immediacy toward institutional embeddedness. It traces four interlinked pathways of transformation: the repositioning from frontline deployment to strategic platform support; the transition from direct intervention to embedded local partnerships; the evolution from emergency relief to community resilience building; and the movement from fragmented projects to regional institutional interfaces. These shifts signal a maturing trajectory toward durable humanitarian governance integration.

20.4.1 From Frontline Action to Strategic Platform Support

The operational environment in Myanmar after the 2025 earthquake demonstrated the limits of physical deployment for foreign social organizations. In response, Chinese humanitarian actors began repositioning themselves not as direct implementers but as infrastructural supporters—focusing on remote sensing, digital platforms, and technical standardization to enhance broader system effectiveness.

As the field became increasingly difficult to access due to administrative delays and security risks, several organizations turned to backend data services. Real-time disaster mapping—generated via Chinese platforms and mobile-based reporting tools—played a critical role in informing international deployment decisions. One example involved integrating geospatial tools with field-based damage reports to produce dynamic rescue maps, offering timely guidance to teams unable to conduct physical reconnaissance (Organisation Z, 2025c).

This shift was not merely technical, but strategic. By building remote coordination interfaces—such as risk dashboards integrating satellite inputs, infrastructure vulnerability data, and border policy alerts—Chinese NGOs reduced their dependence on direct access to contested zones. Such backend systems addressed two core problems: high entry thresholds for field operations, and poor interoperability between domestic platforms and multilateral protocols. Where previous responses had been slowed by data incompatibility, the emphasis on risk modeling and standardized information feeds allowed for more fluid coordination across actors (Laosunthara et al., 2025).

In parallel, attempts were made to position domestic expertise as a global asset. Some organizations began offering modular services, such as early damage estimation algorithms and logistics route optimization models, to regional partners through open-data interfaces. Rather than duplicating frontline tasks, this approach sought to amplify existing efforts by becoming a "strategic multiplier" within broader humanitarian systems—a model better aligned with unstable governance environments where on-ground visibility is limited (Wang et al., 2025).

20.4.2 From Direct Intervention to Embedded Local Partnerships

The 2025 Myanmar Earthquake revealed the growing limitations of a direct intervention model in fragile governance environments. While Chinese social organizations were able to mobilize rapidly, their ability to sustain operations was constrained by legal ambiguity, political sensitivity, and a lack of formal access pathways. These challenges necessitated a strategic reorientation—from standalone deployments to embedded local partnerships that prioritize trust, legitimacy, and continuity.

Rather than functioning as external actors with short-term missions, Chinese organizations began to cultivate embedded cooperation through religious, community-based, and diasporic networks. In particular, partnerships with Buddhist institutions played a vital role in delivering psychosocial support in areas inaccessible to international NGOs due to security restrictions or ethnic tensions. For instance, a joint initiative between Yunnan-based Buddhist groups and Kachin monasteries facilitated culturally sensitive trauma care in displacement camps across northern Shan (Chiu, 2025; Phattharapornjaroen et al., 2025). These collaborations not only enabled entry into otherwise restricted areas but also enhanced local legitimacy through spiritual affiliation and shared cultural values.

Diaspora communities also emerged as crucial intermediaries in the redistribution of aid. In regions such as Mandalay and Yangon, local Chinese business associations and family networks coordinated warehouse access, last-mile logistics, and language translation for Chinese humanitarian teams (Kim et al., 2025). These partnerships helped bridge administrative gaps, reduced coordination frictions, and allowed for more adaptive and localized responses. However, while informal in nature, such networks remained vulnerable to political volatility and lacked the institutional protections needed for long-term engagement.

Recognizing these limits, several Chinese organizations began investing in more formalized liaison mechanisms. Organisation Z, for example, established field contact points and volunteer bases in Lashio and Muse, working with local youth groups and retired teachers to build standby teams for logistics and community support (Organisation Z, 2025d). These semi-permanent presence models allowed for continuity beyond the emergency phase and opened pathways for ongoing recovery and preparedness work.

20.4.3 From Emergency Relief to Community Resilience Building

In the aftermath of the 2025 Myanmar Earthquake, Chinese social organizations demonstrated growing agility in emergency relief. As the operational window closed and international attention waned, the limitations of short-cycle interventions became apparent. In fragile governance settings like Myanmar, where institutional infrastructure is weak and volatility is high, a transition from rapid response to embedded resilience building has emerged as a strategic imperative.

Rather than pursuing large-scale development programs that require stable institutional hosts, Chinese actors increasingly turned to localized, community-based models. These included small-scale Water, Sanitation, and Hygiene (WASH) system restorations in rural areas, where damaged boreholes and latrines posed immediate health risks. In northern Shan State, collaborative efforts with local religious institutions enabled the repair of decentralized water systems, enhancing both health outcomes and trust-building with host communities (Organisation Z, 2025c).

Equally important were initiatives aimed at agricultural support for smallholder farmers. In regions like Magway and Sagaing, where the earthquake disrupted local food systems, temporary agricultural kits—comprising seeds, tools, and technical guidance—helped reestablish household food security. Such efforts not only addressed immediate nutritional needs but also reconnected community members to routine and agency (Organisation Z, 2025d).

Community-based psychosocial interventions also played a stabilizing role. Psychological First Aid (PFA) programs delivered through Buddhist networks and local youth groups helped mitigate collective trauma in displacement camps. Survey data from camps in Myitkyina and Lashio showed measurable improvements in emotional well-being and social cohesion after targeted mental health interventions were implemented (Sigua et al., 2025). These low-cost, high-trust models proved scalable and culturally congruent.

At a strategic level, these practices reflect a shift toward preventative resilience. By investing in decentralized, anticipatory interventions, Chinese organizations are building a platform for future engagement that does not depend on large bureaucratic partnerships. As emphasized in regional analyses, resilience-focused microprojects—such as youth nutrition programs and trauma-informed education in schools—offer long-term returns while reducing future intervention thresholds (Lai et al., 2025).

20.4.4 From Project Logic to Institutional Interface: Toward Durable Regional Mechanisms

The 2025 Myanmar Earthquake highlighted the structural inadequacies of single-mission, project-based humanitarian responses when confronted with fragile governance environments and complex regional dynamics. While Chinese social organizations demonstrated remarkable speed and initiative in the early response phase, the limitations of operating through isolated, short-cycle interventions became increasingly apparent. Sustained engagement in such contexts requires transitioning from project logic to embedded institutional interfaces that align with regional governance frameworks and multilateral coordination norms.

One critical constraint identified during the Myanmar response was the absence of standardized cross-border coordination protocols between Chinese actors and existing regional platforms. Despite the participation of over 30 Chinese social organizations, integration into mechanisms such as the ASEAN Coordinating Centre for Humanitarian Assistance or the UN cluster system remained minimal (Organisation Z, 2025b). This fragmented engagement not only undermined operational efficiency but also limited the strategic visibility of Chinese civil society within broader humanitarian governance structures (Heinkel et al., 2025).

To address this, several Chinese organizations began exploring partnerships to support interoperable systems and joint preparedness mechanisms. For instance, Organization Z initiated informal data-sharing trials with international NGOs and local partners to improve situational awareness during the acute phase, although these efforts were hampered by the absence of formalized agreements or shared data standards (Organisation Z, 2025c). Scholars have since proposed the establishment of a regional coordination interface—anchored in real-time data platforms and risk forecasting tools—that could integrate Chinese NGOs into ASEAN-led disaster management frameworks (Zhao et al., 2026).

Moreover, standard-setting has emerged as a strategic frontier. Field evaluations underscored the need for harmonizing operational protocols across transnational actors, especially regarding WASH, shelter, and psychosocial support delivery. Proposals have been made for China-based NGOs to co-develop modular operating procedures aligned with international guidelines such as those of the Sphere Project or INSARAG, which could facilitate smoother cross-deployment and recognition within regional coalitions (Samad et al., 2025). Collaborative training programs, certification systems, and shared logistical hubs—potentially in partnership with Southeast Asian humanitarian logistics centers—have also been identified as feasible next steps (Heinkel et al., 2025).

20.5 Conclusion

The 2025 Myanmar Earthquake exposed both the operational agility and institutional limitations of Chinese social organizations in transnational humanitarian response. While actors demonstrated unprecedented speed, modularity, and digital coordination, their efforts were constrained by fragmented governance, weak multilateral integration, and unsustainable intervention models. As the chapter reveals, effectiveness in fragile contexts requires more than rapid action—it demands institutional embeddedness, trust-based local partnerships, and alignment with regional governance mechanisms. The Myanmar case thus marks a strategic inflection point: from capacity expansion to systemic anchoring. This shift calls for a redefinition of Chinese civil society's international role—not simply as responders, but as institutional actors capable of shaping, sustaining, and co-producing humanitarian governance in an increasingly unstable global environment.

References

Chiu, T.-L. (2025). From marginalization to localization: Chinese Mahāyāna Buddhism's adaptive strategies in Theravāda Myanmar. *Religions, 16*(11), 1390. https://doi.org/10.3390/rel16111390

Heinkel, S.-B., Miller, C., Thiebes, B., Than, Z. M., Kyi, T. T., Aung, T., Oo, S. S., Mar, W. L., Willkomm, M., Maung, W., Myint, Z. N., Soe, K. K., Spohner, R., & Kraas, F. (2025). Institutional risk and crisis communication on natural hazards and disaster risks in Yangon, Myanmar. *International Journal of Disaster Risk Reduction, 116*, Article 105064. https://doi.org/10.1016/j.ijdrr.2024.105064

Kim, S. Y., Karunathilake, I., & Chiou, H.-Y. (2025). APACPH Statement on the humanitarian crisis following the earthquakes in Myanmar, Thailand, and the Surrounding Region. *Asia-Pacific Journal of Public Health, 37*(5), 512. https://doi.org/10.1177/10105395251360294

Lai, S.-T., Oo, K. M., Htwe, Y. M. M., Yi, T., Than, H. H., Than, O., Min, Z., Oo, T. M., Maung Maung, P., Bindi, D., Cotton, F., Evans, P. L., Heinloo, A., Hillmann, L., Saul, J., Sens-Schönfelder, C., Strollo, A., Tilmann, F., Weatherill, G., Milkereit, C., et al. (2025). Capacity building enables unique near-fault observations of the destructive 2025 M_w 7.7 Myanmar earthquake. *Earth System Science Data, 17*(10), 5149–5164. https://doi.org/10.5194/essd-17-5149-2025

Laosunthara, A., Krutphong, K., Leelawat, N., Wararuksajja, W., Sukulthanasorn, N., Suppasri, A., Thongthip, R., & Chintanapakdee, C. (2025). Initial observations and immediate lessons learned from Thailand's response to the 2025 Mandalay earthquake. *International Journal of Disaster Risk Reduction, 127*, Article 105675. https://doi.org/10.1016/j.ijdrr.2025.105675

Nakasu, T. (2025). Disaster risk assessment in Asian countries: A novel deductive approach. *Discover Environment, 3*(1), 179. https://doi.org/10.1007/s44274-025-00228-4

Organisation Z. (2025a). *Situation report No. 1: Myanmar earthquake (March 28, 2025)*. https://shimo.im/docs/DxBCEccJWnGJgXaK

Organisation Z. (2025b). *Situation report No. 2: Myanmar earthquake (March 28, 2025)*. https://shimo.im/docs/B1AwdwrWKmTDLr3m

Organisation Z. (2025c). *Situation report No. 3: Myanmar earthquake (March 28, 2025)*. https://shimo.im/docs/8Nk6emvyKwigJlqL

Organisation Z. (2025d). *Situation report No. 4: Myanmar earthquake (March 28, 2025)*. https://shimo.im/docs/wV3VMKbVWBfGjZAy

Phattharapornjaroen, P., Burivong, R., & Khorram-Manesh, A. (2025). Myanmar earthquake aftermath—critical update and expanded analysis. *Disaster Medicine and Public Health Preparedness, 19*, Article e125. https://doi.org/10.1017/dmp.2025.112

Samad, M. A., Arifin, K., & Abas, A. (2025). A systematic literature review on the challenges of Southeast Asian countries in natural disaster management. *Cogent Social Sciences, 11*(1), 2435590. https://doi.org/10.1080/23311886.2024.2435590

Sigua, J. A., Caymaz, E., & Castro-Delgado, R. (2025). Emergency medical care provided by humanitarian organizations in response to sudden onset disasters in Southeast Asia: a scoping review. *Disaster Medicine and Public Health Preparedness, 19*, Article e196. https://doi.org/10.1017/dmp.2025.10071

Wang, T., Zhou, Y., Chen, J., Wang, Xiaoqing, Bi, H., & Wang, Xiaoting. (2025). Field survey of building damage at Mandalay during 2025 Myanmar Mw 7.7 earthquake. *Earthquake Engineering and Engineering Vibration, 24*(3), 613–627. https://doi.org/10.1007/s11803-025-2325-5

Zhao, L., Huang, Q., & Wu, C. (2026). Distributionally robust scheduling optimization for pharmaceutical delivery using coordinated mother-end drones under post-earthquake road disruptions. *Transportation Research Part E: Logistics and Transportation Review, 205*, Article 104481. https://doi.org/10.1016/j.tre.2025.104481

Chapter 21
Conclusion: Evolution, Comparative Insights, and Future Potentials

Abstract Bringing together the longitudinal evolution of social organizations, cross-national institutional contrasts, and the forward-looking pathways of institutionalization, professionalization, and digitalization, this concluding chapter reframes disaster governance as a dynamic process of systemic learning and adaptive integration. The comparative analysis demonstrates how different governance logics shape the durability, inclusiveness, and scalability of collaborative arrangements, while empirical evidence highlights the growing interdependence between civil society, state institutions, and digital infrastructures. Building on the theoretical synthesis of collaborative governance, network governance, and co-production, the chapter advances an integrated analytical lens that connects participation mechanisms, organizational relations, performance orientations, and institutional embeddedness. This perspective not only consolidates the book's core findings but also opens avenues for comparative research, policy experimentation, and the continuous co-evolution of resilient disaster governance systems.

Keywords Collaborative governance · Disaster resilience · Civil society engagement · Cross-sector coordination · Institutional transformation

21.1 From Emergency Mobilization to Institutionalized Collaboration: A Comprehensive Review of the Evolution of Social Organizations

Over the past two decades, social organizations have undergone a remarkable transformation in China's disaster governance landscape. Initially emerging as spontaneous responders to catastrophic events, these actors evolved through phases of gradual integration, institutional experimentation, and eventual systemization. Drawing on both Chinese and international cases, this section unpacks four key stages in this trajectory: emergence under crisis, incremental embedding, formal institutionalization, and divergent models of institutional development. These phases illustrate how social organizations have shifted from peripheral improvisation to becoming

Q. Du and R. Shaw, *Social Organizations in Disaster Response and Recovery*, Disaster Risk Reduction, https://doi.org/10.1007/978-981-92-2846-1_21

core components of resilient, digitally enabled, and governance-integrated disaster response systems.

21.1.1 Emergence and Motivation: The Genesis of Emergency-Driven Social Organizations

The genesis of social organizations in disaster governance in China can be traced to the acute emergence of catastrophic events, particularly the 2008 Wenchuan Earthquake and the 2010 Yushu Earthquake. These events acted as critical policy windows that catalyzed spontaneous civic engagement and the rapid mobilization of non-governmental actors in the absence of pre-established formal mechanisms (Alcántara-Ayala, 2025; Clark et al., 2025). In both cases, the overwhelming scale of destruction exposed the limitations of state-centric emergency response systems and created a vacuum that was rapidly filled by social actors—volunteer groups, grassroots NGOs, and informal community networks.

This initial phase of emergency-driven mobilization was characterized by non-institutional spontaneity and low levels of coordination. Social organizations responded in real-time to immediate humanitarian needs, often relying on improvisational tactics and personal networks for logistics and support (Fraser, 2023; Tang & Chen, 2025). Coordination, when present, was largely ad hoc and built through action-based learning rather than predefined protocols. These responses displayed features of what network governance theory defines as "emergent collaboration under pressure"—decentralized, trust-based, and adaptive in nature (Bodin et al., 2022; Emerson et al., 2012).

In Wenchuan, for example, more than 270,000 volunteers participated in response activities, many operating without legal recognition or professional training (Kang, 2020; Menefee & Nordtveit, 2012). Similar dynamics were observed in Yushu, where Buddhist monasteries and local civil actors played essential roles in relief coordination, particularly in the early days when formal aid was delayed (Chu & Huang, 2025; Wei et al., 2013). These cases demonstrate that early mobilization of social organizations in China's disaster landscape was less about formal integration and more about the urgent need to fill response gaps under extreme conditions.

From a theoretical perspective, this phase aligns with the core tenets of collaborative governance that emphasize initial trust-building, recognition of shared purpose, and iterative action as precursors to institutional coordination (Ansell & Gash, 2008; Jagosh et al., 2015). Similarly, in co-production theory, this stage reflects an emergent form of service co-delivery, wherein citizens and informal groups become central to governance outcomes without being formally embedded in state systems (Hilhorst et al., 2020; Osborne et al., 2021).

21.1.2 *Embeddedness and Evolution: The Gradual Institutionalization of Social Organizations*

As China's disaster governance landscape evolved, the role of social organizations underwent a transition from ad hoc mobilization toward more institutionalized participation. This transformation was neither linear nor uniform, but marked by incremental adjustments and policy experimentation—particularly visible during the Lushan (2013), Ludian (2014), and Jiuzhaigou (2017) Earthquakes. In these disasters, social organizations began to move beyond emergent, event-driven actors and gradually embedded themselves within the broader disaster response ecosystem.

Unlike the immediate, decentralized mobilization characterizing earlier events like the Wenchuan and Yushu Earthquakes, subsequent disasters saw the emergence of semi-formal coordination mechanisms. For example, in the Lushan response, local governments supported the establishment of coordination platforms such as the Ya'an Social Organization Coordination Center, which enabled NGOs to interface with government and community actors more effectively (Lu & Li, 2020; Xu et al., 2016). Similarly, during the Jiuzhaigou Earthquake, coordination workstations helped streamline volunteer deployment and resource allocation, showing early signs of structured collaborative governance (Organisation Z, 2017).

These arrangements reflect a gradual policy shift—from tolerance of social actors as auxiliary responders to proactive guidance and incorporation. Local authorities started experimenting with registration systems for NGOs, pre-disaster training, and post-disaster debriefing protocols. Importantly, these steps fostered a platform-based governance structure where multiple actors could negotiate roles and responsibilities without hierarchical imposition (Xu et al., 2018a). Such developments resonate with network governance principles, emphasizing horizontal collaboration, trust-building, and distributed coordination (Emerson et al., 2012; Klijn et al., 2025).

Another notable feature of this period was the increasing professionalization of grassroots actors. In Ludian and Jiuzhaigou, long-term NGOs such as One Foundation and various local civil society groups began specializing in areas like logistics, psychological assistance, and needs assessments. This specialization fostered clearer divisions of labor and reduced role ambiguity (Lu et al., 2023; Wu & Peng, 2017). Over time, these operational refinements contributed to the formalization of best practices and the development of organizational learning platforms.

Theoretically, this stage exemplifies the "emergent institutional hybrid" in collaborative disaster governance—where spontaneous cooperation coexists with incipient institutional rules. It aligns with co-production frameworks that stress iterative learning between state and non-state actors (Osborne et al., 2015) and with adaptive network models that allow for both flexibility and institutional anchoring (Provan & Kenis, 2007).

21.1.3 Stabilized Collaboration and Systemization: Constructing Institutional Disaster Governance

The institutionalization of disaster governance marks a significant transformation in the role of social organizations—from ad hoc participants to embedded actors in formalized, multi-level governance frameworks. This shift is not merely about increased presence, but about their systematic integration through stable coordination platforms, disaster-specific role assignments, and digitalized operational tools.

A representative case is the 2021 Zhengzhou flood, where over 3000 on-site relief actions by social organizations were coordinated within the formal emergency command structure. Unlike earlier disasters where NGOs operated peripherally, in Zhengzhou they participated in simulation exercises, data-sharing protocols, and pre-disaster planning mechanisms, reflecting a new phase of embedded, anticipatory collaboration (Li et al., 2025; Zhao et al., 2023). This indicates a move toward "peacetime–emergency integration," wherein civil actors hold clearly defined responsibilities beyond the immediate crisis.

Parallel developments can be observed internationally. Japan Voluntary Organizations Active in Disaster (JVOAD) and the U.S. National Voluntary Organizations Active in Disaster (NVOAD) serve as enduring coordination hubs, institutionalizing trust-based collaboration through memoranda of understanding, standardized training, and interoperable logistics platforms (JVOAD, 2026b; NVOAD, 2025b). These structures enable both vertical coordination with government and horizontal synergy among NGOs, illustrating the characteristics of what is termed "collaborative governance regimes"—governance systems with consistent decision-making routines, role clarity, and procedural legitimacy (Emerson et al., 2012).

The role of digitalization is particularly pronounced in recent transnational disaster responses. In the 2023 Türkiye–Syria Earthquake, coordination among over 200 actors was facilitated via centralized dashboards and real-time beneficiary tracking, significantly reducing duplication and enhancing targeting efficiency (Schwarz et al., 2025; Tarnas et al., 2023). Similarly, during the 2025 Myanmar Earthquake, social organizations entered coordination early through a pre-configured information-sharing framework compatible with UN cluster mechanisms, allowing seamless integration across sectors (Phattharapornjaroen et al., 2025).

Across these diverse cases, institutional convergence is evident: multi-actor coordination platforms, task specialization, data interoperability, and simulation-based preparedness have become common features. These patterns align with the theoretical benchmarks of mature network governance, particularly those emphasizing high internal legitimacy, node interdependence, and routinized interactions (Bodin et al., 2022; Provan & Kenis, 2007). Importantly, digital platforms now serve not merely as logistical supplements but as foundational infrastructures sustaining cross-sectoral governance capacity.

21.1.4 Divergent Pathways and Theoretical Reflections: Comparing Chinese and International Institutionalization Models

The institutionalization of social organizations in disaster governance has taken divergent forms in China and other international contexts, shaped by differing governance logics and policy environments. In China, institutional engagement often arises in the aftermath of major disasters that create temporary openings for systemic change. The 2008 Wenchuan Earthquake, for instance, marked a moment of spontaneous mass mobilization, but lacked predefined coordination mechanisms. More than a decade later, the 2021 Zhengzhou flood revealed notable progress: local social organizations not only participated in over 3000 on-site actions but were also formally integrated into the municipal emergency command system and involved in pre-disaster simulation planning (Ye et al., 2025; Zhao et al., 2023). Nonetheless, these developments remain largely event-driven and reactive, where collaboration mechanisms are established post hoc, and long-term preparedness structures are still maturing.

In contrast, countries like the United States and Japan have developed anticipatory systems that embed civil society actors into disaster governance through permanent coordination platforms. Organizations such as National VOAD and JVOAD facilitate inter-organizational trust and role alignment through pre-established protocols, memoranda of understanding, and routine joint trainings (JVOAD, 2026b; NVOAD, 2025b). During the 2023 Türkiye–Syria Earthquake response, more than 200 actors coordinated via centralized digital dashboards and real-time needs assessment tools, exemplifying a high degree of interoperability and system readiness (Schwarz et al., 2025).

These contrasting trajectories can be interpreted through different theoretical lenses. Collaborative governance theory, which emphasizes the importance of trust-building, shared vision, and joint action, finds more congruence in systems like U.S. National VOAD and JVOAD, where regular engagement precedes crisis moments (Ansell & Gash, 2008; Emerson et al., 2012). In China, however, network governance theory offers a better analytical fit. Here, evolving institutional arrangements and centralized coordination efforts gradually shape interdependent relationships, even if they originate from fragmented or informal beginnings (Bodin et al., 2022; Provan & Kenis, 2007). Co-production theory bridges both models by highlighting how public value is created jointly by government and civil society, particularly when formal governance capacity is constrained (Brandsen & Pestoff, 2006; Ostrom, 1996).

However, a persistent limitation in the Chinese experience is the limited routinization of pre-disaster collaboration. While simulation-based training and digital platform development are advancing, many coordination processes remain contingent upon high-impact events. By contrast, international frameworks tend to institutionalize roles and data-sharing mechanisms in advance, allowing for rapid and coordinated response regardless of event scale (Phattharapornjaroen et al., 2025).

Ultimately, these differences point to the importance of contextualizing theoretical frameworks. While all three perspectives—collaborative governance, network governance, and co-production—offer valuable insights, their applicability varies depending on the stage of institutional maturity. China's trajectory reflects a meaningful transition from ad hoc mobilization to semi-formal coordination, but further institutional consolidation will depend on enhancing preparedness infrastructure, codifying organizational roles, and fostering long-term legitimacy for social organizations in disaster governance.

21.2 Cross-National Comparison of Disaster Governance: Collaborative Mechanisms Under Institutional Heterogeneity

Rather than treating collaborative governance as a uniform institutional template, this section conceptualizes disaster collaboration as a context-sensitive configuration shaped by political authority, organizational ecology, cultural embeddedness, and hazard characteristics. Through a systematic comparison between China and selected international cases, it identifies how distinct institutional logics generate divergent pathways of coordination, capacity accumulation, and operational stability. The analysis proceeds along four interrelated dimensions: governance structures and institutional foundations, networked organizational configurations, the embedding of cultural resources and social capital, and hazard-specific adaptive dynamics. These dimensions illuminate not only why collaborative mechanisms perform differently across settings, but also how institutional maturity and contextual alignment condition the sustainability and scalability of cross-sector disaster governance.

21.2.1 Governance Structures and Institutional Foundations: Top-Down Versus Decentralized Approaches

The institutional foundations of disaster governance vary markedly between China and international contexts, reflecting divergent paths of state–society relations and modes of coordination. In China, the dominant model is characterized by a top-down, state-centric approach where institutional arrangements often emerge reactively in response to large-scale crises. The 2008 Wenchuan Earthquake and the 2021 Zhengzhou Flood exemplify this pattern. Both events acted as institutional inflection points, catalyzing government-led frameworks for incorporating social organizations. In the aftermath of the Wenchuan disaster, a surge of grassroots mobilization was eventually absorbed into official structures under tightened state regulation (Liu & Palmer, 2021; Lu & Xu, 2015; Tang & Wang, 2020). Similarly, following the Zhengzhou flood, mechanisms such as the inclusion of social actors in emergency command chains and simulation planning were initiated, yet remained largely event-triggered and short-term in orientation (Li et al., 2025; Zhao et al., 2023).

By contrast, international systems such as the NVOAD and JVOAD reflect decentralized, pre-institutionalized collaboration rooted in routine preparedness. These models emphasize the formalization of protocols, role clarity, and platform-based governance long before disasters strike. For instance, NVOAD's structure integrates local, state, and national coordination layers, and relies on predefined standard operating procedures, allowing civil society actors to act autonomously while maintaining alignment with overarching frameworks (NVOAD, 2024; Özkan & Uzun, 2024). JVOAD similarly operates through sector-based working groups and digital coordination platforms that facilitate horizontal accountability among stakeholders (JVOAD, 2025a, 2026a).

The difference in institutional logic is not merely procedural but reflects varying expectations of civil society's role in public governance. In China, civil organizations often require administrative licensing or direct government affiliation to engage in formal disaster response, resulting in limited autonomy and flexibility. This is consistent with the findings that coordination platforms in China tend to be activated "after the fact," shaped by policy responses rather than anticipatory planning (Lu & Li, 2020; Tang & Chen, 2025). In contrast, NVOAD and JVOAD illustrate how proactive institutionalization builds enduring trust and preparedness through routine, pre-disaster engagement.

While both models seek to enhance coordination, their foundations diverge. China's reactive, state-driven institutionalization underscores centralized authority and temporary mobilization, whereas international approaches lean toward distributed governance with durable institutional memory. Bridging this gap will require China to advance from reactive adaptation toward systematic integration and role institutionalization within its disaster governance architecture.

21.2.2 Organizational Ecology and Networked Collaboration Structures

The structural configuration of collaborative networks in disaster governance varies significantly between China and international counterparts, reflecting deeper divergences in organizational ecology, institutional maturity, and capacity development. In China, large-scale disasters often trigger an influx of spontaneous actors and newly formed organizations, producing what can be termed an "explosion–contraction" pattern. Following the Ludian Earthquake, over 130 civil society organizations mobilized for relief efforts within days; however, many exhibited overlapping functions and lacked mechanisms for sustained collaboration, leading to coordination inefficiencies and short-lived alliances (Chen et al., 2020).

This stands in contrast to countries like the United States and Japan, where collaborative networks are shaped by decades of institutional development, functional differentiation, and inter-organizational trust. The NVOAD, for instance, operates through a multi-tiered coordination structure that includes national partners, state chapters, and specialized working groups, each with defined roles and operational protocols (NVOAD, 2025b). This layered framework enables effective division of labor and long-term continuity across disaster cycles. Similarly, JVOAD integrates sectoral coordination, information management, and resource pooling through a central collaborative platform, enhancing both horizontal and vertical alignment among actors (JVOAD, 2025a).

A key differentiator lies in network durability and professionalization. While Chinese civil society demonstrates strong mobilization capacity, particularly in post-crisis contexts, its ecosystem is often fragmented and episodic, constrained by limited inter-organizational linkages and reliance on government facilitation (Xu et al., 2018b; Zhang et al., 2023a). In contrast, NVOAD and JVOAD members routinely engage in joint training, preparedness drills, and pre-disaster planning, fostering relational capital and anticipatory capacity (Everly et al., 2008; Özkan & Uzun, 2024).

Furthermore, international frameworks emphasize interoperability and standardization, which reduce transaction costs and foster adaptive resilience across disaster types. For example, NVOAD's standardized coordination protocols allowed rapid cross-state deployment during concurrent floods and wildfires in 2023, while JVOAD's digital coordination platform enabled real-time information sharing during overlapping typhoon and earthquake responses (JVOAD, 2025c; NVOAD, 2025a).

21.2.3 Cultural Resources and the Embedding of Social Capital

The integration of cultural norms and social capital into disaster collaboration mechanisms significantly shapes the durability and legitimacy of governance systems. In internationally institutionalized models, such as those observed in the U.S. and Japan, cultural and religious networks play a critical role in sustaining engagement across disaster cycles. Organizations affiliated with long-standing faith-based traditions or community volunteerism—such as the American Red Cross or church-based groups within the NVOAD system—benefit from high levels of public trust and continuity. These networks act not only as response actors but also as enduring community anchors that reinforce inter-organizational cooperation and facilitate rapid mobilization during crises (Everly et al., 2008; NVOAD, 2024).

Such embeddedness is fostered through routine participation in preparedness exercises and shared value systems that align humanitarian action with deeply rooted social norms. Trust-based coordination is thus not constructed anew with each event, but instead continually reinforced by repeated interaction and shared cultural infrastructure. The institutional maturity of JVOAD, supported by a culture of civic responsibility and localized volunteerism, similarly demonstrates the resilience of culturally embedded disaster governance (JVOAD, 2025b; Özkan & Uzun, 2024).

In contrast, Chinese social organizations often operate in environments where public trust and organizational legitimacy are still evolving. Case studies from the Lushan Earthquake and the Zhengzhou flood highlight both growing capacities and persistent structural constraints. While community-based organizations played vital roles in immediate relief, their long-term engagement remains highly event-driven and reliant on government authorization or coordination platforms (Zhang et al., 2016; Zhao et al., 2023). In Zhengzhou, for instance, efforts to embed local organizations into formal emergency systems were a significant step forward. However, gaps in institutionalized trust and horizontal linkages across civil actors limited the formation of autonomous, self-organizing networks.

This contrast underscores a key challenge in the Chinese context: the partial disjuncture between emerging civic capabilities and the cultural-institutional frameworks necessary to sustain them. While formal recognition of social actors is increasing, the absence of deeply embedded cultural norms of volunteerism and cross-sector trust restricts the routinization of collaboration. Consequently, Chinese organizations often face difficulties in maintaining continuity and securing stable relational capital between disaster events.

21.2.4 Hazard-Specific Dynamics and the Adaptive Capacity of Collaborative Mechanisms

The nature of a disaster—whether a high-magnitude earthquake or a widespread flood—fundamentally shapes the rhythm, structure, and requirements of collaborative governance mechanisms. Different hazard types entail distinct temporalities of response, resource demands, and organizational configurations, thus placing varying pressures on inter-organizational coordination frameworks.

In seismically induced disasters such as the 2008 Wenchuan and 2010 Yushu Earthquakes, the dominant collaborative logic has been "emergency mobilization"—rapid deployment of search-and-rescue teams, medical units, and logistical support under government command. These events required intensive vertical coordination and immediate access to technical expertise and material resources. In both cases, social organizations emerged spontaneously, yet their roles remained largely reactive and transient. Institutional pathways evolved post-event, as seen in the gradual integration of civil society actors into planning processes only after initial state-led operations (Liu et al., 2011; Tang & Wang, 2020). The lack of pre-disaster agreements or standardized operating procedures constrained early-stage coordination.

Conversely, the 2021 Zhengzhou flood exemplified a different collaborative configuration. As a water-related disaster characterized by urban inundation and extended response periods, it necessitated distributed logistics, shelter management, and continuous public communication. In this context, horizontal coordination among civil society actors—such as mutual aid groups and community-based NGOs—played a more prominent role. The flood response saw experimental inclusion of social organizations in digital coordination platforms, with some participating in early simulation exercises and emergency command systems (Wang et al., 2024b; Yang et al., 2023). Although still event-triggered, this response reflected a shift toward anticipatory alignment.

International cases further highlight adaptive collaborative capacities in cross-hazard scenarios. In Indonesia's 2018 Sulawesi Earthquake-tsunami, hybrid mechanisms emerged to integrate search, medical relief, and psychosocial support in a layered response. Local civil society organizations, international agencies, and military actors operated under a decentralized but interoperable framework (Trias & Cook, 2021). Similarly, in the 2015 Nepal Earthquake, humanitarian coordination mechanisms demonstrated robust risk-type adaptation by integrating shelter, food distribution, and health services into a cohesive response architecture, facilitated by established pre-crisis protocols and civil-military cooperation (Adhikari et al., 2025; Lin, 2021).

These cases suggest that hazard types are not merely triggers but structural variables that shape the institutional ecology of disaster collaboration. Earthquakes, due to their sudden onset and technical complexity, favor centralized, command-based responses, whereas floods and protracted crises offer more scope for decentralized and socially embedded coordination. The international experience shows a higher degree of hazard-specific modularity—where protocols, leadership models, and information systems are tailored in advance. China's evolving system, while still event-responsive, is incrementally moving toward this logic, particularly as simulations, classification standards, and sectoral protocols become more institutionalized.

21.3 Future Pathways of Institutionalization, Professionalization, and Digitalization

As disaster risks grow in complexity, this section explores how civil society organizations can transition from reactive responders to embedded actors through the synergistic advancement of institutionalization, professionalization, and digitalization. It first examines efforts to formalize CSO roles within governance structures, followed by a discussion on building specialized capacities and resilient organizations. It then analyzes the transformative potential of digital platforms in enhancing coordination and adaptive response. Finally, it underscores the strategic convergence of these dimensions as a forward-looking pathway for integrating CSOs into comprehensive disaster governance systems.

21.3.1 Institutionalization and Governance Structure Enhancement

Efforts to institutionalize civil society organizations in disaster governance have gained momentum as states seek to embed non-state actors more deeply into emergency response systems. Rather than relying solely on ad hoc volunteerism, many governments are now designing policy frameworks and legal instruments to formalize CSO engagement. The case of China's post-2021 Zhengzhou flood recovery illustrates this shift. In the aftermath of the disaster, local authorities established structured registration mechanisms for social organizations and introduced cross-sector coordination protocols that improved response speed and transparency (Chen et al., 2022; Li et al., 2025). These mechanisms were later embedded into official contingency planning processes, marking a movement toward codified participation.

Parallel developments can be observed in international contexts. The NVOAD and JVOAD offer long-standing examples of institutionalized CSO networks. These umbrella organizations not only streamline coordination between governmental agencies and nonprofit actors but also maintain formalized standards of cooperation, roles delineation, and communication channels (JVOAD, 2026b; NVOAD, 2023). In both cases, CSO participation is no longer peripheral but integrated into formal disaster governance architecture, including pre-disaster preparedness and post-disaster recovery phases.

One persistent challenge, however, lies in building robust pre-disaster collaboration mechanisms. Across diverse country cases, it remains more common for coordination frameworks to activate during or after disaster onset rather than in anticipation. Despite growing awareness of this gap, institutional inertia and fragmented mandates continue to hinder anticipatory coordination platforms (Kapucu et al., 2024; Tuncalp & Cetindamar, 2026). The importance of establishing joint simulation exercises and cross-organizational training remains under-addressed in many national systems.

Moreover, as CSO engagement becomes more regulated, a new balance must be struck between institutional constraint and operational flexibility. While formalization ensures accountability and reliability, excessive bureaucratization may stifle the agility and local responsiveness that often characterize CSO interventions (Fraser, 2023; McNaught et al., 2024). Effective institutionalization, therefore, requires a governance architecture that supports structured collaboration without undermining innovation and community embeddedness.

21.3.2 Strengthening Professional Capacities and Organizational Resilience

As disasters become more complex and frequent, social organizations are increasingly transitioning from ad hoc actors to critical nodes within disaster governance systems. This evolution requires not only deeper institutional embedding but also the enhancement of professional capacities and organizational resilience. Moving beyond passion-driven volunteering, the shift toward capability- and system-driven response mechanisms has become imperative.

Across both developed and developing contexts, the progression toward professionalization reflects a growing recognition of the need for technical expertise, standardized procedures, and sustained institutional learning. In the United States, the National Voluntary Organizations Active in Disaster network has cultivated a "tiered professional response" model, enabling member organizations to mobilize specialized teams according to disaster scale and typology (NVOAD, 2024). These include trained personnel in logistics, public health, mental health, and housing, operating under standardized protocols and interoperability guidelines (Everly et al., 2008; Özkan & Uzun, 2024). Similarly, JVOAD system fosters sectoral coordination through training and shared accreditation, enhancing consistency across civil society responders (JVOAD, 2026b).

In China, recent disasters such as the 2021 Zhengzhou floods have spurred the emergence of specialized rescue alliances and sector-based mobilization. Empirical accounts from these events document the rising participation of organizations with formal qualifications, pre-deployment training, and digital coordination tools (Li et al., 2025; Zhao et al., 2023). Importantly, these trends demonstrate a departure from spontaneous volunteerism toward structured, role-specific interventions—a development reinforced by evolving regulatory environments and partnership frameworks.

Professionalization is not merely a technical process but also a reconfiguration of identity and accountability. Within the framework of co-production theory, social organizations are no longer peripheral supporters but embedded actors co-producing public value alongside state agencies (Osborne et al., 2021; Ostrom, 1996). This role transformation entails enhanced expectations regarding reliability, procedural transparency, and risk management capabilities. Internationally, credentialing systems and scenario-based training programs are increasingly employed to cultivate adaptive competence and inter-organizational trust (De Sisto et al., 2024; Fraser, 2023).

The road to professionalization is not without challenges. Issues of resource disparity, access to training, and uneven geographic distribution of capacities persist, particularly among grassroots organizations. Therefore, building organizational resilience must accompany skill enhancement—encompassing financial sustainability, internal governance, and digital adaptability (Wang et al., 2024b; Wu et al., 2021).

21.3.3 Empowering Collaboration Through Digital Tools and Platforms

As disaster governance faces growing complexity, digital technologies have emerged as pivotal enablers of coordination, information exchange, and decision-making. In particular, social organizations increasingly rely on digital platforms to enhance their operational agility and strategic alignment during crisis response. The transition toward digital governance is not merely technical; it reshapes the very architecture of collaboration, accountability, and legitimacy in disaster settings.

Digital platforms offer substantial opportunities to overcome traditional coordination challenges. The 2016 Ecuador Earthquake and the 2018 Sulawesi disaster revealed how interoperable tools—ranging from crowdsourced mapping to real-time data dashboards—enabled CSOs, government actors, and international agencies to align actions in rapidly changing environments (Torres & Vaca, 2019; Trias & Cook, 2021). Similarly, Chinese pilot initiatives in "smart emergency" governance, including AI-assisted early warning and multi-platform data integration, illustrate the growing role of technology in facilitating scenario planning and joint decision-making (Mao et al., 2026; Zhao et al., 2023). These digital mechanisms not only streamline logistical operations but also institutionalize collaborative practices, making them less dependent on informal networks or ad hoc coordination.

However, digitalization introduces new layers of governance dilemmas. Issues of data ownership, information asymmetry, and platform governance remain under-addressed. As highlighted in multiple post-disaster evaluations, centralized platforms often struggle to accommodate diverse actors' knowledge systems, leading to marginalization of grassroots inputs (Fraser, 2023; Ye et al., 2025). In addition, the digital divide—both technological and organizational—undermines the inclusivity of digitally mediated collaboration, particularly in rural or under-resourced regions (Higuera Roa et al., 2025; Wang et al., 2024a). These asymmetries risk reinforcing existing hierarchies rather than democratizing response systems.

The shift toward digital governance also carries normative and ethical implications. While algorithmic tools can optimize resource allocation and risk modeling, they may obscure human judgment and local contextual understanding (Liu & Renn, 2025; Schweizer & Renn, 2019). Balancing automation with participatory design thus becomes essential. Some regions have experimented with co-developed platforms that incorporate community feedback loops and multilingual interfaces to bridge this gap (Liang et al., 2019; Xue et al., 2024).

21.3.4 Integrated Pathways: Merging Institutionalization, Professionalization, and Digitalization

Building resilient and adaptive disaster governance increasingly requires a synergistic integration of institutional, professional, and digital capacities. While previous sections have examined these trajectories in isolation, their convergence constitutes the critical pathway through which civil society organizations evolve from ad hoc responders to embedded governance actors.

Institutional frameworks provide the necessary scaffolding for this transformation. Legal mandates, policy guidelines, and coordination protocols not only define the roles of CSOs but also ensure platforms for sustained interaction, accountability, and legitimacy. Internationally, systems like JVOAD and the United States' NVOAD demonstrate how formalized structures support layered multi-actor engagement, encompassing joint preparedness, role allocation, and operational coordination (JVOAD, 2025b; NVOAD, 2025b). In China, developments following the 2021 Zhengzhou flood reflect an emergent institutional shift, with local social organizations participating in emergency simulation drills and assuming formalized roles in municipal response mechanisms (Li et al., 2025; Zhao et al., 2023).

However, institutionalization must be underpinned by the professionalization of CSOs. This includes building specialized competencies, establishing standardized training systems, and promoting sectoral differentiation. In the U.S., NVOAD's certification and credentialing mechanisms ensure operational reliability across state and local levels. In China, the rise of provincial-level rescue alliances—comprised of technically trained volunteers and domain-specific NGOs—illustrates an emerging ecosystem of professional disaster responders. Sustaining this momentum requires long-term investment in capacity-building infrastructures and policy frameworks that recognize the technical value of civil actors (Wu & Peng, 2017; Zhang et al., 2023b).

Digitalization acts as a cross-cutting enabler that links institutional structures with professional practice. The deployment of mobile coordination platforms, real-time data dashboards, and AI-assisted decision tools has revolutionized how disaster actors communicate and collaborate. During the Türkiye–Syria and Sulawesi Earthquake responses, integrated digital systems enabled real-time needs assessments, interoperable logistics, and rapid alignment across CSO-government boundaries (Tarnas et al., 2023; Trias & Cook, 2021). In China, platforms piloted in cities like Zhengzhou and Shenzhen are beginning to embed digital infrastructure into routine emergency workflows, though challenges remain regarding data access, platform interoperability, and ethical governance (Everly et al., 2008; Tang & Chen, 2025).

Moving forward, the integration of institutional, professional, and digital dimensions must be seen as mutually reinforcing rather than sequential. Governments can accelerate this convergence by offering regulatory clarity, fiscal incentives, and digital infrastructure investment. CSOs must commit to internal professionalization and digital adoption. Academic institutions can support through knowledge translation and the co-development of evaluation frameworks for collaborative governance. Such systemic integration offers a viable pathway for reducing fragmentation in disaster governance and embedding civil society actors into the core architecture of twenty-first-century risk management.

21.4 Integrating Theory and Practice: Toward a Knowledge-Informed Model of Collaborative Disaster Governance

Grounded in the empirical insights of the preceding chapters, this section offers a reflective synthesis of how three foundational theories—collaborative governance, network governance, and co-production—have been adapted, challenged, and extended in diverse disaster contexts. Rather than treating these frameworks in isolation, the analysis traces their convergence through evolving governance practices, highlighting how shared processes, structural interdependencies, and co-produced capacities shape disaster resilience. The discussion unfolds across four thematic lenses: the adaptation of collaborative governance under crisis; the evolution of resilient network structures; the expansion of co-production into risk governance; and, finally, the construction of an integrated analytical framework to inform future comparative research and institutional design.

21.4.1 Adaptation and Extension of Collaborative Governance: From Shared Vision to Emergency Resilience

Collaborative governance conceptualizes public problem-solving as a process in which multiple actors build shared purpose, mutual trust, and joint capacity for collective action within institutionalized arenas (Ansell & Gash, 2008; Emerson et al., 2012). In disaster governance, however, collaboration unfolds under conditions of urgency, uncertainty, and asymmetric resources, which requires theoretical recalibration. Evidence across diverse disaster contexts suggests that collaborative arrangements are increasingly shaped not only by deliberative consensus but also by the need for rapid mobilization, operational interoperability, and adaptive learning under stress (McGuire, 2006; Tierney, 2012).

A salient pattern emerging from recent cases is the rise of event-driven "emergency collaboration," whereby extreme events open policy windows for accelerated coordination among governmental agencies, civil society organizations, and private actors. Large-scale disasters in China have demonstrated how institutional constraints can temporarily loosen, enabling social organizations to enter formal response systems through expedited registration, task-based coordination, and platform-mediated deployment (Chen et al., 2022; Li et al., 2025; Zhao et al., 2023). Comparable dynamics are observed in international responses, where multi-actor coordination mechanisms rapidly coalesce around shared operational objectives and resource pooling during acute crises (Tarnas et al., 2023; Trias & Cook, 2021). These patterns extend collaborative governance beyond routine deliberation toward crisis-induced activation and accelerated collective action.

At the same time, the normative orientation of collaboration is shifting from efficiency-driven cooperation toward resilience-oriented governance. Rather than optimizing short-term service delivery alone, collaborative arrangements increasingly aim to enhance systemic robustness, learning capacity, and long-term adaptability across interconnected risk environments (Liu & Renn, 2025; Mark et al., 2024; Schweizer & Renn, 2019). This reframing aligns collaborative processes with broader resilience agendas, emphasizing redundancy, diversity of actors, and institutional memory as essential governance assets.

Another critical evolution concerns the growing institutional embeddedness of collaborative processes. Legal codification, standardized coordination protocols, and formalized partnership platforms are progressively stabilizing collaborative interactions and reducing dependence on ad hoc personal networks (Kapucu et al., 2024; Shan et al., 2025). Such institutionalization enhances predictability and accountability while enabling collaboration to persist beyond episodic crises. Excessive formalization also risks constraining flexibility and local responsiveness, highlighting the need for adaptive institutional design that balances procedural stability with operational agility (Fraser, 2023; McNaught et al., 2024).

21.4.2 Practice-Based Extensions of Network Governance: From Structural Efficiency to Resilient Networks

The evolving landscape of disaster response has challenged traditional notions of network governance, shifting attention from static structural designs to dynamic, resilience-oriented configurations. Rather than focusing solely on efficient resource distribution or formalized partnerships, recent disaster events reveal the growing importance of adaptive networks—those capable of rapid reconfiguration, sustained coordination, and learning in volatile environments.

In post-earthquake contexts across China, including Wenchuan, Lushan, and Jiuzhaigou, decentralized response networks formed spontaneously through digital communication tools and pre-existing interpersonal connections. These temporary, self-organized formations, often involving hundreds of grassroots organizations, mirror what theory describes as "participant-governed networks" (Provan & Kenis, 2007). However, while such emergent formations allowed for rapid mobilization under uncertainty, they frequently encountered bottlenecks in coordination and resource integration during recovery phases—highlighting the limitations of purely informal networks.

To address these limitations, formalized, institutionalized network platforms have gained prominence. The U.S. National VOAD and Japan VOAD exemplify a more centralized "network administrative organization" model, which improves network accountability, clarifies member roles, and standardizes procedures (Klijn et al., 2025). Similar institutionalization can now be observed in Chinese provincial-level alliances that emerged after the 2021 Zhengzhou floods, where government-acknowledged coalitions enhanced coordination among CSOs, professional teams, and state actors.

Beyond structural innovation, a key trend lies in the rise of intermediary platform-type CSOs. These entities serve as relational brokers—facilitating trust, translating knowledge across sectors, and mediating cross-level collaboration. For instance, in both the Ecuador and Sulawesi disasters, CSO platforms enabled timely data exchange and bridged coordination gaps between humanitarian actors (Bodin et al., 2022; Van Der Weert et al., 2022). Such developments point toward a conceptual expansion of network governance from an efficiency-centered paradigm to one oriented around resilience and adaptability.

21.4.3 Practice-Oriented Expansions of Co-production: From Service Delivery to Risk Co-governance

Co-production theory, traditionally rooted in public service delivery, emphasizes joint responsibility, knowledge exchange, and mutual trust between government and non-state actors (Brandsen & Pestoff, 2006; Ostrom, 1996). In disaster governance, this framework has evolved beyond the confines of service efficiency to encompass more dynamic, multi-actor risk governance processes. Evidence from both domestic and international contexts reveals how co-production is increasingly embedded across three critical phases of disaster governance: emergency response, post-disaster recovery, and preparedness.

In the immediate aftermath of disasters, co-production mechanisms enable shared delivery of vital services. During the 2014 Ludian Earthquake, for example, over 100 CSOs and volunteer groups collaborated with local authorities to distribute relief materials, conduct needs assessments, and offer psychosocial support (Chen et al., 2020). Similarly, following the 2021 Zhengzhou flood, spontaneous digital coordination enabled citizens and grassroots actors to co-create rescue maps and organize decentralized aid delivery, bridging the initial vacuum left by formal institutions (Li et al., 2025; Zhao et al., 2023). These practices reflect a shift from hierarchical delegation to horizontal reciprocity in crisis response.

Co-production also becomes salient during post-disaster reconstruction, where cross-sector resource integration is vital. The reconstruction efforts after the 2008 Wenchuan Earthquake involved public–private partnerships, CSO-led community engagement programs, and sustained knowledge exchange between government agencies and civil society actors (Sun, 2017; Xu et al., 2018b). Similar patterns were observed in Ecuador and Türkiye, where platform-based CSOs played brokerage roles in coordinating international and local recovery initiatives, fostering both material restoration and institutional learning (Córdova et al., 2023; Schwarz et al., 2025).

Beyond reactive roles, co-production extends to proactive knowledge generation and capacity-building. The U.S.-based NVOAD exemplifies how civil society actors engage in regular joint training, scenario planning, and technical standard-setting with public agencies, creating a culture of anticipatory governance (NVOAD, 2025b). Such practices embody a broader conceptual shift from service delivery to "relational governance," wherein roles are negotiated, trust is built iteratively, and resilience is co-produced through ongoing engagement (Osborne, 2020; Yumagulova & Vertinsky, 2019).

21.4.4 Integrated Framework Construction: A Synthesis Model for Collaborative Disaster Governance

Building on the empirical insights of this volume, an integrated framework for disaster governance emerges through the synthesis of three foundational theories: collaborative governance, network governance, and co-production. Each offers a distinct analytical lens—focusing, respectively, on interaction processes, structural dynamics, and role constellations—yet all converge in addressing the complexity of contemporary disaster risks.

The first dimension of the proposed model—participation mechanisms—distinguishes between open and closed configurations. Disaster cases reveal that effective governance depends on flexible participation. For instance, China's Earthquake responses (e.g., Wenchuan, Jiuzhaigou, Lushan) demonstrate how spontaneous, open-ended mobilization of grassroots organizations enabled rapid network formation under high uncertainty (Ji & Pang, 2021; Sun, 2017). In contrast, platforms like NVOAD and JVOAD illustrate more structured, closed systems that prioritize standardization and long-term coordination (JVOAD, 2025a; NVOAD, 2024).

Second, the model examines organizational relationships, from flat to hierarchical. Decentralized networks such as temporary CSO coalitions (e.g., in Ecuador and Sulawesi responses) reflect flat, horizontal coordination (Guerrero-Miranda & Luque González, 2021; Trias & Cook, 2021). Conversely, government-sanctioned rescue alliances formed post-Zhengzhou flood show increasing vertical integration and clearer role assignments (Li et al., 2025; Zhao et al., 2023).

The third dimension—performance orientation—maps a spectrum between speed and continuity. While grassroots-led networks offer agility during immediate response (Bodin et al., 2022), they often suffer from fragmentation during recovery phases. Institutionalized platforms provide continuity, enabling trust-building and knowledge retention over time (Van Der Weert et al., 2022).

Finally, institutional embeddedness captures whether governance arrangements remain ad hoc or evolve into normalized systems. Several chapters demonstrate that when collaborative or network-based structures are embedded into legal or administrative routines, they show greater resilience and legitimacy (Kapucu et al., 2024; Shan et al., 2025). This is evident in cases where post-disaster coordination mechanisms transitioned into standing bodies or legal frameworks (e.g., ADRRN, JVOAD).

This four-dimensional framework—participation, relationship, performance, and embeddedness—not only synthesizes diverse empirical patterns but also offers a comparative lens for future disaster governance research. By recognizing the overlapping yet distinct contributions of each theoretical stream, it facilitates a more nuanced understanding of how adaptive, inclusive, and resilient governance can be institutionally realized in risk-prone contexts.

21.5 Synthesizing Insights

Disaster governance has evolved from fragmented emergency responses toward increasingly institutionalized, professionalized, and digitalized systems. Through longitudinal analysis, the chapter traces how social organizations in China transitioned from spontaneous mobilization to embedded actors within formal frameworks. Comparative insights highlight divergent governance logics—China's event-driven integration versus anticipatory systems in countries like the U.S. and Japan. These differences underscore the role of institutional maturity, network structure, and cultural resources in shaping collaborative resilience. Finally, the synthesis of collaborative governance, network governance, and co-production theories culminates in an integrated model that bridges participation, relationships, performance, and institutional embeddedness. Together, these insights chart a pathway for inclusive, adaptive, and future-ready disaster governance across diverse risk environments.

References

Adhikari, P., Upadhaya, B., Adhikari, S. D., Aryal, S., & Wijethilake, C. (2025). The spectacularization of NGOs accountability in disaster situations: Evidence from the 2015 Nepal's earthquakes. *Financial Accountability & Management, 41*(3), 490–507. https://doi.org/10.1111/faam.12427

Alcántara-Ayala, I. (2025). Landslides in a changing world. *Landslides, 22*(9), 2851–2865. https://doi.org/10.1007/s10346-024-02451-1

Ansell, C., & Gash, A. (2008). Collaborative governance in theory and practice. *Journal of Public Administration Research and Theory, 18*(4), 543–571. https://doi.org/10.1093/jopart/mum032

Bodin, Ö., Guerrero, A. M., Nohrstedt, D., Baird, J., Summers, R., Plummer, R., & Jasny, L. (2022). Choose your collaborators wisely: Addressing interdependent tasks through collaboration in responding to wildfire disasters. *Public Administration Review, 82*(6), 1154–1167. https://doi.org/10.1111/puar.13518

Brandsen, T., & Pestoff, V. (2006). Co-production, the third sector and the delivery of public services: An introduction. *Public Management Review, 8*(4), 493–501. https://doi.org/10.1080/14719030601022874

Chen, W., Zhang, H., Comfort, L. K., & Tao, Z. (2020). Exploring complex adaptive networks in the aftermath of the 2008 Wenchuan earthquake in China. *Safety Science, 125*, Article 104607. https://doi.org/10.1016/j.ssci.2020.104607

Chen, Z., Kong, F., & Zhang, M. (2022). A case study of the "7-20" extreme rainfall and flooding event in Zhengzhou, Henan Province, China from the perspective of fragmentation. *Water, 14*(19), 2970. https://doi.org/10.3390/w14192970

Chu, C., & Huang, G. (2025). Hybrid forecasting framework for emergency material demand in post-earthquake scenarios integrating the grey model and Bayesian dynamic linear models. *Sustainability, 17*(15), 6701. https://doi.org/10.3390/su17156701

Clark, N., Boersma, K., Raju, E., Opromolla, A., Orru, K., Hansson, S., Russo, R., Gargiulo, M. V., Duca, G., Capuano, P., Schweizer, P.-J., Cumiskey, L., Steinhausen, M., Branlat, M., Olson, A., Andersen, N. B., Larruina, R., Atun, F., Van Westen, C., et al. (2025). Strengthening all-of-society approaches for disaster resilient societies through competency building: A European research agenda. *International Journal of Disaster Risk Reduction, 121*, Article 105345. https://doi.org/10.1016/j.ijdrr.2025.105345

Córdova, M., Menoscal, J., & Flores, E. M. (2023). Governance and the design of post-disaster policies: A comparative analysis from Latin America and the Caribbean. *Disasters, 47*(3), 766–787. https://doi.org/10.1111/disa.12561

De Sisto, M., Shearing, C., Heffernan, T., & Sanderson, D. (2024). Reshaping disaster management: An integrated community-led approach. *Australian Journal of Public Administration.* https://doi.org/10.1111/1467-8500.12668

Emerson, K., Nabatchi, T., & Balogh, S. (2012). An integrative framework for collaborative governance. *Journal of Public Administration Research and Theory, 22*(1), 1–29. https://doi.org/10.1093/jopart/mur011

Everly, G. S., Hamilton, S. E., Tyiska, C. G., & Ellers, K. (2008). Mental health response to disaster: Consensus recommendations: Early Psychological Intervention Subcommittee (EPI), National Volunteer Organizations Active in Disaster (NVOAD). *Aggression and Violent Behavior, 13*(6), 407–412. https://doi.org/10.1016/j.avb.2008.05.004

Fraser, A. (2023). Reflections from PEARL for an era of systemic risks, uncertain futures and complex governance. *Jàmbá Journal of Disaster Risk Studies, 15*(1). https://doi.org/10.4102/jamba.v15i1.1485

Guerrero-Miranda, P., & Luque González, A. (2021). Social responsibility, sustainability, and public policy: The lessons of debris management after the Manabí earthquake in Ecuador. *International Journal of Environmental Research and Public Health, 18*(7), 3494. https://doi.org/10.3390/ijerph18073494

Higuera Roa, O., Bachmann, M., Mechler, R., Šakić Trogrlić, R., Reimann, L., Mazzoleni, M., Aerts, J. C. J. H., Buskop, F. E., Pirani, A., & Mysiak, J. (2025). Challenges and opportunities in climate risk assessment: Future directions for assessing complex climate risks. *Environmental Research Letters, 20*(5), Article 053003. https://doi.org/10.1088/1748-9326/adc756

Hilhorst, D., Boersma, K., & Raju, E. (2020). Research on politics of disaster risk governance: Where are we headed? *Politics and Governance, 8*(4), 214–219. https://doi.org/10.17645/pag.v8i4.3843

Jagosh, J., Bush, P. L., Salsberg, J., Macaulay, A. C., Greenhalgh, T., Wong, G., Cargo, M., Green, L. W., Herbert, C. P., & Pluye, P. (2015). A realist evaluation of community-based participatory research: Partnership synergy, trust building and related ripple effects. *BMC Public Health, 15*(1), 725. https://doi.org/10.1186/s12889-015-1949-1

Ji, M., & Pang, H. (2021). Survival or crisis? Funding distribution from grant-making foundations to grassroots social organizations in China. *Communist and Post-Communist Studies, 54*(4), 137–156. https://doi.org/10.1525/j.postcomstud.2021.54.4.137

JVOAD. (2025a). *Independent auditor's report for FY2024.* https://jvoad.jp/wp-content/themes/may2024/pdf/kansa2024.pdf

JVOAD. (2025b). *Member organizations.* https://jvoad.jp/member/

JVOAD. (2025c). *Training and capacity-building activities.* https://jvoad.jp/news-cat/event/

JVOAD. (2026a). *About JVOAD and its activities.* https://jvoad.jp/about/activity/

JVOAD. (2026b). *Organization overview.* https://jvoad.jp/about/outline/

Kang, Y. (2020). Dispersed Domination through Patron-Clientelism: The Evolution of the Local State–NGO Relationship in Post-Disaster Sichuan. *Journal of Contemporary China, 29*(124), 598–613. https://doi.org/10.1080/10670564.2019.1677367

Kapucu, N., Parkin, A., Lumb, M., & Dippy, R. (2024). Crisis coordination in complex intergovernmental systems: The case of Australia. *Public Administration Review, 84*(3), 389–399. https://doi.org/10.1111/puar.13830

Klijn, E. H., Koppenjan, J., Spekkink, W., & Warsen, R. (2025). *Governance networks in the public sector* (2nd ed.). Routledge. https://doi.org/10.4324/9781003402138

Li, P., Zhuang, L., Lin, K., She, D., Chen, Q., Wang, Q., & Xia, J. (2025). New perspectives on urban stormwater management in China, with a focus on extreme rainfall events. *Natural Hazards, 121*(4), 3745–3774. https://doi.org/10.1007/s11069-024-06994-3

Liang, Y., Cheng, J., Ruzek, J. I., & Liu, Z. (2019). Posttraumatic stress disorder following the 2008 Wenchuan earthquake: A 10-year systematic review among highly exposed populations in China. *Journal of Affective Disorders, 243*, 327–339. https://doi.org/10.1016/j.jad.2018.09.047

Lin, P. (2021). China's evolving humanitarian diplomacy: Evidence from China's disaster-related aid to Nepal. *Asian Journal of Comparative Politics, 6*(3), 221–237. https://doi.org/10.1177/20578911211019255

Liu, H., & Renn, O. (2025). Polycrisis and systemic risk: Assessment, governance, and communication. *International Journal of Disaster Risk Science, 16*(4), 526–549. https://doi.org/10.1007/s13753-025-00636-3

Liu, J., Fan, Y., & Shi, P. (2011). Response to a high-altitude earthquake: The Yushu earthquake example. *International Journal of Disaster Risk Science, 2*(1), 43–53. https://doi.org/10.1007/s13753-011-0005-8

Liu, Q., & Palmer, D. A. (2021). Chinese NGOs at the interface between governmentality and local society: An actor-oriented perspective. *China Information, 35*(2), 158–178. https://doi.org/10.1177/0920203X20942094

Lu, Y., & Li, R. (2020). Rebuilding resilient homeland: An NGO-led post-Lushan earthquake experimental reconstruction program. *Natural Hazards, 104*(1), 853–882. https://doi.org/10.1007/s11069-020-04194-3

Lu, Y., Wang, Y., & Zhan, C. (2023). Building rural community disaster resilience in developing countries: Insights from a Chinese NGO's Safe Rural Community programme. *Disasters, 47*(4), 1090–1117. https://doi.org/10.1111/disa.12586

Lu, Y., & Xu, J. (2015). NGO collaboration in community post-disaster reconstruction: Field research following the 2008 Wenchuan earthquake in China. *Disasters, 39*(2), 258–278. https://doi.org/10.1111/disa.12098

Mao, J., Cao, X., Zhang, R., Tang, J., Zhao, P., & Qin, H. (2026). Characterising the high spatial heterogeneity of urban flood resilience under extreme rainfall using mobile signal data. *Geomatics, Natural Hazards and Risk, 17*(1), 2605116. https://doi.org/10.1080/19475705.2025.2605116

Mark, S., Holder, S., Hoyer, D., Schoonover, R., & Aldrich, D. P. (2024). Understanding polycrisis: Definitions, applications, and responses. *SSRN Electronic Journal*. https://doi.org/10.2139/ssrn.4593383

McGuire, M. (2006). Collaborative public management: Assessing what we know and how we know it. *Public Administration Review, 66*(s1), 33–43. https://doi.org/10.1111/j.1540-6210.2006.00664.x

McNaught, R., Nalau, J., Hales, R., Pittaway, E., Handmer, J., & Renouf, J. (2024). Innovation and deadlock in governing disasters and climate change collaboratively—Lessons from the Northern Rivers region of New South Wales, Australia. *International Journal of Disaster Risk Reduction, 105*, Article 104366. https://doi.org/10.1016/j.ijdrr.2024.104366

Menefee, T., & Nordtveit, B. H. (2012). Disaster, civil society and education in China: A case study of an independent non-government organization working in the aftermath of the Wenchuan earthquake. *International Journal of Educational Development, 32*(4), 600–607. https://doi.org/10.1016/j.ijedudev.2011.10.002

NVOAD. (2023). *2023 triennial review*. https://www.nvoad.org/wp-content/uploads/Triennial-Review-2023.pdf

NVOAD. (2024). *Strategic plan 2024–2026*. https://www.nvoad.org/wp-content/uploads/Strategic-Plan-2024-2026.pdf

NVOAD. (2025a). *National VOAD guidance for disaster response and recovery for renters and rental properties*. https://www.nvoad.org/wp-content/uploads/Rental-Guidance-Document-2025.pdf

NVOAD. (2025b). *State of disaster recovery: Expending assistance*. https://www.nvoad.org/wp-content/uploads/State-of-Disaster-Recovery-Expediting-Assistance-Report.pdf

Organisation Z. (2017). *Analytical situation report: Jiuzhaigou earthquake (August 8, 2017)*.

Osborne, S. P. (2020). *Public service logic: Creating value for public service users, citizens, and society through public service delivery* (1st ed.). Routledge. https://doi.org/10.4324/9781003009153

Osborne, S. P., Nasi, G., & Powell, M. (2021). Beyond co-production: Value creation and public services. *Public Administration, 99*(4), 641–657. https://doi.org/10.1111/padm.12718

Osborne, S. P., Radnor, Z., Kinder, T., & Vidal, I. (2015). The SERVICE framework: A public-service-dominant approach to sustainable public services. *British Journal of Management, 26*(3), 424–438. https://doi.org/10.1111/1467-8551.12094

Ostrom, E. (1996). Crossing the great divide: Coproduction, synergy, and development. *World Development, 24*(6), 1073–1087. https://doi.org/10.1016/0305-750X(96)00023-X

Özkan, O., & Uzun, A. (2024). The democratic touch of disasters: The perception of non-governmental organizations in public institutions in the context of disaster governance. *SSRN*. https://doi.org/10.2139/ssrn.4989771

Phattharapornjaroen, P., Burivong, R., & Khorram-Manesh, A. (2025). Myanmar earthquake aftermath—Critical update and expanded analysis. *Disaster Medicine and Public Health Preparedness, 19*, Article e125. https://doi.org/10.1017/dmp.2025.112

Provan, K. G., & Kenis, P. (2007). Modes of network governance: Structure, management, and effectiveness. *Journal of Public Administration Research and Theory, 18*(2), 229–252. https://doi.org/10.1093/jopart/mum015

Schwarz, A., Sellnow, D. D., Diers-Lawson, A., & Sellnow, T. L. (2025). The crisis response to the Türkiye-Syria earthquakes in 2023: Instructional emergency risk communication by the government, the national disaster management agency, and NGOs via press releases and microblogs. *International Journal of Disaster Risk Reduction, 129*, Article 105770. https://doi.org/10.1016/j.ijdrr.2025.105770

Schweizer, P.-J., & Renn, O. (2019). Governance of systemic risks for disaster prevention and mitigation. *Disaster Prevention and Management: An International Journal, 28*(6), 862–874. https://doi.org/10.1108/DPM-09-2019-0282

Shan, H., An, Y., Bai, H., & Shi, J. (2025). Intergovernmental collaborative governance of emergency response logistics: An evolutionary game study. *Natural Hazards, 121*(1), 705–730. https://doi.org/10.1007/s11069-024-06785-w

Sun, T. (2017). Earthquakes and the typologies of state–society relations in China. *China Information, 31*(3), 304–326. https://doi.org/10.1177/0920203X17720337

Tang, G., & Chen, B. (2025). Unpacking the dynamics of nonprofit purpose-oriented networks in disaster relief: A tale of two cases. *Public Administration and Development*. https://doi.org/10.1002/pad.70044

Tang, G., & Wang, F. (2020). What factors contribute to nonprofit collaboration? An analysis of response and recovery efforts after the 2008 Wenchuan earthquake, China. *Safety Science, 125*, Article 104624. https://doi.org/10.1016/j.ssci.2020.104624

Tarnas, M. C., Karah, N., Almhawish, N., Aladhan, I., Alobaid, R., & Abbara, A. (2023). Politicization of water, humanitarian response, and health in Syria as a contributor to the ongoing cholera outbreak. *International Journal of Infectious Diseases, 131*, 115–118. https://doi.org/10.1016/j.ijid.2023.03.042

Tierney, K. (2012). Disaster governance: Social, political, and economic dimensions. *Annual Review of Environment and Resources, 37*(1), 341–363. https://doi.org/10.1146/annurev-environ-020911-095618

Torres, J., & Vaca, C. (2019). Cross-lingual perspectives about crisis-related conversations on Twitter. In *Companion Proceedings of The 2019 World Wide Web Conference* (pp. 255–261). https://doi.org/10.1145/3308560.3316799

Trias, A. P. L., & Cook, A. D. B. (2021). Future directions in disaster governance: Insights from the 2018 Central Sulawesi Earthquake and Tsunami response. *International Journal of Disaster Risk Reduction, 58*, Article 102180. https://doi.org/10.1016/j.ijdrr.2021.102180

Tuncalp, D., & Cetindamar, D. (2026). Grassroots digital humanitarianism and regenerative synergy in the 2023 Türkiye earthquake. *Technology in Society, 85*, Article 103174. https://doi.org/10.1016/j.techsoc.2025.103174

Van Der Weert, G., Burzynska, K., & Knoben, J. (2022). An integrative perspective on interorganizational multilevel healthcare networks: A systematic literature review. *BMC Health Services Research, 22*(1), 923. https://doi.org/10.1186/s12913-022-08314-6

Wang, J., Feng, J., Sun, B., Guo, J., Hu, S., & Zhang, M. (2024a). Characterization of dam failure flooding in an urban reservoir under different rainfall conditions. *Scientific Reports, 14*(1), 27560. https://doi.org/10.1038/s41598-024-78968-1

Wang, W., Zhu, X., Lu, P., Zhao, Y., Chen, Y., & Zhang, S. (2024b). Spatio-temporal evolution of public opinion on urban flooding: Case study of the 7.20 Henan extreme flood event. *International Journal of Disaster Risk Reduction, 100*, 104175. https://doi.org/10.1016/j.ijdrr.2023.104175

Wei, B., Su, G., & Liu, F. (2013). Public response to earthquake disaster: A case study in Yushu Tibetan Autonomous Prefecture. *Natural Hazards, 69*(1), 441–458. https://doi.org/10.1007/s11069-013-0710-2

Wu, F., & Peng, L. (2017). The rise of non-governmental disaster relief in China. *SSRN Electronic Journal*. https://doi.org/10.2139/ssrn.2916221

Wu, M., Gao, X., Cao, M., Papa, E., & Zhu, X. (2021). The changes of intergovernmental collaboration dynamic in post-disaster destination management: Network analysis. *Journal of Hospitality and Tourism Management, 48*, 32–45. https://doi.org/10.1016/j.jhtm.2021.05.006

Xu, D., Hazeltine, B., Xu, J., & Prasad, A. (2018a). Public participation in NGO-oriented communities for disaster prevention and mitigation (N-CDPM) in the Longmen Shan fault area during the Wenchuan and Lushan earthquake periods. *Environmental Hazards, 17*(4), 371–395. https://doi.org/10.1080/17477891.2018.1491382

Xu, J., Xu, D., & Lu, Y. (2016). Resident participation in post-Lushan earthquake housing reconstruction: A multi-stage field research method-based inquiry. *Environmental Hazards, 15*(2), 128–147. https://doi.org/10.1080/17477891.2016.1142417

Xu, J., Xu, D., Lu, Y., & Wang, Q. (2018b). A bridged government–NGOs relationship in post-earthquake reconstruction: The Ya'an service center in Lushan earthquake. *Natural Hazards, 90*(2), 537–562. https://doi.org/10.1007/s11069-017-3056-3

Xue, R., Luo, J., Zuo, G., & Yang, X. (2024). Dynamic evolution analysis and risk assessment of drought processes: A case study in Shaanxi, China. *Ecological Indicators, 169*, Article 112913. https://doi.org/10.1016/j.ecolind.2024.112913

Yang, Y., Yin, J., Wang, D., Liu, Y., Lu, Y., Zhang, W., & Xu, S. (2023). ABM-based emergency evacuation modelling during urban pluvial floods: A "7.20" pluvial flood event study in Zhengzhou, Henan Province. *Science China Earth Sciences, 66*(2), 282–291. https://doi.org/10.1007/s11430-022-1015-6

Ye, Q., Ma, Y., Xu, S., Sotelo, M. Á., & Li, Z. (2025). Understanding urban mobility responses to extreme precipitation events: A case study of Zhengzhou, China. *IET Intelligent Transport Systems, 19*(1), Article e70045. https://doi.org/10.1049/itr2.70045

Yumagulova, L., & Vertinsky, I. (2019). Moving beyond engineering supremacy: Knowledge systems for urban resilience in Canada's Metro Vancouver region. *Environmental Science & Policy, 100*, 66–73. https://doi.org/10.1016/j.envsci.2019.05.022

Zhang, H., Zhang, X., Comfort, L., & Chen, M. (2016). The emergence of an adaptive response network: The April 20, 2013 Lushan, China earthquake. *Safety Science, 90*, 14–23. https://doi.org/10.1016/j.ssci.2015.11.012

Zhang, Q., Li, R., Sun, J., Lu, F., Xu, J., & Zhang, F. (2023a). A review of research on the record-breaking precipitation event in Henan Province, China, July 2021. *Advances in Atmospheric Sciences, 40*(8), 1485–1500. https://doi.org/10.1007/s00376-023-2360-y

Zhang, X., Wang, C., Wang, W., & Zhang, H. (2023b). Inter-organizational collaboration after institutional reform in China: A perspective based on the revision of the emergency plan. *International Journal of Disaster Risk Reduction, 98*, Article 104084. https://doi.org/10.1016/j.ijdrr.2023.104084

Zhao, X., Li, H., Cai, Q., Pan, Y., & Qi, Y. (2023). Managing extreme rainfall and flooding events: A case study of the 20 July 2021 Zhengzhou flood in China. *Climate, 11*(11), 228. https://doi.org/10.3390/cli11110228

MIX
Papier | Fördert gute Waldnutzung
FSC® C083411

Kolibri 360 GmbH
Ferdinand-Jühlke-Straße 7
99095 Erfurt, Deutschland
produktsicherheit@kolibri360.de